D1620914

Strategies for Protecting National Critical Infrastructure Assets

A Focus on Problem-Solving

John Sullivant

WILEY-INTERSCIENCE
A John Wiley & Sons, Inc., Publication

Published by John Wiley & Sons, Inc., Hoboken, New Jersey.
Published simultaneously in Canada.

For general information on our other products and services or for technical support, please contact our Customer Care Department within the United States at (800) 762-2974, outside the United States at (317) 572-3993 or fax (317) 572-4002.

Wiley also publishes its books in a variety of electronic formats. Some content that appears in print may not be available in electronic format. For information about Wiley products, visit our web site at www.wiley.com.

Wiley Bicentennial Logo: Richard J. Pacifico

Library of Congress Cataloging-in-Publication Data:

Sullivant, John.
  Strategies for protecting national critical infrastructure assets / John Sullivant.
    p. cm.
  ISBN 978-0-471-79926-9 (cloth)
1. Infrastructure (Economics)—Security measures—United States. 2. Infrastructure (Economics)—Risk assessment--United States. 3. Terrorism—United States—Prevention. I. Title.
  HV6432.S85 2007
  363.325'720684—dc22   2007026434

Printed in the United States of America.

10 9 8 7 6 5 4 3 2 1

Strategies for Protecting National Critical Infrastructure Assets

BICENTENNIAL
1807
⊕WILEY
2007
BICENTENNIAL

THE WILEY BICENTENNIAL—KNOWLEDGE FOR GENERATIONS

*E*ach generation has its unique needs and aspirations. When Charles Wiley first opened his small printing shop in lower Manhattan in 1807, it was a generation of boundless potential searching for an identity. And we were there, helping to define a new American literary tradition. Over half a century later, in the midst of the Second Industrial Revolution, it was a generation focused on building the future. Once again, we were there, supplying the critical scientific, technical, and engineering knowledge that helped frame the world. Throughout the 20th Century, and into the new millennium, nations began to reach out beyond their own borders and a new international community was born. Wiley was there, expanding its operations around the world to enable a global exchange of ideas, opinions, and know-how.

For 200 years, Wiley has been an integral part of each generation's journey, enabling the flow of information and understanding necessary to meet their needs and fulfill their aspirations. Today, bold new technologies are changing the way we live and learn. Wiley will be there, providing you the must-have knowledge you need to imagine new worlds, new possibilities, and new opportunities.

Generations come and go, but you can always count on Wiley to provide you the knowledge you need, when and where you need it!

WILLIAM J. PESCE
PRESIDENT AND CHIEF EXECUTIVE OFFICER

PETER BOOTH WILEY
CHAIRMAN OF THE BOARD

In memory of

Peter Sullivant
Matina Sullivant
Jerry Sullivant

And to Maia, my granddaughter of three years rapidly going on 30.
A genius on resetting my TV and computer controls, and challenging
her mom to pay attention to her while working on this book.

Papou

ACKNOWLEDGMENTS

There are many professional people I want to thank for the opportunity to write this book. Special appreciation is extended to Roger Woodson, Virginia Howe, and Victoria Roberts for their guidance and encouragement from conception to publication.

My career has been influenced by many professional associates, and each in their own way contributed to my education, philosophy, and vision. I will always be grateful for their praise, criticism, and advise. Singly and collectively they encouraged me to bring these thoughts to print. The list goes far beyond the writing of this book–it spans four decades of friendship and working together with great professionals throughout the security industry.

Marko Bourne	Thomas Kitchen	Robert Schultheiss
James Broder	Paul Magallanes	Chuck Sennewald
George Campbell	James Malloy	Bruce Seymour
Larry Cassenti	Wayne Messner	Dan Sughrue
Thomas Dexter	Bonnie Michelman	Jerry Sutherlen
Dennis Dwyer	Gerald Orozco	John Sutton
Jack Gundrum	Charlie Patterson	Ronald Thomas
Richard Grassie	Jack Publicover	Lou Tyska
Jack Hurley	Red Robidas	John Wrona

To those not listed and to everyone who ever said to me, "You should write a book…," I say, here it is. Thank you for your confidence.

To all my siblings: Jerry, George, Nonda, and Stella who supported my career. To my daughters Tina, Nika, and Pamela whose love and support brought this book to life.

But no one deserves my appreciation more than my daughter Nika. I am grateful for her hard work and dedication to researching material and collating data, and her leadership in keeping me focused. She took advantage of her unique writing expertise in theater and film to make this complicated work user friendly and understandable. Throughout this work I relied heavily on her wise counsel, and I am grateful.

The list would not be complete without thanking the editors of John Wiley for their faith in me and the potential of this book. This is my first book and they were taking a chance on its success. They deserve credit for having the vision to see the benefits of this work to the industry.

It is my fondest wish that security professionals everywhere embrace these techniques and that this book helps bring much needed consistency to the industry.

ABOUT THE AUTHOR

John Sullivant, CSC, CHS-III, CPP, is a senior security executive consultant and president/owner of S^3E–Sisters Three Entrepreneurs Security Consultants Company. S^3E is a Small Business and Disabled Veteran Business Enterprise home based in West Hollywood, CA with a global reach that specializes in security risk management, vulnerability and threat analysis consultation, security strategic planning, and development and implementation of security programs.

Mr. Sullivant is an industry leader in antiterrorism planning, preparedness, and response with more than four decades of experience formulating and addressing strategy, policy, and initiatives important to complex security challenges. He is recognized as one of the nation's most distinguished and foremost security experts in the industry. Areas of expertise include:

- **Concepts of Operations:** Operations research and behavioral; analysis; organizational development; design and implementation; performance metrics and evaluation design; interoperability assessments; benchmarking programs and operations; and program evaluation and redesign.
- **Threat Characterization:** Threat and risk assessment; vulnerability analysis and consequence analysis; business and economic impact analysis; design threat profile development; target analysis and scenario development; exercise design; and red-teaming and analysis.
- **Programmatic Strategic Planning:** Mission analysis; requirements generation; policy analysis; standards development and implementation; investment decision-making; security planning, emergency preparedness and continuity planning; acquisition and resource allocations; administration, operations, and program management.
- **Systems Analysis:** Security technology analysis; architecture development and systems engineering; integrated system testing and assessment; test-bed planning; and program management.

Mr. Sullivant had a distinguished 24-year career with the United States Air Force serving in a multitude of sensitive assignments, ranging from Vietnam to the Pentagon. Many of these involved the development and implementation of comprehensive global security programs surrounding highly sensitive security matters.

Appointments called for leadership in a rapidly changing, complex environment containing ambiguity under uncertain and highly visible circumstances.

In the private security sector since 1982, Mr. Sullivant cultivates and maintains strong and productive business relationships with corporate executives by: providing advise, strategic direction and foundation for policy and strategic goals; translating them into meaningful action plans; coordinating across multiple organizational lines and remote sites to ensure the effective implementation of action plans and making mid-course corrections; driving cross-functional change to manage complex, long-term projects; evaluating the consistency and application of best practices; establishing action plans, timetables, and outcome measurements; and developing resources.

Mr. Sullivant's experience spans the national critical infrastructure sector:

Transportation: Civil Aviation, Railroads, Port Authorities. **Energy:** Nuclear Power, Electric and Water Utilities. **Public Health:** Pharmaceutical Laboratories, Medical Research Centers and Hospitals. **Manufacturing:** Production and Food Processing, Distribution. **Services:** Banking and Finance, Telecommunications, Shopping Centers and Malls, Entertainment and Special National Events. **Government:** Intelligence, RDT&E, C3 Centers, Military Installations, Law Enforcement and Corrections.

He is a seasoned speaker and has given numerous exacting presentations to corporate executives, program officials, and governing bodies in industry and government.

For more information:

www.S3EConsultants.com
JohnSullivant@S3EConsultants.com

CONTENTS

Chapter 7 Exhibits

Chapter 11 Exhibits

Chapter 12 Exhibits

Chapter 14 Appendix

15 THE AGRICULTURE AND FOOD SECTOR   421

Chapter 15 Exhibits

Chapter 15 Appendix

16 THE BANKING AND FINANCE SECTOR   451

Chapter 16 Exhibits

Chapter 16 Appendix

PREFACE

I founded Sisters Three Entrepreneurs [S^3E] Security Consultants Company in 1990. S^3E has always been a topic of conversation, so I must share the origin of its name. It represents my three daughters [Tina, Nika, and Pamela] and at the time, it was my intent to create a Small Business Women-Owned Enterprise. However, over time the girls had other aspirations as all children do, and they went their separate ways, developing their own interests and skills, establishing their independent careers, and doing quite well.

In deciding to write this book I began to research and categorized ideas in what appeared to be a logical sequence or common grouping. It became very clear early on that the broad span of the national critical infrastructure sectors and their complexity, and the volume of information collected and sorted proved logic perhaps did not prevail.

Whatever the original book outline was at the start of the project, it has taken on several new faces, and that gives me great comfort now that the writing process is complete. I am thankful for the creativity, flexibility, and optimism that emerged early in the project and lasted throughout the outcome. The experience was invaluable to tailoring my security assessment model to the individual infrastructure sectors.

My overriding task then became the promise of sharing four decades of knowledge and experience with other professionals, peers, and co-workers by presenting a very comprehensive and complicated process, in pieces sufficiently small enough to "chew on" without need for further coaching. In sharing this knowledge and experience, I hope to raise the industry standard of professionalism and quality service to all clients–they deserve nothing less.

At a time when the United States government and corporate America are spending billions of dollars performing security assessments, this book offers a new systematic integrated cost-effective approach to investigating the high-risk national critical infrastructure environment worthy of acceptance by all those charged with the awesome responsibility of homeland security. It is my fondest wish that this contribution will become a cornerstone in protecting the Nation's Critical Infrastructure Assets.

Part I

UNDERSTANDING THE ENVIRONMENT

Part I will guide the reader through background information that triggered America's response to the horrific events of September 11, 2001, to an understanding of the threats that pose significant impact. Emphasis is placed on the greatest national threats facing America's infrastructure. References to and acknowledgments of excellent works completed by others are given where appropriate. Chapters 1 and 2 offer a clear and distinct appreciation of the need to declare risk assessment as the vital first link in solving a national problem and of the management actions that must follow after the risk assessment is completed.

Chapter 1

INTRODUCTION

WHAT THIS BOOK IS ABOUT

Before making assumptions about security assessments based on past experience, consider first that you are about to be introduced to an entirely different perspective on the subject in this book. To begin, it is helpful to understand the environment that shapes not only our daily routine but our strategic thinking as well. And in doing so we will embrace the ability to "think outside the box."

Assessing vulnerability and risk has certainly been part of our experience since the beginning of humankind. It was evident when the leader of the pack told his most trusted kin to "go live among them and report back all you see and hear" what the task was, what was its importance to the survival of the tribe, and what was the standard of performance expected. As tribal cultures developed into more sophisticated civilizations, pharaohs, kings, emperors, prime ministers, and presidents of nations built and maintained their dynasties and countries by exploiting the strengths and weak-

3

nesses of both friend and foe and in particular members of their immediate and extended family. Agents dispatched throughout the region and foreign lands became "trusted envoys." They swore to observe the societal lifestyle and economy of the targeted population; determine attitude, morale, and loyalty; assess the will of the people and the strength of the military; discover the makeup of the government and its short- and long-term intentions; and secretly report their observations, conclusions, and recommendations to no one but the leader who sent them.

Due to the well-executed strategies and intelligence gathered in these types of missions, wars have been won or lost, territories and countries defeated or conquered, people enslaved or freed, political aims achieved or failed, and elections or appointments gained or lost. History offers us an opportunity to learn about the nature of these past assessments—why they succeeded, why they failed—and hopefully make us wiser in the process. This includes adapting our own methodologies to meet the needs of modern times and modern threats.

WHY THIS BOOK IS IMPORTANT

This book helps with these objectives by presenting a comprehensive methodology for conducting security assessments of the nation's critical infrastructures. **The S^3E Security Assessment Methodology focuses on clearly identifying, measuring, and prioritizing security risk for high-threat environments.** For too long many executive managers and security directors have been intimidated by this task and either taken investigative shortcuts or postponed the security assessment for another day. Conversely, many managers and organizations complete their assessments but still do nothing by way of strategic security planning. They report that embracing a comprehensive security strategy is not feasible due to impending obstacles, or they refute the security issues identified, believing they are not at risk. Others are very much aware of their facility's shortcomings but lack either the knowledge of how to proceed or the security budget to move forward. **The security-assessment methodology presented here provides a detailed roadmap that can help enterprises that are in denial of risk or lack an influential security director who can raise security standards and adopt protection planning.** Otherwise, in today's world of increasing vulnerability, such organizations are liable to become bad risk investments. Their business ratings and insurance premiums no doubt reflect this corporate culture or soon will.

The U.S. government does not perform comprehensive security assessments of all the national critical infrastructure sectors and key assets. Typically the owners or operators of infrastructure enterprises perform these assessments and mostly look to professional security consultants to carry out the task. Such assessments are important from both a local and a national planning perspective in that they enable authorities to evaluate the potential effects of a terrorist attack or other emergency on a given facility or sector and then invest accordingly in protection measures. Security assessments also serve as the building blocks for threat-vulnerability integration, allowing authorities to determine which facilities and sectors are at most risk. This aids local and national planners in developing thresholds for future standards for preemptive or protective action and setting priorities for changes to facility designs.

Enterprises [corporations, business organizations, government agencies, etc.] need to undergo systematic changes in their security posture as modern threats diversify and intensify and as the technical sophistication of terrorism increases with the availability of knowledge and materials to carry out acts of violence. World events have put terrorism at the forefront of the American psyche. These events and others have shown that terrorism in the U.S. is not about to end. In fact, it is only just beginning.

Various security-assessment models are currently in use, but many simply do not address the demands placed on business management. Some go partway but tend to introduce their own drawbacks and difficulties. They can be generic in nature or limited in scope and may only offer elementary guidance to a general audience. Others are neither consistent nor comparable in their methodology, thereby complicating protection planning and resource allocation.

Those sources that do offer a degree of useful detail are either government documents or private industry works with controlled distribution. For consultants who provide confidential security services to an array of clients, they can only access and use these sensitive materials under strict supervision and control and only when under contract for a specific project. In other instances, engineering firms, system integrators, and consultants have found it necessary to expand on works created by various professional associations and develop their own approaches. At best the data is scattered throughout the industry, forcing many security practitioners to research extensively to collect and assemble vital information. The most successful models are those that are performance-based, but many lack agreement on a best formula and outcome. **This book is an attempt to establish a much needed industry standard for generating solid and thorough security assessments for a varied clientele.**

Also of importance is the environment in which security assessments are conducted. If not performed by qualified individuals with a full understanding of analytical and security principles, they can do more harm than good. Poorly conducted assessments are a waste of time and resources and can lead enterprises to take action that is ineffective and may give them a false sense of security.

This book proposes to ease the research burden, develop investigative protocols for infrastructure assets, and pull together baseline data into a comprehensive and practical guide to help the serious reader understand advanced concepts and techniques of security assessment, with an emphasis on meeting the security needs of the National Critical Infrastructure. At a time when the U.S. government and corporate America are spending billions of dollars on performing risk assessments, establishing an innovative, acceptable, and proven methodology is vital for the national-critical-infrastructure environment. This methodology expands on the work of others, bringing together the best methods, techniques, measurements, and collective expertise of many colleagues and other professionals in an effort to raise the performance bar for conducting security assessments.

In summary, this book is a critical contribution to the field of security analysis in several ways. It offers a series of integrated strategies to evaluate the effectiveness and efficiency of an entity's security program. It pulls together user-friendly data into a comprehensive and practical guide to help the reader understand and apply advanced analytical techniques. And it provides a proven methodology applicable not only to infrastructure assets but to other organizations as well.

WHO CAN BENEFIT FROM THIS BOOK

This book will be an excellent guide for serious security practitioners, specifically:

- CEOs, presidents, and VPs of critical infrastructure businesses
- Homeland Security officials
- Directors of security and security managers
- Business-continuity managers and Risk-management managers

- Facility, traffic-management, and warehouse managers
- Human-resources managers
- Safety managers and quality-assurance managers
- Architects and engineers
- Judicial and correctional professionals in responsible leadership positions
- IT professionals responsible for computer security
- Security-system integrators
- Security consultants
- Managers and supervisors with operational responsibility for critical infrastructure businesses
- Government and military personnel with security and intelligence responsibilities
- Professionals responsible for security operations, emergency services, and public relations
- Law-enforcement professionals
- Researchers and investigators in the behavioral sciences
- Forensic examiners in the fields of medicine, psychology, criminology, accident reconstruction, crime-scene reconstruction, criminal investigation, engineering, and risk analysis

The book will also serve as a valuable reference in the academic world for:

- Educators, professors, and instructors in relevant fields
- Curriculum-development professionals
- Upper-division undergraduate and graduate students pursuing security-analysis expertise
- Security-training academies

HOW TO USE THIS BOOK

The text is divided into five major sections presented in a logical learning sequence. It is best to approach the book in the order established, as each chapter presents essential principles necessary to understanding subsequent material.

Part 1. Understanding the Environment

Part I guides the reader through the purpose and use of this book and the specific elements of western cultures that attract terrorism.

Chapter 1, Introduction, sets the stage by discussing:

- What this book is about
- Why this book is important
- Who can benefit from this book
- How to use this book

In **Chapter 2, Environments that Influence the Security Assessment: Threats, Western Values, and the National Critical Infrastructure Sectors**, American values are contrasted with the terrorists' ideology. This chapter discusses:

- Western social values: strengths, weaknesses, fears, and aspirations
- Safeguarding western values, safeguarding American values, the American population
- America's Inherent Vulnerabilities: American Values in Contrast with Tyranny's Oppression
- The Importance of the National Critical Infrastructure Sectors
- The protection challenge
- The nation's most direct threats and consequences

Chapter 2 offers an understanding of the greatest current threats to America's economy and security. It presents the clear and distinct need for security assessments as the vital first link of effective protection and the management actions that must follow to enhance enterprise security. Without a strong understanding of environments that influence the security assessment, it is a useless tool in guarding against terrorist attacks, major disasters, and other emergencies.

Part II. Understanding Security Assessment

Part II guides the reader through a comprehensive description of the S^3E **Security Assessment Methodology**. This model was developed and improved over many years of research and experience. It has been successfully used in industry and government domestically and internationally across the entire spectrum of the infrastructure sectors, including some of the nation's most highly classified one-of-a-kind national experimental resources; military air, land, and sea assets; and national intelligence facilities. Over 3,000 security assessments have been completed in 30 countries and 5 continents using this approach.

Chapter 3,The Security Assessment: What, Why, and When, introduces and discusses:

- Why perform a security assessment?
- What is the scope of a security assessment?
- When should a security assessment be performed?
- Which security-assessment model is best?

Chapter 4, Proven Security Assessment Methodology, introduces the S^3E **Security Assessment Methodology**'s architecture and elements, defines its purposes and objectives, describes the behavioral and physical sciences at play and the techniques employed in the process, and addresses the standards adopted to evaluate and measure success. This chapter discusses:

- The security-assessment challenge
- A proven security-assessment model and methodology
- The security-assessment methodology as a system-level performance-based approach to problem solving
- Distinct benefits of the S^3E Security Assessment Methodology
- Enterprise key security strategies
- Enterprise security performance strategies
- Enterprise key security operational capabilities
- Security-assessment measurement criteria

Chapters 5 through 10 identify how the security-assessment methodology is performed. The process consists of six independent and separate tasks that are bound together by significant interrelationships and dependencies.

Chapter 5, Task 1, Project Strategic Planning: Understanding Service Requirements, discusses:

- Project mobilization and start-up activity
- Site-investigation preplanning
- Planning, organizing, coordinating, project-kickoff meeting

- Attending and co-chairing project-kickoff meeting
- Reviewing available project information and conducting workshops
- Conducting interviews
- Documenting the entire security-assessment process

Chapter 6, Task 2, Critical Assessment: Understanding the Service Environment, discusses:

- Site and facility mission and services
- Facility configuration and layout
- Asset and resource identification and criticality
- Documenting the characterization process

Chapter 7, Task 3, Identify and Characterize Threats to the Service Environment, discusses:

- Developing a design-basis threat statement
- Identifying adversarial groups and their capabilities
- Identifying the range and levels of threats
- Assigning likelihood ratings to ranges and levels of threats to assets
- Assigning ranges of malevolent acts to adversary attractiveness
- Determining loss consequences and probability of occurrence

Chapter 8, Task 4, Evaluate Program Effectiveness, discusses:

- Evaluating the status of operational systems, functions, processes, and protocols
- Evaluating the status of the overall security program including physical, operations, information, personnel security, and training
- Measuring program effectiveness

Chapter 9, Task 5, Program Analyses, discusses:

- Finalizing and refining the design-basis-threat profile
- Assessing vulnerability
- Determining and finalizing rank order for protection
- Developing workable solutions

Chapter 10, Task 6, Reporting Security Assessment Results, discusses:

- Developing enterprise-security strategies
- Developing mitigation solutions and cost estimates to implement program enhancements
- Presenting the report to executive management
- Presenting findings and recommendations to governing authorities

Part III. Tailoring the S^3E Security Assessment Methodology to Specific Critical Infrastructures

Part III provides the critical link for applying the methodology in diverse environments. The security-assessment methodology introduced in Part II is presented with many faces and an equal amount of applications tailored to specific infrastructure sectors. It therefore relies heavily on the reader's comprehensive understanding and ability to successfully employ the basic security-assessment principles presented in Part II in an effective and efficient manner.

Chapters 11 through 21 focus on particular infrastructure sectors. The template initially introduced in Part I is reintroduced but tailored to address these infrastructures:

- **Chapter 11. The Water Sector**
- **Chapter 12. The Energy Sector**
- **Chapter 13. The Transportation Sector**
- **Chapter 14. The Chemical Industry and Hazardous Materials Sector**
- **Chapter 15. The Agriculture and Food Sector**
- **Chapter 16. The Banking and Finance Sector**
- **Chapter 17. The Telecommunications Sector**

Chapters 11 through 21 are designed to stand alone, permitting the security analyst to concentrate on a particular sector with minimal reference to any previous chapter in Part II. To my knowledge, introducing such an integrated approach under one cover sets a new standard. It offers the diversified consultant, the security practitioner with multidiscipline responsibilities, and the academic the availability of a quick, reliable, and practical "briefcase" reference to use in the office as well as on the road.

Chapter 2

ENVIRONMENTS THAT INFLUENCE THE SECURITY ASSESSMENT:

Threats, Western Values, and the National Critical Infrastructure Sectors

One cannot reap the full benefits of this book without first having a comprehensive understanding of the environments that influence and surround the conduct of a security assessment. These environments are: threats, western values, and the makeup of the National Critical Infrastructure Sectors.

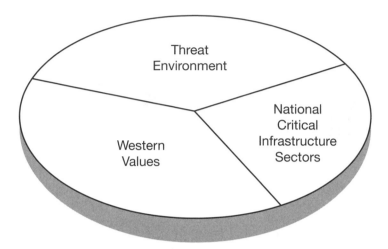

Exhibit 2.1 Environments that influence the security assessment

UNDERSTANDING THE THREAT ENVIRONMENT

The Birth of Modern-Day Terrorism

Terrorism is as old as civilization and throughout history has taken on various forms and guises. Understanding the numerous threats of our time and knowing the aims and deceptions of terrorists are critical to developing a strategy of protective measures necessary to defeat the threat.

The "act of terrorism" or "terrorist activity" has its roots in the beginning of modern civilization. The first recorded terrorists were holy warriors who killed civilians in first-century Palestine. Jewish zealots would publicly slit the throats of Romans and their collaborators. In seventh-century India, the Thuggee cult would ritually strangle passersby as sacrifices to the Hindu deity Kali. In the eleventh-century Middle East, the Shiite sect known as the Assassins would eat hashish before murdering civilian foes. According to the Council on Foreign Relations, historians can trace recognizably modern forms of terrorism back to such late-nineteenth-century organizations as Narodnaya Volya ["People's Will"] in Russia. One particularly successful early case of terrorism was the 1914 assassination of Austrian Archduke Franz Ferdinand by a Serb extremist—an event that helped trigger World War I. [1]

The term "terrorism" was first coined during France's Reign of Terror in 1793-94. Originally the leaders of this systematized attempt to weed out "traitors" among the revolutionary ranks praised terror as the best way to defend liberty, but as the French Revolution soured, the word soon took on grim echoes of state violence and guillotines.

The Psychology of Terrorism

Terrorists seek to demoralize free nations with dramatic acts of murder and destruction. They seek to wear down resolve and will by killing the innocent and spreading fear and anarchy. They operate as a network of many violent extremist groups around the world, striking separately and at times in concert but retaining their separate identity and quest for recognition. Terrorism, and in particular the suicide bomber, is meant to produce psychological effects that reach far beyond the immediate horror of the attacks. The strategy of suicide terrorists is to make people paranoid and xenophobic, fearful of venturing beyond their homes even to shop, go to school, or attend events. Terrorists hope to compel societies to acquiesce or surrender to stated demands. The fundamental characteristics of suicide bombings and the strong attractions for a terrorist organization behind them are universal: they are inexpensive and effective, and the organization has nothing to lose. They are less complicated and compromising than other kinds of terrorist operations. They guarantee media coverage and tear at the very fabric of trust that holds societies together. Suicide bombers are responsible for almost half of all attacks and on average kill four times as many people as other terrorist acts [2].

In the past terrorist groups were mainly motivated by political ideology, usually avoiding mass-casualty attacks for fear of alienating their political constituencies and potential recruits. Today, religiously motivated terrorist groups—or those that hide behind religion—exhibit fewer self-imposed constraints. They believe in annihilating their enemies, thus the increase in more dramatic terrorist acts involving a higher kill ratio than in the past [3]. A defining moment came during the late 1960s with the hijacking of commercial airlines. Later, radical Islam caught the attention of the world in 1979 when the late Ayatollah Khomeini assumed power and established the Islamic Republic of Iran [4]. The subsequent hostage crisis at the American Embassy in Tehran highlighted the dangers radicalism posed. Today, the psychology of terrorism is to "think bigger." We are seeing a trend in which terrorists now target a nation's critical infrastructure to achieve three general types of effects:

DIRECT INFRASTRUCTURE EFFECTS:	Cascading disruption or arrest of the functions of critical infrastructures or key assets through direct attacks on a critical mode, system, or function.
INDIRECT INFRASTRUCTURE EFFECTS:	Cascading disruption and financial consequences for an enterprise through public and private-sector reactions to an attack.
EXPLOITATION OF INFRASTRUCTURE:	Exploitation of elements of a particular infrastructure to disrupt or destroy another target.

Exhibit 2.2 Terrorists' long-term strategic objectives

As a nation we must protect the critical infrastructure and key assets from acts of terrorism that would:

- Impair an enterprise's ability to perform essential services and ensure the general public's health and safety
- Undermine the enterprise's capacity to deliver minimal essential public services
- Damage the enterprise's ability to function
- Undermine the community's morale and confidence in the enterprise's capability to deliver services

The Many Faces of Terrorism

The war on terrorism is not a war of religions or a clash between civilizations. For example, the civilization of Islam, with its humane traditions of learning and tolerance, has no place for these violent sects of thugs, killers, assassins, and tyrants. The faith of Islam teaches men and women moral responsibility and forbids the shedding of innocent blood. Those who believe in Jihad—the holy war against all nonbelievers—practice a fringe form of Islamic extremism that has been rejected by Muslim scholars and the vast majority of Muslim clerics as a movement that perverts the peaceful teachings of Islam. These extremist terrorists follow a doctrine that commands them to kill Christians, Jews, and Americans and to make no distinction among military and civilians, including women and children. They believe that the Middle East must fall under the rule of radical governments, moderate Arab states must be overthrown, nonbelievers must be expelled from Muslim lands, and the harshest practice of extremist rule must be universally enforced. Islamic extremists claim that the United States is a target because it is a supporter of reactionary, nonprogressive regimes; the United States leads the anti-Islamic movement; the United States is the home of Jewish-American-Israeli capital; and attacks within the United States present an element of tactical surprise.

International Terrorism
The enemy has no rules—only the determination to instill fear, control behavior, and kill. What drives current international terrorism is the belief and duty to fight the modern, usually Western, world. In this vision books are burned, terrorists are sheltered, women are whipped, and children are schooled in hatred, murder, and suicide. International terrorism has long been recognized as a serious foreign and domestic security threat. A *modern trend* in terrorism is toward loosely organized, self-financed international networks of terrorists. *Another trend* is toward terrorism that is religiously or ideologically motivated. Radical Islamic fundamentalist groups, or groups using religion as a pretext, pose terrorist threats of varying kinds to United States interests and to friendly regimes. A *third trend* is the apparent growth of cross-national links among different terrorist organizations, which may involve combinations of military training, funding, technology transfer, or political advice.

Once considered insignificant and a local law-enforcement nuisance, terrorism in general remained relatively unchallenged until recently. Today we witness it standing fully erect, exposing itself to the entire world, and we recognize the madness for what it really is—misguided hatred and misconceived intervention in the affairs of nations, the practice of religion, and the behavior of individuals across all aspects of life. For terrorists, killing people has been the dramatic backdrop to guarantee worldwide attention with a single horrific act. Notwithstanding the terrible loss of life, the real targets of most terror campaigns have not been the unfortunate victims but their survivors and the captive world audience. [5]

The end of the Cold War and the demise of the Soviet Union dramatically rearranged the thrust and complexity of international terrorism. The restraining influences of this former superpower on terrorist groups have vanished, rendering old strategic-threat assumptions obsolete. What we have witnessed with the collapse of Soviet control in the region is the diffusion of power across the globe and the cancerous growth of a new threat. The independence gained by the former Soviet states has had a price: regional and global instability as these countries try to establish themselves and maintain order, an increase in organized and petty crime, the proliferation and sale of weapons of mass destruction to rogue states, and new safe havens for terrorists. The corrupting elements in these new regimes pose a substantial challenge to the United States and world stability. [6].

In the midst of the race to gain recognition for their respective causes, political and religious terrorists seek to assert themselves in rapid succession in a number of countries including the United States, giving rise to various terrorist subgroups. In particular, the threat posed by radical Jihadists has metastasized into something more widespread, and perhaps more lethal, than other global terrorist organizations that are developing and managing terror plots.

It is essential to understand the nature of terrorism and its causes. The main thrust of the current terrorist threat in the West comes from an ideological war within Islam. A radical Islamic faction is attacking moderate Muslims no less viciously than Westerners. Conventional security thinking was mostly geared to catching or preventing criminals motivated by greed working in deviant subcultures or those using their psychopathic and antisocial behavior for personal gain. In contrast, the motives of the modern foreign terrorist focus on real socioeconomic and political issues and defy all the normal rules of governance, including rules of war and combat. In fact, their aim is precisely to show how powerless governments are by using unconventional means that instill insecurity and fear in the population and doubt about the government's ability to control the situation. Terrorists do not focus on personal gain but rather work for a higher calling or cause, such as Jihad.

The culture of the West, and in particular of the U.S., functions in spans of time dictated by elections or appointments. Democratic societies want quick action and look for instant poll results. Elected or appointed governments do not usually have the years needed to mount lengthy campaigns. But if they do nothing for a period, the opposition party will use this inaction as a spear to wage its own internal criticism of the incompetence of the existing governing party. Conversely, terrorists are not constrained by time, conditions, elections, or the rule of laws. They do not need to win "the war" to have immediate effect—they need only win a particular battle at a particular time. A perfect example is the Madrid train bombing in April 2004 and the Spanish government's withdrawal of its military forces from Iraq soon afterward [7].

Terrorists are strategic actors. Terrorism is theater at its finest hour, and the television networks are the box office. Politically and ideologically motivated terrorist groups thrive on publicity and aim at producing immediate dramatic effects. Well-publicized attacks attract worldwide attention to the terrorists and their causes, forcing governments, in the eye of the public, to direct unprecedented resources to counter the threat. This threat is achieved by attacking highly symbolic and vulnerable targets—a "shock and alarm" tactic oriented toward causing heavy property damage rather than mass casualties [not both?]. Targets are deliberately chosen based on weaknesses observed in defense and preparations. They weigh the difficulty in successfully executing a particular attack against the magnitude of its impact. They monitor media accounts, listen to policymakers as nations discuss how to protect themselves, and adjust their plans accordingly. Where nations insulate themselves from one form of attack, terrorists shift and focus on another exposed vulnerability [8].

In a post-Cold War environment, terrorist activity has moved away from government targets and placed its gun-sight on commercial enterprise targets. The Department of Transportation and Commerce is replacing the Department of Defense as the latest symbolic icon of the American establishment. **Infrastructure assets are much softer targets than government facilities and have therefore become more attractive to the terrorist.** Moreover, the economic and psychological impact of destroying the infrastructure is more devastating than damaging a single military facility. Large numbers of people and business establishments become the victim of a single attack, and the media does the rest of the intended damage by reporting pre-determined misinformation offered by an "unknown" or "confidential" source. [9]

Economically related targets are chosen for almost two out of three terrorist attacks, and except for specific retaliatory purposes business targets appear to be better suited to meeting terrorist objectives than government targets. Organizations that have national or international notoriety or symbolic significance are extremely vulnerable. Terrorist groups are now turning their attention to more critical, vulnerable targets such as bridges and tunnels; telecommunications and computer networks; power, oil, and gas systems; transportation; and drinking water [10]. Evidence obtained in Afghanistan, Iraq, and other countries indicates that terrorists consider these assets as viable targets for a dramatic entrée onto the stage, even greater than the events of 9/11. Several foiled terrorist plots confirm this determination.

Specifically, the following plots come to mind:

- LONDON, ENGLAND–Plot to bomb underground subway and bus discovered [2005]

- NEWARK, NJ–Lot to buy "Dirty Bomb" and smuggle shoulder-held missiles into the United States discovered. [2005]
- NEW YORK, NY–Plot to blow up subway station at Herald Squared uncovered. [2004]
- NEW YORK, NY–Plot to use tourist helicopters in terrorist attacks uncovered. [2004]
- BRITIAN, UNITED KINGDOM–Authorities discover plot to attack Heathrow International Airport. [2004]
- MADRID, SPAIN–Plot unfoiled to blow up high speed train. [2204]
- MADRID, SPAIN–Police foil attempt to blow up city's main subway station. [2002]
- LOS ANGELES, CA–Plot to blow-up LAX Airport during the millennial discovered. [1999]
- MANILA, PHILIPPINES–Plot uncovered to blow up 12 U.S. aircraft over the Pacific [1995]
- PARIS, FRANCE–Plot uncovered to fly jet liner into Eiffel Tower. [1994]
- NEW YORK, NY–Plot uncovered to blow up Lincoln and Holland Tunnels, and other tunnels and bridges throughout Manhattan. [1993]

Terrorists are also masters of publicity. They use the news media for maximum effect. The media brings the world into the living room of every household, constantly increasing the awareness of potential threats and actual attacks. This magnifies the problem, giving the terrorists the publicity they crave and instilling fear into the population. Since terrorism relies on the media to present dramatic news, each subsequent terrorist attack needs to be more newsworthy than the last. As such, they are looking to more sophisticated technology—possibly dirty bombs, chemical and biological weapons, and even crude nuclear devices—to deliver their next evening newscast [11]. While global terrorist organizations may continue to focus on spectacular iconic targets, independent terrorist cells will likely narrow in on the water and power, communications, public-health, and rail and mass-transit systems.

International Terrorist Cells
International terrorist cells exist in most major American cities. Findings in Afghanistan confirm that most of the 19 men who hijacked the planes on September 11 were trained in secret camps. So were tens of thousands of others. These dangerous killers, schooled in methods of murder and destruction, often supported by hostile regimes, are now spread

throughout the world—including the United States—like time bombs set to go off without warning. According to government officials, presently within the United States there are representatives from state sponsors of terrorism, members of international terrorist groups, as well as violent radical Islamic fundamentalists:

- These include representatives from Libya, Iran, Iraq, Syria, Sudan, North Korea, and Cuba. Many have resided in America since 1992. [12]
- Every major terrorist organization in the Middle East has created significant infrastructure networks across America. Intelligence reports indicate that the Abu Nidal Organization [AND], the Islamic Resistance Movement in Palestine [HAMAS], the Popular Front for the Liberation of Palestine [PFLP], Hezbollah, Al Gama al Islamiyyia [IG], Palestinian Islamic Jihad [PIT], Al Furgra, and al Qaeda have established operational networks in the U.S., Puerto Rico, Canada, and Mexico. Radical Islamic fundamentalist groups are established and operating in over 40 major U.S. cities and over 60 countries. The activities of these groups include recruiting, weapons training, bomb making, robbery, extortion, counterfeiting money and documents, and other criminal acts. [13]

Attempts to sabotage New York City's federal buildings, the United Nations complex, and both the Lincoln and Holland Tunnels, in addition to a series of independent political assassinations and kidnappings of public figures, highlight an unprecedented conspiracy for large-scale urban terrorism. Other unprecedented terrorist offensives against European targets and the United States include bombing landmark buildings and lacing a large city's water supply with a deadly botulism toxin developed in Iranian biological-warfare laboratories in Esfahan. These activities demonstrate the intent and capability of international terrorists to conduct attacks on a scale not previously seen. [14]

We also know for certain that enemies of the United States posing as visitors and invited guests will take advantage of the constitutional liberty and personal freedoms we enjoy. Without hesitation, they will attempt to violate the very foundation of our laws to inflict damage, death, and destruction upon our nation, our communities, and our people. Other discoveries in Afghanistan, Iraq, and the Philippines confirm that the depth of their hatred is equaled only by the madness of destruction that they design. Diagrams of American nuclear power plants and water-supply systems, detailed instructions for making chemical and biological weapons, and surveillance maps of American cities provide thorough descriptions of threats to landmarks in America and throughout the world.

There also remains an underlying potential threat from those posing as foreign students and immigrants who may have entered the United States to support internal terrorist activities prior to the September 11 attacks. Although most foreign students possess legal status and contribute to America's diversity and economy, there are over four million persons residing illegally in the country. A flow of illegal immigrants across our porous borders makes it virtually impossible to exclude all foreign terrorists.

This new threat environment has increased the number of terrorist cells operating on their own, relatively independent of their leaders. The United States-led war on terror has disrupted the global terrorist organizational structure and shifted responsibility for initiating and executing attacks to local terrorist cells, like those responsible for recent attacks in London, Madrid, Tokyo, and Malaysia.

Domestic Terrorism
The most difficult terrorism threat to contain comes from citizens of the United States within our borders. Over the past 10 years, the number of actual and attempted bombings in the United States has quadrupled to more than 3,000 annually according to government reports. Bombings, even more than other security problems including a bomb hoax, can empty a building or entire business district for days. They can stop the flow of commerce on land, air, and sea and bring industry and government to a standstill. The continued sabotage of government and industry facilities and public areas and the threat of individual mail bombs and other bomb-threat situations clearly demonstrate the general vulnerability of a free society to terrorism.

Domestic terrorism is a separate and independent force that is difficult to distinguish from the foreign threat. Most domestic terrorists are "special-interest extremists" that may operate alone or as members of a small unit. Extremist and anarchist groups include militias, separatists, and numerous antigovernment organizations. Most encourage the massing of weapons, ammunition, and supplies to prepare for a confrontation with law enforcement. **Some of these groups even view the overthrow of the federal government as their constitutional obligation.** Independent groups further provide a formula for the spread of white supremacy, while others pursue antigovernment sentiments stemming from environmental issues, right to life, and the freedom of religion, speech, and the press regardless of whose inalienable rights they crush. [15]

Militia Groups

Militia groups pose a significant threat to governance. They mistrust the federal government, are against the United Nations, the World Bank, and taxes, and exhibit racial and religious bigotry. While the majority of militia members are law-abiding citizens who do not pose a direct threat, there is an ongoing focus on radical elements that are capable and willing to commit violent acts against law enforcement, government agencies, and military, civilian, and international targets. The militia movement has increased significantly over the past few years. Hate philosophies of varying degrees have begun to creep into the growing militia movement. At the same time, many groups choose to remain autonomous, depending on their extremist view of a particular issue.

Extremist Groups

Extremist groups in the United States mostly consist of those focusing on the right to life, animal rights, and saving the environment:

- Anti-abortion activists, made up primarily of extremist right-wing Christians, have demonstrated against clinics, bombed facilities, and killed employees and patients.
- Animal-rights terrorism is still a foreign concept to most Americans who hear only of the occasional raid on a research lab. Such attacks are more common than people know and are increasing in frequency and violence. The activists' main objective is to stop the use of animals for experimentation, sport, or food. Their strategy is mainly economic sabotage. The primary organization leading this movement is the Animal Liberation Front [ALF], a left-wing extremist group [16]. Attacks in 2001 included the fire-bombings of a primate lab in New Mexico and a federal horse corral in California.
- Radical left-wing environmental groups target logging facilities, real-estate developers, and other organizations perceived to engage in activities detrimental to the environment. In 1998 the Earth Liberation Front [affiliated with the ALF] fire-bombed a ski resort in Vail, CO, destroying $12 million in property. In 2003 they targeted car dealerships in CA, damaging and destroying 125 SUVs.

The 1995 bombing of the Murrah Federal Building in Oklahoma City and the 1996 Olympic bombing in Atlanta highlight the threat of domestic terrorist acts designed to achieve mass casualties. Moreover, the U.S. government averted seven planned terrorist acts in 1999—two potentially large-scale, high-casualty attacks being organized by domestic extremist groups. These and other incidents have placed a new dimension on domestic terrorist attacks.

The Changing Face of Terrorism

International Proliferation of Weapons of Mass Destruction

New and emerging dangers are on the horizon. Looming over the entire issue of international terrorism is a trend toward proliferation of weapons of mass destruction. Currently, Iran is seen as the most active state sponsor of terrorism and is aggressively seeking nuclear-arms capability. North Korea has admitted to having a clandestine program for uranium enrichment and nuclear weapons. Terrorists have attempted to acquire the means to make weapons of mass destruction through their own resources and underground connections.

Lethal chemical and biological agents and radiological materials released into a city center or residential neighborhood could take the lives of thousands and render an entire region useless for months if not years. Such infrastructure attacks could instantaneously cripple the economy or even place the nation under siege. [17]

Vulnerability of Weapons Stockpiles Abandoned by the former Soviet Union

The lack of adequate controls on biological, chemical, and nuclear technologies within the former Soviet Union has resulted in a flood of buyers eager to purchase lethal materials from an expanding black market. Added to this volatile mix are scientists and technicians—at times out of economic desperation—who are prepared to sell their skills to the highest bidder [18]. It is understandable why national-security analysts are concerned about the ramifications of nuclear materials in the hands of terrorists. Four significant trends illustrate this concern [19]:

- The security of nuclear-storage sites and accountability of weapons-grade nuclear materials within the territories of the former Soviet Union and within Russia have lacked positive controls for some time. The Russian Mafia is known to have masterminded the theft, diversion, transportation, and distribution of nuclear materials outside the country. The threat is fueled by widespread corruption, lack of law enforcement and security, severe fiscal constraints, and the breakdown of military discipline and morale. Russian interior-ministry officials reported 54 cases of theft of nuclear materials in 1993 and 1994, all by insiders at government and military facilities, research institutes, fabrication plants, and transportation facilities.

- Security agents in Turkey, Morocco, and South Africa have seized weapons-grade nuclear materials belonging to Russia. Potential buyers have included states such as Libya, Iraq, Iran, North Korea, nuclear extortionists, terrorist organizations, and rogue military elements. While the availability of nuclear devices to terrorists is still a matter of debate among the intelligence communities, the use of nuclear material with a conventional explosive charge to introduce a contamination scenario is very real, as is the use of biological and chemical agents.

- Since the movement of terrorist groups from country to country has been made easier by the lifting of border controls in most European countries, the movement of nuclear materials from the former Soviet Union through these nations is also easier to accomplish and more difficult to detect.

- Hospitals in Eastern Europe have reported an alarming number of radiation-poisoning cases. Apparently, smugglers trafficking in plutonium, strontium-90, cobalt-60, and cesium-137 from the former Soviet Union do not realize that sophisticated protective measures are necessary to handle and transport such deadly materials and are unaware of the serious health hazards that can result from their improper handling.

Weapons of Mass Destruction

A new and more dangerous threat is here. We are now witnessing terrorist groups implementing unconventional methods and tactics never before used: [20]

- Indications have surfaced that al Qaeda has attempted to acquire chemical, biological, radiological, and nuclear weapons. As a result, stakes in the war against international terrorism are increasing, and margins for error in selecting appropriate risk-mitigation steps to prevent, prepare for, and respond to terrorist attacks are diminishing.

- The Aum Shinrikyo cult was able to procure technology and blueprints for producing sarin, a deadly nerve gas, through official contacts in Russia in the early 1990s. The gas was subsequently used in an attack on the Tokyo subway in March 1995 that killed 12 people and injured 5,000.

The Chemical Weapons Threat

Chemical weapons are extremely lethal and capable of producing tens of thousands of casualties. These are poisonous gases, liquids, or solids that have toxic effects on people, animals, and plants. They are character-ized by the rapid onset of medical symptoms and easily observed signa-tures, such as colored residue, dead foliage, pigment odor, and dead ani-mals and insects. Most chemical weapons cause serious injuries or death. The most dangerous chemical agents are sarin, VX, and mustard gas. They are relatively easy to manufacture using basic equipment, trained person-nel, and precursor materials that often have legitimate dual uses. As the 1995 Tokyo subway attack revealed, even sophisticated nerve agents are within the reach of terrorist groups. [21]

SARIN
Sarin is one of the world's most dangerous chemical-warfare agents. Experts say that it is more than 500 times as toxic as cyanide:

- Sarin was found in artillery rounds on the battlefield near the Baghdad International Airport in 2003.
- Iraq began producing sarin in 1984 and admitted to possessing 790 tons in 1995. Saddam Hussein used it on the Kurds in northern Iraq during the 1987-88 campaign known as the Anfal.
- The Aum Shinrikyo Japanese doomsday cult used a low-lethality batch of sarin in 1995 in a terrorist attack on the Tokyo subway system that killed 12 and sent more than 5,000 people to hospitals. They left punc-tured packages of liquid sarin in subway cars and stations, which gave officials time to seal off the affected areas. If purer sarin had been released, particularly as an aerosol, the attack might have been much worse. A year earlier, the cult killed seven people in a sarin-gas attack in the central Japanese city of Matsumoto.

VX
VX is the deadliest nerve agent ever created. VX can kill by severely disrupting the nervous system. It is unlikely that terrorists could use VX, as synthesizing it is complicated and extremely dangerous. It requires the use of toxic and corrosive chemicals and high temperatures in a sophisti-cated chemical laboratory:

- The Japanese doomsday cult Aum Shinrikyo, which recruited trained chemists from Japanese universities, managed to synthesize small quantities of VX to use for assassinations. In December 1994 and January 1995, Aum Shinrikyo used VX injections in three assassina-tion attempts on enemies of the cult. One person died.
- Syria has reportedly successfully produced VX or a similar agent and tested missile warheads armed with VX.

- Although not verified, it is reported that Saddam Hussein used VX against Iranian forces in both the 1980-88 Iran-Iraq War and in the 1988 chemical attack on Iraqi Kurds in the town of Halabja. That massacre reportedly killed 5,000 people and created serious health problems for thousands more.

Terrorists lacking access to trained organic chemists might be more likely to steal munitions containing VX from a poorly guarded chemical-weapons depot in a country such as Russia.

MUSTARD GAS

Mustard gas is a potentially deadly chemical agent that attacks the skin and eyes. It is one of the best known and most potent of chemical weapons and causes severe blisters and, if inhaled, damages the lungs and other organs. It is usually disabling—sometimes gruesomely so—but not fatal: [22]

- Unconfirmed reports indicate that groups linked to the al Qaeda network tried to obtain the ingredients to make mustard gas in Afghan labs.
- Saddam Hussein used mustard gas on Kurds in northern Iraq during the 1987-1988 campaign known as the Anfal. The worst attack occurred in March 1988 in the Kurdish village of Halabja, where a combination of chemical agents including mustard gas, sarin, and VX killed 5,000 people and left 65,000 others facing severe skin and respiratory diseases, abnormal rates of cancer and birth defects, and a devastated environment.

Mustard gas is less likely to kill large numbers of people than such nerve agents as sarin and VX. However, depending on the level of exposure, it could leave victims with injuries more lasting than those from other nerve gases.

The Biological Weapons Threat

According to the Centers for Disease Control and Prevention, the bacteria that cause plague and tularemia, the toxin that causes botulism, and hemorrhagic-fever viruses [HFVs] such as ebola and Marburg—along with the anthrax bacteria and the smallpox virus—pose the greatest hazard to public health based on their death rates, ease of dissemination and transmission, and the potential to inspire public panic. [23]

Biological weapons, which release large quantities of living, disease-causing microorganisms, have extraordinary lethal potential. These are organisms or toxins that can cause deadly disease in people, livestock, and crops. There are different types, including bacteria, viruses, and toxins, and are often colorless and odorless. The size of these agents—5 microns, or less than one-fifth the width of human hair—means that they can remain airborne for hours and up to a day or more. They can bypass the filtering mechanism in a human's upper-respiratory system and enter the lungs and bloodstream. **Like chemical weapons, biological weapons are relatively easy to manufacture, requiring straightforward technical skills, basic equipment, and a seed stock of pathogenic microorganisms.** Biological weapons are especially dangerous because we may not know immediately that we have been attacked, allowing an infectious agent time to spread. Moreover, they can serve as a means of attack against humans as well as livestock and crops, inflicting casualties as well as economic damage. **Biological weapons are potentially the most dangerous weapons in the world and the most difficult to protect against.**

ANTHRAX

Anthrax is not easy to acquire, make, or deploy. It is an acute infectious disease caused by the spore-forming bacterium *Bacillus anthracis.* Anthrax most commonly occurs in wild and domestic cattle, sheep, goats, camels, antelopes, and other herbivores. It can also occur in humans when they are exposed to infected animals or tissue from infected animals:

- In March 2002 the U.S. military reported the discovery of a partly constructed laboratory in Afghanistan, in which al Qaeda may have planned to develop anthrax and other biological assets.
- In October and November of 2001, anthrax was distributed in envelopes through the U.S. mail system, causing 5 deaths and 21 injuries.
- The Japanese cult Aum Shinrikyo tried to spread anthrax from its Tokyo office building in 1993 and failed dismally.

SMALLPOX

Smallpox can spread like the common cold through person-to-person contact and through the air. It is one of the most devastating diseases known to mankind and has killed an estimated 300 to 500 million people in the twentieth century alone. As recently as 1967, 15 million contracted the disease, of which 2 million died. The World Health Organization officially declared smallpox eradicated in 1979. The last known human case of smallpox contracted naturally was in Somalia in

1977, and the very last known human case—caused by a laboratory accident in Birmingham, England—occurred in 1978.

In America in 1763 there was fighting between the British and the Indians for lands west of the Appalachians. The British general ordered the distribution of blankets from the smallpox hospital to the Indian chiefs with whom they were negotiating. An epidemic soon spread and decimated the Indian population. [24]

The world community cannot assume that smallpox viruses are not available to rogue nations or terrorist groups. Samples of the virus remain in labs in the Centers of Disease Control and Prevention [CDC] in Atlanta and in labs in Russia. [25]

BOTULISM

Botulism is a potentially fatal but noncontagious disease marked by muscle paralysis. It is caused by exposure to nerve toxin, the single most toxic substance known to science. Its extraordinary potency has made it one of the most widely researched bioweapons: a single gram of botulinum toxin would theoretically be enough to kill more than a million people if a lethal dose were administered to each person individually. *Clostridium botulinum*, the bacterium that produces botulinum toxin, can be found naturally in soil and is sometimes present in undercooked food or improperly canned goods. Botulinum toxin is also the first biological toxin to be approved as a medical treatment: it is used to treat neuromuscular disorders, lower back pain, and cerebral palsy, and, marketed as Botox, is injected to temporarily eliminate wrinkles by paralyzing the facial muscles.

PLAGUE

Plague is a lethal illness caused by the bacterium *Yersinia pestis*. It takes two main forms:

- Bubonic plague, the more common and less deadly variant, is carried by rodents and transmitted to humans via flea bites. It cannot spread from person to person. This form caused the Black Death that devastated China, the Middle East, and Europe in the fourteenth century, killing a larger proportion of the world population than any single war or epidemic since.

- Pneumonic plague, which infects the lungs, travels through the air and is highly contagious. It's also rarer and more lethal than bubonic plague. If those infected do not receive treatment, their mortality rate can approach 100 percent.

TULAREMIA

Tularemia is one of the world's most contagious diseases. Also known as rabbit fever, it is caused by *Francisella tularensis*, a rare bacterium carried by small mammals including rabbits and squirrels. Humans typically contract the disease from contact with the tissues or body fluids of infected animals or from the bites of infected insects. Tularemia cannot spread from person to person. Experts say that if tularemia bacterium were used as a weapon, it would probably be aerosolized, thereby causing an especially serious inhaled form of the disease. Although tularemia is less lethal than some other agents, death rates for those infected with the inhaled form can still climb as high as 30 to 60 percent if left untreated.

HEMORRHAGIC-FEVER VIRUSES

Hemorrhagic-fever viruses are a deadly and gruesome class of viruses that produce high fever and leakage from blood vessels, ultimately causing bleeding from internal organs as well as the eyes, ears, nose, and mouth. Animals in Africa, Asia, the Middle East, and South America carry HFVs. They can be transmitted from person to person via contact with blood or other body fluids. There are four HFV families of which filoviruses, including the notorious ebola and marburg viruses, and arenaviruses, such as the lassa virus, are considered the most serious bioterror threats. There is no treatment for ebola, and up to 90 percent of those who contract the disease die.

The Dirty-Bomb Threat

Radiological weapons, or "dirty bombs," combine radioactive material with conventional explosives. Such a bomb could produce instant hysteria around the world. The devise is designed to injure or kill by creating a zone of intense radiation that could extend several city blocks. People in the immediate area would be killed, and radioactive material would be dispersed into the air and reduced to relatively low concentrations. Such bombs could be miniature devices or as big as a truck bomb:

- According to a UN report, Iraq tested a one-ton radiological bomb in 1987 but gave up on the idea because the radiation levels it generated were not deadly enough.
- There is information that Osama bin Laden attempted to purchase several nuclear "suitcase" bombs from contacts in Chechnya and Kazakhstan and that about 100 of these dirty bombs disappeared from Russian stockpiles in the early nineties. A former U. S. intelligence officer believed at the time that several were stored in an

al Qaeda complex in Kandahar, Afghanistan. However, there are no reports that any have been found in al Qaeda facilities captured and searched.

- In January 2003, British officials found documents in the Afghan city of Heart that led them to conclude that al Qaeda had successfully built a small dirty bomb. During this same period, British authorities reportedly disrupted a plot to use the poison ricin against personnel in England. [26]

- In late December 2003, the *Washington Post* reported that Homeland Security officials worried that al Qaeda would detonate a dirty bomb during New Year's Eve celebrations at college-football bowl games. According to government officials, the Department of Energy sent scores of undercover nuclear scientists with radiation-detection equipment to five major U.S. cities to look for such bombs.

The use of nuclear material with a conventional explosive charge to introduce a contamination scenario is very real, as is the use of biological and chemical agents: [27]

- Easily introduced into the environment through a conventional explosion, special nuclear materials could contaminate and deny access to business areas and financial districts for several years.

- Lethal chemical and biological agents released into the center city or residential neighborhoods could take the lives of thousands and render entire regions of the country useless for months.

- Pound for pound, poison gas and such deadly germs as anthrax, ricin, and sarin can have the same mass-killing power as a nuclear bomb. During the Gulf War, Iraq got as far as filling warheads with deadly germs such as sarin and cancer-causing aflatoxin. At the end of the war, hundreds of tons of biochemical arsenal were destroyed by United Nations observers [28], while 150 tons remain unaccounted for to this day. Most recently on the Iraq battlefield, several artillery rounds containing sarin were fired at U.S. military forces.

The relative ease of constructing a dirty bomb makes it a particularly worrisome threat. Even so, expertise matters. Not all dirty bombs are equally dangerous: the cruder the weapons, the less damage caused. We don't know if terrorists could handle and detonate high-grade radioactive material without fatally injuring themselves first. Depending on the sophistication of the bomb built, wind conditions, and the speed with which the area of the attack was evacuated, the number of deaths and injuries from a dirty-bomb explosion might not be substantially greater than from a conventional bomb. However, panic over potential radioactivity and the chaos of evacuation measures could snarl a city. Moreover, the area struck would be off-limits for a least several months—possibly years—during cleanup efforts, which could paralyze a local economy and reinforce public fears about being near a radioactive area.

The Nuclear-Weapons Threat

Nuclear weapons have enormous destructive potential. Terrorists who seek to develop a nuclear weapon must overcome two formidable challenges. *First,* acquiring or refining a sufficient quantity of fissile material is very difficult, though not impossible. *Second,* manufacturing a workable weapon requires a very high degree of technical capability—though terrorists could feasibly assemble the simplest type of nuclear device. To get around these significant though not insurmountable challenges, terrorists could seek to steal or purchase a nuclear weapon:

- As early as 1993 al Qaeda agents tried repeatedly and without success to purchase highly enriched uranium in Africa, Europe, Russia, and in particular from Pakistani scientists and a nuclear-power plant in Bulgaria. In November 2001 Osama bin Laden announced that he had obtained a nuclear weapon, but U.S. intelligence officials dismissed his claims. Documents recently discovered in Afghanistan also describe al Qaeda's nuclear aspirations.

- In 1981 Saddam Hussein's drive to obtain nuclear weapons was set back when Israeli jets destroyed his Osiraq nuclear reactor outside Baghdad. When U.N. inspectors dismantled a revitalized Iraqi nuclear-weapons program in 1991, they found that Iraq was probably within two to three years of having enough highly enriched uranium to build a bomb.

The Most Dangerous Terrorist Group to the World Community and the United States

While many terrorist groups currently exist throughout the world, the most notorious and feared terrorist group of our time is al Qaeda, headed by Osama bin Laden. Al Qaeda was established in the late 1980s to bring together Arabs who fought in Afghanistan against the Soviet

Union. Its current goal is to establish a pan-Islamic caliphate throughout the world to overthrow regimes it deems "non-Islamic" and expel Westerners and non-Muslims from Islamic countries, particularly Saudi Arabia. In February 1998, Osama bin Laden issued a statement under the banner of the "World Islamic Front for Jihad Against Jews and Crusaders," saying it was the duty of all Muslims to kill U.S. citizens—civilian or military—and their allies everywhere.

Al Qaeda is part of a dangerous trend toward sophisticated terrorist networks spread across many countries, linked together by information technology, enabled by far-flung networks of financial and ideological supporters, and operating in a highly decentralized manner. Unlike traditional adversaries, these terrorist networks have no single "center of gravity" whose destruction would entail the defeat of the entire organization. While Afghanistan has been incapacitated as a safe haven for al Qaeda, unrest in politically unstable regions continues to create an environment conducive to the operations and sanctuary of terrorist groups. Particularly, the Taliban is developing a strategy to return to Afghanistan once again. Homegrown extremists, inspired by al Qaeda, are at work in many nations. While thousands of al Qaeda operatives have been captured or killed, many remain at large, including leaders working to reconstitute the organization and continue operations. Their operatives and cells continue to plan attacks against high-profile landmarks and critical infrastructure targets in Europe, the Middle East, Africa, Southeast Asia, and the United States.

Al Qaeda is only part of a broader threat that includes other international terrorist organizations with the will and capability to attack the United States. The most dangerous of these groups are associated with religious extremist movements in the Middle East and South Asia. Hezbollah was formed in 1982 in response to the Israeli invasion of Lebanon and is dedicated to liberating Jerusalem, ultimately eliminating Israel. It has formally advocated the ultimate establishment of Islamic rule in Lebanon. Until September 11, Hezbollah was responsible for more American deaths than all other terrorist groups combined, including those killed in the 1983 bombing of the U.S. Marine Corps barracks in Lebanon. Hezbollah has never carried out an attack within the United States but could do so if the situation in the Middle East worsens or the group feels threatened by U.S. actions.

Other terrorist groups, from Hamas to the Real Irish Republican Army, have supporters in the United States. To date, most of these groups have largely limited their activities in the United States to fundraising, recruiting, and low-level intelligence, but many are capable of carrying out terrorist acts within our borders. Al Qaeda and the groups in its network have taken their terrorist war home to more than 60 countries, including Afghanistan, Algeria, Bali, Bosnia, Britain, Chechnya, China, Egypt, India, Iraq, France, Germany, Malaysia, Pakistan, the Philippines, Saudi Arabia, Somalia, Spain, Sudan, Syria, the United States, Uzbekistan, Yemen, and other countries. The greatest impact during the decade of the nineties was felt in Algeria where 85,000 people were killed during horrific massacres carried out by the Islamic Group. This organization was finally driven out of Algeria, and its members joined al Qaeda. **It is estimated that over 70,000 hard-core terrorists live in more than 60 countries and are recruiting from their own nations and neighborhoods.** Thousands have been brought to camps in several Middle Eastern and North African countries, where they have been trained in the tactics of terror. They then return back to their homes or are sent to hide in countries such as England, Spain, Germany, Canada, the Philippines, and the United States until they are called upon to execute their missions. [29]

Osama bin Laden's declared "Holy War" against free countries of the world has led to the exposure of a terrorist conspiracy against the United States government and United States interests worldwide. Al Qaeda has issued statements under the banner of "The World Islamic Front for Jihad Against the Jews and Crusaders" warning that it will attack United States and Israeli targets. It has threatened to force the closure of American and Israeli businesses and economics, bring down their aircraft, prevent the safe passage of their ships, and occupy their embassies. Intelligence developed in late October 2000 uncovered a plan to attack United States interests in Turkey, Bahrain, and Qatar. Confidential intelligence sources have advised that the Islamic militant terrorist groups in the Middle East will eventually attempt to strike targets in the United States. Intelligence experts see the concept of American targets as a natural progression of the hatred that exists against the United States and the maniacal and fanatical determination on the part of the fundamentalist terrorist to wreak death and destruction against all perceived enemies of Islam.

In the aftermath of 9/11, more than 7,000 al Qaeda operatives and associates have been arrested in over 100 countries, and over $150 million in financial assets have been seized from terrorist accounts worldwide.

WESTERN SOCIAL VALUES: STRENGTHS, WEAKNESSES, FEARS, ASPIRATIONS

Safeguarding Western Values

The nature of free societies greatly enables terrorist operations and tactics and at the same time hinders a nation's ability to predict, prevent, or mitigate the effects of terrorist acts. [30] Attacks in America and Britain and the infiltration of terrorist cells in Germany are recent examples. Western values and influence are attacked in the East as well—in Iraq, Egypt, Bali, and Japan, to name a few countries. Terrorists use fundamentalist religion, racism, and ignorance in their quest to subjugate their people, oppress free thought, and halt any outside influences from reaching their societies. Today the terrorist threat to Western civilizations takes many forms, has many hiding places, and is often invisible. Terrorists find Western ideals a threat to their way of life, their control of ideas, and the power they hold over their people.

In the new century we have witnessed firsthand the emergence of threats from a wide range of terrorist groups. The age of international terrorism has converged and exploded into an era of changing risks and escalated fears that have created a whirlwind of change at all levels of industry and government throughout the world.

Safeguarding American Values

The tragic events of September 11, 2001, and other global attacks have become the reference point of a new era to many government and enterprise executives and shareholders. Over the past six decades America has sought to protect its sovereignty and independence through a strategy of strong global political, economic, military, and cultural presence and engagement. In doing so it has helped many other countries and peoples advance along the path of democracy, open markets, individual liberty, and peace with their neighbors. Yet there are those in different parts of the world who are threatened by this western influence. They use terrorism against European nations, the United States, and their allies because they cannot defeat these countries through traditional battle. Their strategy has therefore become: to take advantage of western freedom and openness by secretly inserting terrorists into a country to attack the homeland, to attack interests abroad, and to assault Western allies. Terrorists have been successful in each of these endeavors.

DEMOCRACY: The American system is a form of government in which state governments share power with federal institutions. It preserves a series of checks and balances, a distribution of state and federal rights, and an affirmation of the rights and freedoms of individuals. The system requires that all actions adhere to the rule of law. It relies on the stability and continuity of the government to exercise constitutionally prescribed procedures and powers. America has more than 87,000 different jurisdictions.

- America has a vision of openness and public debate.
- America chooses lawful change and civil disagreement.
- Terrorists seek to impose a grim vision in which dissent is crushed, and every man and woman must think and live in colorless conformity.
- Terrorists seek to create coercion, subversion, and chaos in overthrowing governments and destabilizing entire regions.

LIBERTY: The freedom of expression, religion, and movement, property rights, and freedom from unlawful discrimination are rights guaranteed to all Americans.

- America believes that every person has the right to speak his or her mind.
- America respects people of all faiths and welcomes the free practice of religion.
- For terrorists, free expression is a threat rather than an advantage to society.
- Terrorists want to dictate how people think and worship. Terrorists may hide behind a peaceful faith but at the same time murder innocent men, women and children.

CULTURE: America is an open, pluralistic, diverse society that engages in dialogue rather than the dogmatic enforcement of any one set of values or ideas. The American culture also is characterized by compassion and strong civic engagement.

- America values education.
- Terrorists do not believe women should be educated, have health care, or be able to leave their homes.

SECURITY: America enjoys great security from external threats, has no hostile powers adjacent to its borders and is insulated from attack by two vast oceans. Externally, the United States seeks to protect its interests through a strategy of global presence and engagement. Internally, it relies primarily on law enforcement and the justice system to provide for domestic tranquility.

- America continually strives to uphold freedom, democracy, and equality. Americans are a strong and determined people with a steady resolve to defend basic American values.
- Terrorists believe that free societies are essentially corrupt and decadent, and with a few hard blows will collapse in weakness and panic.

ECONOMY: The U.S. economy is a free market system predicated on private ownership of property and freedom of contract, with limited government intervention. People work for their individual prosperity and government ensures that all have equal access to the marketplace.

- America values the dignity of life.
- Terrorists choose to destroy dignity.

Exhibit 2.3 America's values in contrast with tyranny's oppression

- An estimated 284.8 million people lived in the United States on July 1, 2001.

  Source: U.S. Department of Commerce

- 54.2% of the nation's population lives in ten states–three in the Northeast, three in the Midwest, three in the South, and one in the West.

- The average population density within the United States is 79.2 people per square mile.

- The average population density in coastal metropolitan areas is 320.2 people per square mile.

- Over 225 million Americans live in metropolitan areas.

- Nearly 85 million Americans live in metropolitan areas of 5 million people or more.

- Each year, the United States admits 500 million people, including 330 million non-citizens, through our borders.

  Source: 2000 Census

- Over 4 million people were processed through security at the last Olympics, over 85,000 at the last Super Bowl, and approximately 20,000 each at the Republican and Democratic National Conventions.

- In 2007, the country's urban population will surpass the rural population for the first time.

  Source: U.S. Secret Service

Exhibit 2.4 Demography, *The American Population, National Strategy for Homeland Security–July 2002*

After gaining a comprehensive understanding of the basics of terrorism, its psychological impact on American culture is pertinent to performing any homeland-security assessment. Of particular significance to the current international extremist threat is contrasting American values to the objectives of fundamental terrorism.

America is Inherently Vulnerable
The American Population

The U.S. population is large, diverse, and highly mobile, allowing terrorists to live and travel among us and hide within our midst until they are ready to attack our homes, places of business, recreation areas, and governance. [31]

Much of America lives in densely populated urban areas, making major cities conspicuous targets. Americans congregate at schools, sporting arenas, malls, concert halls, office buildings, high-rise residences, and places of worship—presenting targets with the potential for many casualties.

The Importance of the National Critical Infrastructure Sectors

Infrastructure security is critically important to the nation's economy, but until 9/11 it had largely been ignored. According to Michael A. Gips, the National Critical Infrastructure was established by the President's Commission on Critical Infrastructure Protection (PCCIP) in 1999 and initially included water services, electric power, gas and oil, transportation, banking and finance, emergency services, and government continuity. After the devastating events of September 11, 2001, two executive orders added information systems, facilities that use nuclear material, agriculture, and special events of national significance such as the Super Bowl, World Series, Olympics, and national conventions, to name a few. The National Strategy for Homeland Security, issued in July 2002, added the chemical industry, the defense industrial base, and the postal and shipping sectors. In February 2003 the National Strategy for the Physical Protection of Critical Infrastructures and Key Assets identified national monuments, symbols, and icons; dams and nuclear-power plants; and large gathering sites such as stadiums, shopping malls, amusement parks, recreation areas, and office buildings as key assets. [32]

The National Critical Infrastructure Sectors provide the foundation for our national security, governance, economic vitality, and way of life. Their continued reliability, robustness, and resiliency create a sense of confidence and form an important part of our national identity and purpose. They frame our daily lives and enable us to enjoy one of the highest overall standards of living in the world. [33]

While many use the term "national critical infrastructure," few really understand its makeup and significance to the national economy. **The companies that own and operate these enterprises provide more than 85 percent of America's economic strength.** The critical infrastructure sectors are particularly important because of the products and services they provide and because they are complex systems.

Because the American economy is dependent on networks of physical infrastructure such as energy and transportation systems and virtual networks such as information systems, an attack against one or more of these may disrupt an entire system and cause significant damage to the nation. Defining and bisecting the infrastructure is a major step of the security-assessment process to make us more secure from future terrorist attacks as well as reducing our vulnerability to natural disasters, organized crime, and computer hackers.

It's important to recognize that the U.S. government does not have a monopoly on insight and ingenuity. The establishment of the Commission of Critical Infrastructure Protection and the Critical Infrastructure Assurance Office relies on a voluntary public-private partnership involving corporate and nongovernment organizations to offer advice and expertise to the government. The leadership of these sectors also possesses a wealth of information about both vulnerabilities and the most practical protective measures to save lives and prevent social and economic disruption.

Consultants who perform security assessments provide a valuable first link in supporting the long-term U.S. strategies to defeat terrorism and ensure the longevity of our values and way of life. By understanding the critical-infrastructure work environment, its mission and services, contribution to the community and the nation, and the intentions and capabilities of terrorist groups to inflict harm, security assessments provide the basic foundation for protective measures.

The Protection Challenge

The Patriot Act defines the nation's critical infrastructures as "systems and assets, whether physical or virtual, so vital to the United States that the incapacity or destruction of such systems and assets would have a debilitating impact on security, national economic security, national public health or safety, or any combination of those matters."

America's critical infrastructure encompasses a large number of sectors possessing assets, systems, and functions vital to the economy, public health and safety, national morale, governance, and national security. The list of critical infrastructures will likely continue to evolve over time.

Given the immense size and scope of theses enterprises, the protection strategy to be employed must be based on a thorough understanding of industry complexities.

There are 50 states, 4 territories, and 87,000 local jurisdictions that comprise this nation, and each has an important and unique role to play in the protection of critical infrastructure and key assets.

While the federal, state, and local governments provide guidance and information to key industry decision-makers on the nature of the threat and assist in preparing for a catastrophic terrorist attack, it is industry with the support of security consultants that must plan, develop, and implement prudent and reasonable security programs to respond to the wide range and levels of threats.

Agriculture and Food	1.9 million farms 87,000 food-processing plants 1 million restaurants & food service outlets $991.5 billion revenues; $140 billion exports
Municipal Water System	160,000 public water systems 1,800 federal reservoirs 19,500 municipal wastewater facilities 800,000 miles of wastewater sewer lines 300 million people served Includes 75% of U.S. population
Public Health	5,800 registered hospitals
Emergency Services	87,000 U.S. localities
Defense Industrial Base	250,000 firms in 215 distinct industries
Telecommunications	2 billion miles of cable 20,000 switches, access tandems
ENERGY	
Electricity	2,800 power plants
Oil and Natural Gas	Produces 20% of world's supply 300,000 producing sites; 7,500 bulk stations 153 refineries; 4,000 off-shore platforms 600 natural gas processing plants 1,400 product terminals 160,000 miles of crude oil pipelines 278,000 miles of natural gas pipelines 1,119,00 miles of distribution pipelines
TRANSPORTATION	
Aviation	19,000 general aviation airports 450 major commercial airports 10,000 FAA facilities
Border	5,525 miles of Canadian border 1,989 miles of Mexican border $1.35 trillion in imports $1 trillion in exports
Passenger Rail and Railroads	120,000 miles of major railroads 30 million tons of freight, daily 1.1 billion passenger trips, daily (continued on next page)

Exhibit 2.5 The protection challenge

Highways, Trucking, and Busing	590,000 highway bridges
Pipelines	2 million miles of pipelines
Maritime	95,000 miles of shoreline and waterways
	3.4 million square mile economic zone
	361 seaports; 16 million cargo containers
	50% of U.S. imports arrive in containers
Mass Transit	500 major urban public transit operations
Banking and Finance	26,000 FDIC insured institutions
Chemical Industry and Hazardous Materials	60,000 chemical plants
	Nation's top exporter
	$400 billion revenue
Postal and Shipping	127 million delivery sites
	$200 billion revenue
KEY ASSETS	
National Monuments and Icons	5,800 historic buildings
Nuclear Power Plants	104 commercial nuclear power plants
Dams	80,000 dams
Government Facilities	3,000 government owned/operated facilities
Commercial Assets	460 skyscrapers

Source: National Strategy for Homeland Security – July 2002 National Strategy for the Physical Protection of Critical Infrastructures & Key Assets – February 2003.

Exhibit 2.5 The protection challenge *(continued)*

A review of the National Security Strategy for Homeland Security and the National Strategy for the Security of the United States of America provides valuable background information to the diversified security consultant, the security practitioner with multidiscipline responsibilities, and the academic to develop operational and security requirements and protocols for any infrastructure sector and its respective assets.

The Importance of Key Assets
Key assets and high-profile events are individual targets whose attack—in the worst-case scenarios—could result in not only large-scale human casualties and property destruction but also profound damage to our

national prestige, morale, and confidence. These key assets include symbols or historical attractions such as prominent national, state, or local monuments and icons; schools; courthouses; museums; entertainment parks; national conventions and celebrations; and sporting events such as the Olympics, Super Bowl, and World Series. They represent the natural and cultural grandeur of our country. They celebrate American ideals and way of life and therefore present attractive targets for terrorists, particularly when coupled with high-profile events and celebratory activities that bring together significant numbers of people. Individually key assets do not endanger vital systems. However, a successful strike against such targets may result in a significant loss of life and property in addition to long-term adverse public health and safety consequences. [34]

Conclusion

In tandem with understanding western and American culture and their social values, equal importance needs to be placed on understanding terrorism, its psychological impact on a society, the specific terrorist groups in place today, and the changing face of terrorism. Collectively this knowledge is vital to conducting successful security assessments in today's dynamic threat environment, and in particular to formulating workable and cost-effective solutions to maintain the strategic integrity of America's economy and security for future generations.

The citizens of the United States face increased threats of harm since the terrorist attacks of September 11, 2001, and the delivery of anthrax-contaminated letters later that year. These threats related not just to individuals but also to the country's vital institutions, systems, and infrastructure. The possibility that terrorist groups may introduce a new type of terror by employing a mix of conventional explosives with radioactive particles or the use of chemical or biological contaminants is very real. Such a threat could produce horror on a scale never before imagined. [35] The events of September 11, 2001, and the subsequent passage of the Bioterrorism Act of 2002 have pushed security to the top of the agenda of many executives. Although great strides have been taken by many business owners to protect assets, much work remains to be done. While the American infrastructure has generally been immune to terrorist attacks in the past, post-9/11 intelligence indicates now more than ever that this gap is quickly diminishing. Documents captured in raids within the United States, Asia, Europe, and the Middle East and other intelligence information analyzed reveal that al Qaeda and its affiliates as well as other terrorist groups have both the capability and desire to attack western interests.

Security guards and employees across the country have observed and continue to report to local and federal authorities suspicious people conducting preoperational surveillance and taking photographs of facilities and surrounding areas. Terrorists continue to scheme, organize, and carry out preplanning activities as individuals or in groups. Attack plans are very detailed, frequently rehearsed, and at times able to be carried out at a moment's notice. [36]

Success in guarding against these threats will be achieved and maintained by those entities willing to implement security strategies, tools, and methods developed to service a world under a state of heightened alert. Insurance companies are increasingly forcing substantial management and operational changes on commercial enterprises with inefficient security programs. Another catastrophic event may need to occur before entities that have not advanced their security posture understand the lessons others have learned.

We now know for certain that the American infrastructure vulnerability continues to gain the increased attention of terrorists, and targets of the future will be cities, utility companies, technology companies, and high-profile corporations. Secondary explosions—attacks against our critical infrastructure assets—have the potential to reach volcanic proportions. Against this prognosis, there is no such thing as absolute safety, and the American infrastructure will continue to remain a preferred target for a number of international and domestic terrorist groups.

NOTES

1. Terrorism: Questions and Answers: The Markle Foundation and the Council on Foreign Relations (April 2004)

2. Bruce Hoffman, "The Logic of Suicide Terrorism," *The Atlantic Monthly*, June 2003

3. Peter S. Probst, "How Can We Tackle Tomorrow's Terrorists?", *Security Management* (January 1996)

4. Robert M. Jenkins, "The Islamic Connection," *Security Management* (July 1993)

5. Frank J. Gaffney, Jr., Director of The Center for Security Policy, quoted in "The Terrorists Among US" by Nathan M. Adams, *Reader's Digest Special Report* (December 1993)

6. ibid.

7. *INTERSEC*, Vol. 13 (May 2004)

8. ibid.

9. *The National Strategy for Homeland Security* (July 2002)

10. Peter S. Probst, "How Can We Tackle Tomorrow's Terrorists?", Security Management (January 1996)

11. *INTERSEC*, Vol. 13 (May 2004)

12. Nathan M. Adams, "The Terrorist"

13. *Patterns of Global Terrorism 1999*, U.S. Department of State (April 2000)

14. Nathan M. Adams, "The Terrorist"

15. *Patterns of Global Terrorism 2003*, U.S. Department of State (April 2004)

16. *The National Strategy for Homeland Security* (July 2002)

17. 17. Terrorism: Questions and Answers: The Markle Foundation and the Council of Foreign Relations (June 2004)

18. *Patterns of Global Terrorism 2003* U.S. Department of State (April 2004)

19. *Terrorism and National Security: Issues and Trends,* Congressional Research Services, Library of Congress (October 2, 2003)

20. *Patterns of Global Terrorism 2003* U.S. Department of State (April 2004)

21. Council of Foreign Relations (June 2004)

22. ibid.

23. ibid.

24. Howard Zinn, *A People's History of the United States,* (New York: Harper Collins, 2003)

25. Centers of Disease Control and Prevention

26. *Terrorism and National Security: Issues and Trends* U.S. Department of State (2003)

27. ibid.

28. Albert E. Snell and Edward J. Keusenhathen, "Mass Destruction Weapons Enter Arsenal of Terrorist," *National Defense* (January 1995)

29. *Patterns of Global Terrorism 2003*

30. *The National Strategy for Homeland Security* (July 2002)

31. ibid.

32. *National Strategy for the Physical Protection of the Critical Infrastructure and Key Assets* (February 2003)

33. ibid.

34. ibid.

35. *The National Strategy for Homeland Security* (July 2002)

36. *Terrorism and National Security: Issues and Trends* U.S. Department of State (2003)

Part II

UNDERSTANDING SECURITY ASSESSMENTS

Chapter 3 introduces what, why, and when a security assessment should be performed. Work completed by others is also highlighted to give the reader an overview of various approaches.

Chapter 4 introduces the proven S^3E Security Assessment Methodology and guides the reader through a step-by-step discussion.

Chapters 5 to 9 guide the reader through a comprehensive description of the security-assessment process by defining its purpose and objectives, the behavioral and physical sciences at play, the techniques employed in the process, and the measurement and evaluation tools and standards used to perform an objective risk assessment.

Chapter 10 addresses the importance of preparing a comprehensive and factual security-assessment report.

Chapter 3

THE SECURITY ASSESSMENT: WHAT, WHY, AND WHEN

WHY PERFORM A SECURITY ASSESSMENT?

Security is About Minimizing Risk

It's not a question of "if" but a matter of "when, where, and how." Forget the conviction that it will not happen at your company, on your watch.

The security assessment should produce a professional, candid, independent, and objective analysis of enterprise security vulnerability to measure the effectiveness of existing protective measures, evaluate the current status of the security program, and identify gaps in the security process.

A comprehensive, thorough, and useful security assessment helps executive management determine means and ways to increase the capability and effectiveness of the enterprise to prevent the damage or destruction of its assets and resources or the disruption of operations. It identifies organization or agency strengths and weaknesses and presents recommended mitigation actions to reduce or illuminate security vulnerability. While a desirable goal of the security assessment is to achieve a perfect and absolute risk-free environment, the reality is that this is unattainable, and executive management and stakeholders must agree on what is or is not an acceptable risk.

A comprehensive and thorough assessment not only enables decisive near-term action but guides the rational long-term investment of effort and resources. For example, a comprehensive assessment can help determine whether to invest in permanent, physical "hardening of an asset" or in maintaining a reserve of personnel and equipment that can meet a temporary "surge"-protection requirement during periods of heightened security.

The security assessment identifies specific initiatives to drive protection priorities. More importantly, it establishes a foundation for building and fostering a cooperative environment in which all elements of an enterprise can carry out their respective protection responsibilities more effectively and efficiently.

The definition of what constituted adequate security changed on September 11, 2001, and so did the definition and makeup of a security assessment. History has a tendency of repeating itself, and one of the benefits of a well-designed and -executed security assessment is to offer a "lessons-learned strategy" to bridge the past to the present and hopefully the present to the future.

For example, in reviewing cases dating back three decades, the seasoned security consultant and security analyst can show that corporate America has often based its security strategy on "work politics," on an incorrect understanding of the threat, and on the misapplication of adopted protective measures. The results of these efforts more often than not have led to a false sense of security and expensive liability suits. Until recently enterprises have conducted security assessments with little thought or adequate emergency preparedness for meeting the current dynamic threat environment.

Today the adversary is highly skilled, raising the bar of vulnerability to new heights. In today's corporate world security threats are so diverse that they pose one of the most serious challenges facing executive management, making the corporate security organization a dynamic entity constantly responding to changing priorities and competitive demands. Traditional law-enforcement investigating techniques have served America well in the past, but they do not work adequately in today's threat environment. As a nation we have made significant advancements to meet the changing threat. As such, the conduct of security assessments must also take on a new, dynamic structure and meaning. No longer can we be content with focusing our security assessments strictly on cameras that seldom monitor some distant perimeter. Other operational capabilities are in play that cannot be ignored.

Risk concerns an event that has potential negative impact. It represents the possibility that such an event will occur and adversely affect an enterprise's activities and operations, as well as the achievement of its mission and strategic objectives. It encompasses identifying vulnerabilities, threats, and consequences. Enterprises typically define risks as events related to terrorism, criminal activity, natural disasters, and other emergencies with a quantitative and/or qualitative measurement involving such variables as a predetermined level of financial damage or loss to property and/or potential for injury or loss of life. Ranges and levels of significant risk include:

- Vandalism
- Breaking and entering
- Robbery
- Employee theft
- Workplace violence
- Arson
- Loss of intellectual property
- Insider financial fraud and deception
- Ethical breaches
- Disruption of business operations
- Natural disaster
- Tampering with or destroying equipment
- The diffused nature of international and domestic terrorism in its various disguises

Corporate America faces new challenges posed by the unexpected, such as fears of a new terrorist attack, a new outbreak of anthrax, the use of ricin or sarin, water contamination, and the impact of other unpredictable events. The vulnerability of America's critical infrastructure never has been greater or more complex than it is today, and the risks are high. Without a doubt, traditional crime pales in comparison to the devastating cultural, emotional, and financial effects of terrorist attacks. But these criminal activities have also had great influence on corporate executives, ensuring that preventable foreseeable events don't wreak havoc on business goals and objectives.

The risk of vulnerability in a changing environment will continue to increase. **No two enterprises are the same, and neither are their security requirements.** While the range and level of risk may differ from one critical infrastructure sector to another, from one enterprise to another and between industry and government, the need for prudent and responsible action remains the same distinct challenge. Concluding thorough and comprehensive security assessments translates threat information into protective measures.

These threats are rupturing the very fiber of the American economy. They are eroding consumer confidence. For instance, the short-term consequences of the 9/11 horrendous loss of life, destruction of property, and apprehension in the stock market are well documented. According to a report issued by Congress's Joint Economic Committee, several studies have attempted to put a price tag on it. The immediate loss of "human and nonhuman capital" has been estimated at $20 to $60 billion. Short-term lost economic output has been estimated at $47 billion, with another $1.7 trillion in lost stock-market wealth. The long-term costs of 9/11 are inconclusive, with no realistic forecast. The full costs of terrorism including the effects of biological, chemical, or even cyber attacks are almost impossible to estimate.

As these threats continue and new threats emerge, vulnerabilities in the changing infrastructure environment will continue to increase. No doubt, expenditures to combat terrorism across national critical infrastructure assets will continue to divert more funds from other activities. Conservative analysts place the cost at $400 billion annually. Corporate profits are threatened, and no doubt terrorism will continue to take its toll.

The Changing Threat Environment

Since the beginning of the 20th century, terrorism has migrated in both scope and scale, drawing its strength from continuous media coverage and silent supporters. The more familiar form of terrorism first appeared on

July 22, 1968, when the Popular Front for the Liberation of Palestine undertook the first terrorist hijacking of a commercial airplane. Since then terrorism has taken on many faces—among them international terrorism, domestic terrorism, and independent crusaders with a particular cause. Terrorism has impacted governments, industries, and private citizens. It has destroyed, damaged, and disrupted:

- Government and commercial facilities and operations
- Oil pipelines and production capabilities
- Aircraft and sea vessels including air and maritime operations
- Olympic Games and other events
- Railroads and mass-transportation systems
- Telecommunications and transmission lines
- Construction projects
- Private property

Corporate America is Adjusting to the Changing Threat Environment

The attacks on the World Trade Center and the Pentagon on September 11, 2001, and the subsequent anthrax mail attacks have been viewed as the single most defining moment in the world of corporate security. American business CEOs now face challenges more difficult than organized crime and other traditional criminal elements. There are over 130 major extremist groups and terrorist organizations that oppose America in some way. Their new target is corporate America.

Conducting security assessments makes increasing sense. Western nations are awakening to new and emerging dangers potentially involving the world's most destructive weapons.

The knowledge, technology, and materials needed to build weapons of mass destruction are increasingly available all over the world. These capabilities have never been more accessible than now. Terrorists may conceivably steal weapons of mass destruction, weapons-usable fissile materials, or related technology from states with such capabilities, create them with their own resources, or obtain them through connections. Several state sponsors of terrorism already possess or are working to develop weapons of mass destruction and could provide materials or technical support to terrorist groups. Once terrorist groups acquire these weapons and the means to deliver them, the potential consequences could be more devastating than any attack suffered to date. Evaluating potential security risks to workers and the public posed by uncontrolled release of and exposure to toxic or hazardous substances; release of biological, chemical, or radiological weapons; or interruption of drinking-water supplies enhances safety and public-health protection. A comprehensive security assessment helps determine whether to invest in permanent physical hardening, the isolation of facilities, or other preventive and protective measures.

Enterprise fraud, waste, and abuse committed by corporate senior executives also has become a national investigative priority. The passage and implementation of the Sarbanes-Oxley Act of 2002 set into motion corporate governance functions and strict auditing procedures including adding a more complicated layer of a diverse array of responsibilities. Such investigations and audits have prompted an increase in forensic-accounting services, and auditors must now take on a broader scope of responsibility respective to the insider fraud threat. It makes good sense to be diligent, including identifying vulnerabilities against profit loss, corruption, and other white-collar crimes.

A solid business case exists to conduct enterprise-continuance assessments. While most enterprises are not direct targets of terrorist attacks, the greatest threat to many involves collateral damage brought about by a nearby attack involving conventional explosives against other high-risk corporate entities. The consequences of such events and their resultant effects in terms of human injury or death are dire. The partial or total destruction of facilities, the shutdown of operations and loss of enterprise leadership, the disruption of vehicular traffic and mass transportation, the demand on lost utilities, the availability of emergency-services responders, and the closure of streets and establishment of security corridors around a large area would directly affect enterprise operations for an indefinite period. The demand on emergency-service responders to work in harm's way would also impact the community.

Continuance planning increases a corporation's probability of survival. It represents the wise choice among many costly alternatives to guide the long-term investment of effort and resources. Building on the efforts of the security assessment, its results provide direction to enhance the enterprise. In this respect, the security assessment belongs and applies to the total enterprise, not just to the security organization.

WHAT IS THE SCOPE OF A SECURITY ASSESSMENT?

The scope of a security assessment is usually determined by the statement of work issued by the enterprise. Most security assessments are limited in nature for a variety of reasons. For instance, investigative work may already have been performed by the client or another consultant and perceived to be adequate. Budget and time constraints often control the assessment as well.

The security assessment evaluates in-place protective measures for **detection, deterrence, delay, assessment,** and **response** to identify threats and security operational and physical vulnerabilities and to offer solutions to increase the security capabilities and effectiveness of the corporation to combat terrorism and prepare for other potential emergencies.

WHEN SHOULD A SECURITY ASSESSMENT BE PERFORMED?

The security assessment is not a one-time task. It must be performed periodically in order to remain relevant to changes in business objectives and operational processes, technological advances, modifications to facilities, the construction of new structures, or the relocation of operations. A security assessment should also be performed when threat conditions and circumstances significantly change, emerging threats present a clear and present danger, and after a major terrorist attack, natural disaster, or other emergency.

WHICH SECURITY ASSESSMENT MODEL IS BEST?

Numerous security-assessment models exist, which provide a foundation for selecting and implementing actions to reduce the risk associated with current or anticipated threats. The best models are those that focus on performance-based results. James F. Broder, CPP, security consultant, lecturer, and author of *Risk Analysis and the Security Survey,* suggests that the success and creditability of a security assessment depend on four factors:

- What is it that needs to be done?
- What is the performance capability or expectation for accomplishing the security mission?

- What performance-based evaluation criteria are used to measure that level of performance?
- How strongly does senior management support and approve of security at the outset?

Other elements indispensable to a successful security assessment include the following:

- A user-friendly process that enables the enterprise staff to understand its structure and process without special knowledge of security analytical skills.
- A methodology that encourages the management staff to place security high on the agenda and promote security awareness across the enterprise.
- An approach that identifies the areas of greatest vulnerability to the enterprise as a whole, promotes better decision-making across divisional lines, and helps avoid excessive or unnecessary expenditures.

CONCLUSION

A comprehensive security assessment identifies initiatives to drive protective measures. It increases the enterprise's probability of continuance in the market place and provides a direction to prioritize management actions, and the allaction of resources and budget.

Security assessments performed on a periodic basis keep pace with changes in business objectives and practices, threat conditions and circumstances, and market changes.

Chapter 4

A PROVEN SECURITY ASSESSMENT METHODOLOGY

THE SECURITY-ASSESSMENT CHALLENGE

There can be no argument that given the current threats, conducting a security assessment is more dynamic now than ever before. Threats to the nation's infrastructure sectors are much higher and the consequences of loss much greater than those posed to other business enterprises. These targets directly affect the economy and security of the entire country. Some examples include: the events of September 11, 2001, and the subsequent anthrax attacks in October and November of the same year in the U.S.; attacks in Madrid and Malaysia in April 2004; the London bombings in July 2005; and the Tokyo attacks in January and March of 1995.

Many new requirements for assessing security have been introduced since September 11, 2001. These new parameters come from the U.S. and other governments, international organizations such as the International

Civil Aviation Organization [ICAO] and the International Maritime Organization [IMO], and professional institutions such as the American Society of Industrial Security [ASIS] International, American Water Works Association [AWWA], the National American Electric Reliability Council [NAER] for Utilities, the Center for Chemical Process Safety [CCPS], and the National Institute for Standards and Technology [NIST].

An accurate security assessment begins by selecting a proven methodology, but selecting the correct one can be confusing. Government and industry continue to perform extensive research in an attempt to develop specific methodologies appropriate to each infrastructure sector. Several security-assessment models are in existence now with many more on the drawing board. Many of these are excellent tools, but others miss the mark in one or more areas of analysis. The industry is inundated with pertinent information, which can lead to contradictory guidance. At best the data is scattered throughout the industry, forcing many security practitioners to research extensively in order to collect needed data and assemble relevant facts. Aligning disparate assessment models presents a significant challenge to the security practitioner and the security consultant. In many cases they are neither consistent nor comparable, thereby complicating protection planning and resource allocation across the board.

In order to be successful, the design of any methodology must be based on a solid platform. This includes a set of acceptable performance-based measurement and evaluation standards and strong evidence of reliability and credibility. Too many of the available methodologies, including those modified by design and engineering firms, lack sufficient guidance on how to exercise a systematic approach to security problem-solving and do not have definitive performance-based standards and measurement criteria to support judgments made during the decision-making process. Models that are vague and ambiguous tend to be confusing and sometimes frustrating to implement. In addition, many of the various models available for use today are inconsistent in their design, purpose, and application. They lack a uniform platform and must be tailored for a specific use. This often ventures into a major revision of acceptable investigative practices.

Given a clear and distinct platform, the execution of a security assessment should fundamentally be the same irrespective of the infrastructure sector, the asset being protected, or the environment in which the assessment is conducted. Only the investigative methods, techniques, and tools should vary slightly to meet unique requirements or competitive demands. No justification exists to alter the structure of the assessment process itself.

ANALYSIS OF SEVERAL INDUSTRY MODELS

Now more than ever, national security requires that security professionals have at their disposal the benefit of the best assessment tools available to combat terrorism and other criminal activity. However, not every tool is useful, and some tools may even be counterproductive. Particularly suspect are those approaches that have failed in the past to produce acceptable results and those that require consistent modification from one assessment to another in their basic structure.

This can best be depicted by a comparative analysis of several methodologies. In concert two viewpoints must also be made. One is put forth by Michael Goldsmith, Executive Director of NWTC, Inc., who suggests that all methodologies should have specific elements in common, or what is repeatedly referred to here as the assessment platform or foundation. The second viewpoint is the premise of this book—a series of integrated and indispensable components that formulate a comprehensive security assessment.

Each of the models illustrated in the exhibit has particular constraints, either by design or default. Many are generic in nature or limited in scope, only offer elementary guidance to a general audience, and fall short of offering a "best practices" formula. Those that do offer a degree of useful detail are either government or private-industry models with controlled distribution. By design, many of these documents are confidential in nature, making the sharing of information difficult at best. For consultants who provide confidential security services to an array of clientele, they can only access and use these sensitive materials under strict supervision and control and only when under contract for a specific project. Michael Goldsmith suggests that it has been industry practice that his list of common elements is tailored for individual industries. Exhibit 4.1 supports his theory, but other forces come into play. For instance:

- What do you tailor and to what degree?
- How do you tailor to be effective without losing sight of objectivity?
- What do you tailor and why is it necessary?
- What errors, omissions, and misapplications begin to take form?

Little has been written on this complicated transition process, and attempts to tailor a given methodology to a particular industry consistently received mixed reviews among colleagues.

Michael Goldsmith's Suggested Common Elements for All Security Assessment Models	
♦ Characterize Mission and Objectives	♦ Identify Consequences
♦ Assess Likelihood of Event Occurrence	♦ Review Existing Security Measures
♦ Prioritize Critical Assets	♦ Develop Prioritized Action Plan
Examples of Industry Security Assessment Models *(Structural processes that have been tailored by each infrastructure sector or professional security organization.)*	

AMERICAN SOCIETY FOR INDUSTRIAL SECURITY GENERAL RISK ASSESSMENT MODEL [1]	CENTER FOR CHEMICAL PROCESS SAFETY MODEL	AMERICAN WATER WORKS ASSOCIATION MODEL [2]
♦ Identify Assets ♦ Specific Loss ♦ Frequency of Events ♦ Impact of Events ♦ Mitigation Options ♦ Feasibility of Options ♦ Cost Benefit Analysis ♦ Decision ♦ Reassessment	♦ Screen Vulnerabilities ♦ Characterize Facility ♦ Derive Severity Levels ♦ Assess & Prioritize Threat ♦ Site Analysis & Site Survey ♦ System Effectiveness ♦ Analyze Risks ♦ Risk Reduction ♦ Action Plan	♦ Characterize Mission, Objectives, System ♦ Identify, Prioritize Adverse Consequences ♦ Determine Critical Assets ♦ Assess Likelihood or Occurrence ♦ Assess Existing Measures ♦ Analyze Risk ♦ Develop Prioritized Action Plan
NATIONAL AMERICAN ELECTRIC RELIABILITY COUNCIL MODEL [3]	DEFENSE SECTOR MODEL [4]	GENERAL PORTS MODEL [5]
♦ Identify Assets & Loss Impact ♦ Identify & Analyze Vulnerability ♦ Assess Risk & Determine Assets ♦ Identify Countermeasures ♦ Identify Cost Trade-offs ♦ Associated Dependencies	♦ Identify Assets ♦ Identify Threats ♦ Identify Consequences ♦ Determine Vulnerability ♦ Determine Readiness ♦ Assess Mitigation ♦ Identify Residual Vulnerability ♦ Develop Action Plan	♦ Identify & Evaluate Assets ♦ Identify Threats, Likelihood of Occurrence ♦ Select & Prioritize Countermeasures ♦ Determine Level of Effectiveness to Reduce Vulnerability ♦ Identify Vulnerability
FOOD & AGRICULTURE SECTOR MODEL [6]	TRANSPORTATION SECTOR [7]	NATIONAL INSTITUTE OF STANDARD TECHNOLOGY MODEL [8]
♦ Assess Criticality ♦ Determine Recognizability ♦ Assess Accessibility ♦ Assess Vulnerability ♦ Assess Likelihood of Threat ♦ Determine Severity to Public Health ♦ Determine Shock Value of Attack ♦ Determine Recuperability	♦ Assess Threat Environment ♦ Assess Vulnerability ♦ Assess Consequences ♦ Mitigate Vulnerability ♦ Consequence Reduction	♦ Identify Hazards ♦ Profile Hazards and Vulnerability ♦ Inventory Assets ♦ Estimate Loss
	SOUTH CAROLINA SECURITY ADVISORY COMMITTEE [9]	
	♦ Identify Risks ♦ Prioritize Risks ♦ Risk Assessment ♦ Risk Control and Mitigation ♦ Risk Monitoring	*(continued on next page)*

Exhibit 4.1 Comparison of selected security assessment models

The exhibit also clearly illustrates many inconsistencies among various models. In perusing the representative samples, one has to look hard to understand the rationale for the sequencing of some of the tasks. Notwithstanding these variances among the respective models, inadequacies in implementing a given model within a particular sector exist. The driving force behind this is the lack of a solid platform and a systematic, integrated approach to each of the steps.

All these models place emphasis on the physical security aspects of the security program. They have typically been confined to examining

Author's Presentation of Indispensable Components of a Comprehensive Security Assessment
(Only aspect customized is the criticality and prioritization of an enterprise's systems, functions, assets, and resources. Structural methodology remains intact.) Refer to Exhibit 4-2 for detailed methodology.

- Characterize Mission, Objectives, Culture and Political and Social Dependencies
- Characterize Physical Structure, Configuration of Facility, and Physical Security Features
- Characterize Operational Systems, Functions, Processes, Protocols, and External & Internal Dependencies & Influences
- Identify and Categorize Assets as Critical or Non-critical
- Define Range and Levels of Threats and Consequence of Events Against Asset Attack
- Assess Vulnerability to Operational Systems, Functions, Processes, Protocols, and External & Internal Dependencies
- Evaluate Effectiveness of In-place Protective Measures and External and Internal Dependences and Influences
- Evaluate Effectiveness of Security Strategies, Security Organization, and Security Operations
- Evaluate Effectiveness of Proposed Protective Measures and Cost Benefit Analysis
- Identify Residual Vulnerability, Acceptable Risks, and Compensatory Measures
- Quality Assurance Review of Assessment Process, Findings, Observations, Conclusions, and Report
- Report Assessment Results and Prioritized Action Plan, Milestone Schedule, and Budget Estimate

Legend:

[1]: The American Society for Industrial Security [ASIS] has had an effective generic model for the past two decades that is commonly used throughout the commercial sector. It is periodically evaluated and updated by its national committee and has recently gone through another upgrade.

[2]: The American Water Works Association, in conjunction with Sandia National Laboratory, developed this model in 2002, patterned after one adopted by the Nuclear Regulatory Commission used for commercial nuclear power facilities.

[3]: In 2002, the North American Electric Regulatory Council issued this voluntary guidance for conducting security assessments within the utility industry.

[4]: The Defense Model was first developed in the late 1970's with major revisions to the process in the 1980's. It has consistently been updated and revised to meet the threat demands imposed on the defense establishment.

[5]: The General Ports Model was issued in 2002 and is currently under evaluation for improvements.

[6]: The Food & Agriculture Sector Model is in its infancy stage.

[7]: The Department of Transportation has had its assessment model for many years and has possibly been used on more occasions than any other, except for the ASIS model.

[8]: Several private firms and government agencies have partnered to develop risk assessment models. While some information is available to security consultants, most information is restricted for security reasons.

[9]: In May 2002, the Governor's Workplace Security Advisory Committee developed this risk assessment model for state agencies to use. Since then, other states have adopted similar models.

Exhibit 4.1 *(continued)*

security fences, gates, and doors; installing security systems; and expanding the security system to meet emerging security needs. They fail to recognize other aspects of the security program such as cybersecurity, information and personnel security, communications, and security protocols. Moreover, little or no attention is typically given to business operations, processes, protocols, and external and internal dependencies. The security organization and security management have largely been treated as separate issues and investigated in a piecemeal fashion. They are seldom integrated into a much broader strategic security vision. Several circumstances foster this situation. Among them are the constraints and limitations imposed on the security consultant by management, lack of understanding security principles and security operational capabilities by the enterprise, and the orientation and background of security-assess-ment-team members who are not security professionals.

To make the best use of the information, alternative approaches to threat assessments are needed. Analysis without access to classified or proprietary data needs a framework to define procedures to make "best-guess" assess-ments of those characteristics and prediction of probabilities using the infor-mation that is available. In this book, the author hopes to bridge the gaps and advance the "best-guess" method to "best professional expertise."

Because security requirements vary from one infrastructure sector to another and from one time to another, a generic approach to conducting risk assessments lacks uniformity and applicability. Security threats from terrorists are a relatively new phenomenon for many of the infrastructure sectors, and as a result standard protocols are in their developmental infancy. Determining successful protective measures for these sectors is extremely difficult, as many are being implemented for the first time and their effectiveness is yet to be revealed. In addition, for these industries assets and operations typically are dispersed—often over hundreds or even thousands of square miles—so standard approaches to security, developed for enterprises with highly centralized assets and clearly defined bound-aries, are difficult to apply to this other dynamic environment.

SELECTED INDUSTRY REFERENCES

In addition to the models illustrated, other industry guidance that con-tributes to the successful security-assessment analysis and decision-mak-ing process is available for use by a security-assessment team, a design and engineering team, or an enterprise staff. Among them are:

- FEMA *Antiterrorism Design Criteria*
- Department of Defense *Uniform Facilities Criteria*
- R.S. Means *Building Security: Strategies and Cost*

FEMA Antiterrorism Design Criteria

FEMA has developed a series of guidance manuals to assist business own-ers and state and local communities in preparing for risk mitigation. These manuals address the need for risk assessment for a variety of hazards.

FEMA 426, *Reference Manual to Mitigate Potential Terrorist Attacks Against Buildings,* provides guidance to architects and engineers on how to reduce physical damage to buildings, related infrastructure, and people caused by a terrorist attack. The manual presents incremental approaches that can be implemented over time to decrease the vulnerability of buildings to terrorist threats.

FEMA 427, *Primer for Design of Commercial Buildings to Mitigate Terrorist Attacks,* introduces a series of concepts that can help building designers, owners, and state and local governments mitigate the threat. The manual contains extensive qualitative design guidance for limiting or mitigating the effects of terrorist attacks, focusing primarily on explosions but also addressing chemical, biological, and radiological attacks.

FEMA 428, *Primer to Design Safe School Projects in Case of Terrorist Attacks,* provides the design community and school administrators with the basic principles and techniques to design or remodel a school that is safe from terrorist attack.

FEMA 429, *Building Insurance, Finance, and Regulatory Communities,* introduces the issues of terrorism risk management in buildings and the tools currently available to manage these risks.

Department of Defense Uniform Facilities Criteria

The Department of Defense *Uniform Facilities Criteria* [UFC], published in July 2002 for unrestricted distribution, sets standards for site planning, structural design, architectural design, and electrical and mechanical design that play a role in protecting buildings from explosive threats. DOD's focus on minimum setback distances as the primary approach separates it from GSA and Department of State, which place more emphasis on building hardening.

R. S. Means

In 2003, R. S. Means published *Building Security: Strategies and Cost* to assist building owners and facility managers to assess risk and vulnerability to their buildings, develop emergency-response plans, and make choices about protective measures and designs. *Building Security* includes pricing information for several security-related components, systems, and equipment as well as the labor required for installation.

THE S^3E SECURITY ASSESSMENT MODEL AND METHODOLOGY

A clear distinction needs to be made between the sample models presented and the S^3E **Security Assessment Methodology** and its security strategies proposed in this book. The developed design methodology provides a fresh, strategic, systematic approach to security problem-solving that embraces all aspects of business and security-related operations, placing them into perspective. Expanding on the works of others, the framework of the methodology incorporates the very best components from various risk-assessment models and research papers. These components have been integrated into a forensics investigative process that can be effectively applied to any environment.

The second layer, the S^3E **Security Assessment Methodology** (see Exhibit 4.2), shows the detailed breakdown of the process by specific task and the associated interfaces of each program element to be explored. Each component of the model is transferred into a performance-based, task-driven activity based on the proposition that a threat to a vulnerable asset results in security risk and consequences. Here, the strategic-planning architecture [Elements of Tasks 1, 2, and 3] is customized to meet the dynamic needs of an enterprise, thereby giving the methodology local meaning to stakeholders.

The S^3E security-assessment methodology is a systems-level performance-based approach to security problem-solving. In an increasingly complex and dangerous world, enterprises and government agencies of all types require solutions to protect operations, employees, facilities, and proprietary information. The methodology presented offers an explicit approach for achieving these objectives.

The S^3E **Security Assessment Methodology** is a process that encompasses a series of mitigating actions that permeate an enterprise's activities and reduce the likelihood of an adverse event occurring and having a negative impact. In general, the security assessment is a portfolio that addresses enterprise-wide vulnerability and risk. It addresses "inherent," or preaction, vulnerability [a weakness that would exist absent any mitigating action] and "residual," or postaction, vulnerability [that weakness that remains even after mitigating actions have taken place].

The S^3E **Security Assessment Methodology** represents a comprehensive security-management action plan that focuses on evaluating and enhancing an enterprise's security program from a systems-level perspective:

- It employs a timely approach and a wide variety of complementary strategies, keyed to the security program's **prevention, control, detection,** and **intervention** functional objectives.
- It integrates physical measures, cybersecurity, processes, information, people, facilities, and equipment as well as their internal and external dependencies and the relationship of such dependencies to other critical program elements.

Because it is a performance-based security-assessment methodology, it blends the evaluation and measurement of a mix of facility and land use, operations and processes, procedures and techniques, personnel and efficiencies, and technology effectiveness critical to the security program. The methodology aims at minimizing exposure to risk and loss in the areas of corporate policies, human resources, security technology, and physical or architectural barriers:

- It is a powerful tool for identifying, evaluating, and controlling risk.
- It permits the consultant assessment team to fold the results of analysis within a larger context of requirements, bringing protective measures into direct alignment with the corporation's business-continuity-planning initiatives, including emergency-preparedness planning and response and recovery activities.

The *S^3E* **Security Assessment Methodology** consists of the following indispensable components: strategic planning, program effectiveness, program analysis, and reporting and implementation plan. We'll discuss each of them in turn.

Strategic Planning

Strategic planning itself has several components.

The Internal (or operational) enterprise environment is the institutional "driver" of the security-assessment process. It includes the enterprise's organizational and management structure, processes that provide the framework for executing services, and the control and monitoring of activities.

Criticality assessment is an asset's relative importance based on a variety of factors such as mission or function; the extent to which systems, functions, facilities, and resources are at risk; and significance in terms of enterprise security, economic security, or public safety. Criticality assessment is important because it provides, in combination with other factors, the basis for prioritizing those assets that require greater or special protection relative to finite resources and funding.

The threat (or undesirable event) assessment is an assessment of any event that may disrupt, damage, or destroy enterprise systems, functions, and facilities or injure or cause the death of human resources. Threats are acts of terrorism, criminal enterprises, lone wolves, industrial accidents, natural disasters, and other events that would impact business or government operations.

Strategic Planning
Tasks 1, 2, & 3
Operational Environment
Critical Assessment
Threat Assessment

Program Effectiveness
Task 4
Infrastructure Interdependency
Evaluation [P^{E1}]

Operational Environment
Task 1
Enterprise Goals
Corporate Image, Core Values
Areas of Responsibility
Performance Objectives
Business Incentives, Investment

Commitments
Federal Statutes, State Laws, Municipal
Codes, Confidentiality Laws, Civil
Common Law, Willful Torts, Contractual
Obligations, Agreements, Governing
Authorities, Voluntary Commitments

Characterization
Mission, Location, Facility Layout
Facility/Land Use, Neighboring
Facilities, Terrain, Climate
Resources, Operations, Processes

Operations – Logistics
Visitors, Customers, Vendors; Mail and
Material Delivery, Receiving, Shipping;
Staging, Storing, Distribution &
Delivery; Time-Sensitive Processes;
Communications

Critical Assessment
Task 2
Resources, Operations, Functions
Administrative, Aircraft/Fleet Facilities
Board/Conference Rooms
Cable/Communications Rooms
Electrical/Mechanical Rooms
Physical Infrastructure
Telecommunications Centers
IT Network Centers
Training Facilities
Warehouses, Loading Docks
SCADA & Security Systems
Infrastructure Sharing
Critical Interdependencies
Primary, Backup, Redundant Systems

Threat Assessment
Task 3
International, National, Regional, Local
Range & Levels of Threats
Design Basis Threat Profile, Probability
Threat Categories, Capabilities,
Methods & Techniques

Enterprise
Roles & Responsibilities
Operations, Processes, Protocols
Functional Relationships
Security Awareness
Reporting Security Incident
Alert Notification System

Security Capabilities
Management Team
Organization & Composition
Staffing, Skills, Experience
Qualifications & Training
Operational Capabilities
Dependency Programs

Barriers & Delay Measures
Terrain, Approaches, Fences
Barrier, Walls, Gates

Detection Measures
Interior/ Exterior Sensors
Special Purpose Sensors
Tamper Systems

Access Control Measures
People/Vehicle/Deliver Controls
Screening, Verification

Assessment Measures
Perimeter Security, Lighting
Posts and Patrols, CCTV

Communications Media
Radio, Telephone, PA,
Hardwire, Wireless
Fiber Optic
Information Network

Security Control Center[s]
Annunciation & Display
Event Storage, Reports
Alarm Integration
Information Networks
Infrastructure
Primary, Secondary, Backup,
Redundancy

Response & Recovery
Tactics, Techniques
First & Second Responders
Integrated Responders
Closeout Actions

Quality Assurance/Quality Control Process

Exhibit 4.2 The S^3E security assessment methodology

Program Analyses
Task 5
Mitigation Evaluation (P_{E1})
Risk Characterization (C)
Constraints, Trade-offs

Reporting &
Implementation Plan
Task 6
Risk Mitigation
Mitigation Selection (P_{E2})
Monitoring Mitigation

Security Strategies
Vulnerability Analysis
Loss Consequence
Rank Order Facilities & Assets
Priorities, Protective Measures
Day-to-day Planning
Emergency Preparedness Planning
Business Continuity Planning
Performance Expectations
Hiring Practices

Business Losses
Theft of Trade Secrets
Loss of Competitiveness
Shrinkage, Vandalism
Shelf Life, Spoilage
Damage to Critical Assets
Production Shutdown
Service Contamination
Business Continuity

Business Disruptions
Disasters & Accidents
Work Place Violence
Bomb Threats
Demonstrations
Unidentified Materials
Specified Threats
Inefficiencies & Waste
Sudden Unexpected Layoffs
Insensitive Terminations
Ignorance or Indifference

Business Liability
Insurance Premiums
Exposure to Shareholders
Employees, Suppliers
Carriers, Customers
Unsafe Work Environment
Neighborhood Social Order
Ineffective Security Plan
Ineffective Safety Plan
Ineffective Parking Plan
Loose Visitor Controls
Personnel, Vehicle Screening
Qualifications & Training
Supervision

Integrated
Transformation
Solution

Action Plan

Reduce Vulnerability
Reduce Liability
Enhance Work Place
Enhance Best Practices

Strategies

Strategic Vision
Resources
Priority Initiatives
Milestones
Budget

Deliverables

Progress Reports
Schedules
Assessment Report
Presentations

Partner Organizations
Federal Agencies
State Agencies
Municipal Agencies
Trustees
Board of Governors/Directors
Commissions
Other Governing Bodies

Exhibit 4.2 *(continued)*

Program Effectiveness

The components of program effectiveness are discussed below.

The security vulnerability assessment is an assessment of any inherent state—physical, technical, or operational—of a system, function, facility, or resource that can be exploited by an adversary to cause harm, damage, or destruction. Vulnerability is an inherent state to the extent that it is susceptible to exploitation relative to the effectiveness of existing protective measures $[P_{E1}]$. Residual vulnerability is the inherent state that remains when the mitigation-selection $[P_{E2}]$ process is completed.

Security-risk assessment is a qualitative and/or quantitative determination of the probability of occurrence $[P_A]$ of an undesirable event. It includes scenarios under which two or more risks interact, creating greater or lesser impacts.

Program Analysis

Following are the elements of program analysis.

Security-risk characterization involves the application of a graded scale of risk $[P_A]$ and the severity of its consequences [C]. Security-risk characterization is the crucial link between security-risk assessment, mitigation evaluation, and mitigation selection, recognizing that not all risks can be addressed as resources and funding are inherently scarce. Accordingly, security-risk characterization forms the basis for deciding which actions are best suited to mitigate the assessed vulnerability and risk.

Security-mitigation evaluation assesses the efficacy of mitigation alternatives relative to their likely effect on reducing risk, effectiveness of performance, reliability and dependability, and cost-effectiveness.

Reporting and Implementation Plan

Security-risk mitigation is the implementation of prioritized mitigation actions in priority order, commensurate with assessed risk. Depending on risk tolerance, no action may be taken. This is characterized as **risk acceptance**. If the enterprise does choose to take action, such action falls into three categories: **[1] risk avoidance**—existing activities that expose the enterprise to risk; **[2] risk reduction**—implementing actions that reduce the likelihood of impact or risk; and **[3] risk sharing**—implementing actions that reduce the likelihood or impact by transferring risk to or sharing it with affiliated enterprises and external dependencies. In each category, the enterprise implements actions as part of an integrated "systems" approach, with built-in redundancy that addresses

residual risk [the risk that remains after actions have been implemented]. The systems approach then consists of taking actions in personnel [e.g., training, deployment], processes [e.g., operational procedures], technology [e.g. software or hardware], infrastructure [e.g. institutional or operational configurations], and governance [e.g., management and internal control and assurance]. In selecting actions the enterprise assesses their benefits and costs [where the amount of risk reduction is weighed against the cost involved] and identifies potential financing operations for the actions chosen.

Security-mitigation selection involves a management decision as to which mitigation alternatives should be implemented among the possibilities, taking into account risk, the effectiveness of mitigation alternatives, and costs. Selection among mitigation alternatives should be based upon established criteria. Mitigation selection does not necessarily involve prioritizing all resources to the highest-risk area but attempts to balance overall risk and available resources. However, there are as of yet no clearly preferred selection criteria for most infrastructure sectors, although potential factors might include risk reduction, net benefits, equality of treatment, or other stated values.

Monitoring and evaluation of security-risk mitigation entails testing, evaluating, and validating the effectiveness of implemented actions against the established strategic objectives and performance measures to ensure that the entire process remains current and relevant, reflecting changes in the mission and operations, processes and people, and threats. Monitoring and evaluation includes, where and when appropriate: peer review, testing, validation, evaluation of the impact of the actions on future operations, and identification of unintended consequences that in turn would need to be mitigated. The process requires frequent review, restarting the "loop" of quality-assurance assessment, mitigation, and monitoring and evaluation.

The S^3E **Security Assessment Methodology** has successfully been used and refined over the last three decades in over 3,000 security assessments, domestically and internationally. This approach has been exceptionally well received by those enterprises that shun models that are either generic in nature or too confusing to understand. Its thoroughness and specificity help the enterprise better understand and relate to the process and engender a high level of confidence in the professionalism and quality of the approach. The methodology involves working hand-in-hand with enterprise executive management and their staffs to establish the rapport and trust that are vital in performing a security assessment. In the end, the effort produces highly positive analyses that lead to clear, defendable, and workable solutions. This approach is well-suited for large multidiscipline corporations. For smaller enterprises, a scaled-down version of the methodology can be just as effective.

The *S³E* **Security Assessment Methodology** helps the enterprise gain a greater awareness of threats and impacts on business operations while formulating a framework to establish a sound security program. It blends the fields of organizational sciences, management theories, operational- and training-need analysis, and the application of security and information technologies into a flexible, responsive, integrated, top-down-bottom-up approach to security problem-solving. These disciplines are folded into the process because feasible solutions must be based on a series of integrated functional operational needs measured against a series of collective risks—otherwise, there is no relationship between identified problems and proposed solutions and therefore no workable solution.

The *S³E* **Security Assessment Methodology** determines program status and guides the development of workable solutions. It provides an adaptable capability to determine an enterprise's critical infrastructure and assets and to examine the strength and weaknesses of the systems, functions, and processes. This is accomplished by combining an analysis of asset functions and their relationships to operations with an assessment of vulnerability and threat against an evaluation of in-place protective measures and by measuring required or discretionary risk-mitigation alternatives that may be implemented to reduce vulnerability and threat to acceptable levels. A flexible six-step process captures the fundamental elements of the methodology. Results are reinforced in the conclusions of a formal report that is management-oriented and designed to prioritize the application of recommended protective measures to reduce vulnerability and threat and improve program effectiveness. The methodology guides the development of well-designed program solutions that are palatable to the culture and operation of an enterprise. It is exceptionally helpful in identifying the enterprise's security-program strengths and weaknesses to determine whether performance requirements have been met or whether consequences will have to be mitigated or the effectiveness of security increased.

The *S³E* **Security Assessment Methodology** presents five distinct benefits:

- First is the flexibility of the methodology—its greatest strength. The experienced investigator is able to adjust the focus of assessment parameters to meet enterprise needs without jeopardizing the integrity of the overall process.

- Second, the objective investigator recognizes that solutions are not always hardware- or software-driven. Experience tells us that creative management measures, progressive business practices, and a motivated workforce can often reduce the scope and cost of scientific and engineering solutions.
- Third, a key value-added benefit of the methodology is how the experienced investigator can apply a four-dimensional perspective to the evaluation of alternatives by bringing together a delicately balanced, integrated solution composed of **resources, processes, facilities,** and **technology** into the decision-making process.
- Fourth, when evaluating security risk and business exposure, the methodology does not distinguish between safety issues and security issues. Peter E. Tarlow argues that any problem that affects the wellness of the corporation or its resources is one and the same and that any distinction between these two sides of the coin is mere academic sophistry. Taking it further, the methodology considers safety issues and security issues as integral elements of a larger whole—**liability.**
- Last, the methodology's "built-in" quality-assurance review offers the enterprise and the security-assessment teams the immediate opportunity to evaluate judgments and to consider new circumstances and conditions that may evolve or influence judgments already reached in a previous step that now require further investigation. Quality-assurance reviews may occur at any point during the assessment and analysis process.

The methodology is effective and produces best results when the security-assessment team exercises objectivity through independence of thought and action and zeros in on the problem. The methodology involves the discreet execution of a deliberate, comprehensive, performance-based, forensic investigative technique. It requires a clear understanding and application of several disciplines: risk analysis, business analysis, improvement analysis, process analysis, time-sensitive analysis, operational analysis, performance measurement, testing, auditing, inspection, interviewing, cost analysis, and an extensive background in asset-protection measures including a thorough understanding of the national strategies for protecting the nation's infrastructure.

The **S^3E Security Assessment Methodology** rests on three key security strategies:

- Enterprise performance standards to accomplish the security mission
- Enterprise security operational capabilities to implement requirements and expectations

- Enterprise security program performance-based standards and metrics to measure the performance effectiveness of the overall security program

Enterprise Performance Standards to Accomplish the Security Mission

When enterprise performance strategies are embraced by executive management, staff, and employees, the process effectively identifies and tears down barriers and obstacles including those that foster "stove pipe" thinking. Enterprise performance strategies are designed to ensure the establishment of indispensable enterprise security operational capabilities and security program performance-based standards and metrics to measure achievement.

The *S³E* **Security Assessment Methodology** has as its foundation those indispensable corporate performance strategies that make up the enterprise infrastructure and give it its purpose. To provide the greatest value to the enterprise, the *S³E* **Security Assessment Methodology** embraces six critical and inseparable corporate performance strategies that are the foundation of corporate operations. These strategies, shown in Exhibit 4.3, contribute directly to identifying the strengths and weaknesses of the enterprise's security program. They provide the foundation for the analysis process and in the selection and implementation of clear, defendable solutions to reduce security risk associated with current, anticipated, or emerging threats.

Enterprise Security Operational Capabilities to Implement Requirements and Expectations

The *S³E* **Security Assessment Methodology also identifies six strategies are indispensable to the effective performance of a security mission.** An effective security program has all six capabilities: **deterrence, delay, detection, assessment, response,** and **recovery.** These capabilities permeate all aspects of an enterprise. The significance and interrelationship of these security operational strategies are discussed in Chapter 8, "Evaluate Program Effectiveness." For the time being, it is only important to know that these elements play a crucial role in preventing a terrorist attack, criminal activity, or other emergency.

Creditability of the *S³E* **Security Assessment Methodology is established in its measurement criteria.** To be meaningful, security operational performance capability must have some level of expectation. To be objectively measured, performance must have a clear beginning, a visible process, and a recognizable closure or specific end. The measurement criteria used must be distinct between degrees of strengths and weaknesses in

ENTITY MISSION STRATEGY	Enterprise goals, objectives, and culture represent its contribution to the business world, its' standing within the community, and its acceptance by employees.
BUSINESS OPERATIONS STRATEGY	The processes, techniques, and best practices used by the enterprise to deliver services and products in quality fashion, timely pursuit, and at reasonable market value reflect its image and mirror its social value within the community.
PERSONNEL SECURITY STRATEGY	Protecting "people assets" from harm is critical to the enterprise's reputation, creditability, and integrity. Protecting employees includes a set of protocols, policies, and best practices such as pre-employment screening, background investigations, testing, codes of conduct, ethics, performance standards, professional development, certification, policies and procedures, awards, and disciplinary actions. It also includes maintaining a safe work environment from terrorism, assaults, kidnappings, theft, and other crimes, as well as protective measures against explosive blast effects and chemical, biological, and radiological exposure.
PHYSICAL SECURITY STRATEGY	The characterization of enterprise facilities, appearance and layout, and their accessibility sets the tone for social behavior, standard of character, and integrity. Characterization encompasses a system of tangible and intangible protective measures working in harmony to prevent, respond to, and recover from unwarranted interruption of business operations and services to the community.
INFORMATION SECURITY STRATEGY	Proprietary, intellectual data, and confidential business information formulate the foundation of competitive information which contribute to business success. It includes the protection of service or product information, research and development, trade secrets, customer lists, marketing plans, financial data, legal records, and personnel files. Information security addresses physical, technical, and legal considerations.
CYBER SECURITY	In today's business environment more than 90 percent of data is processed and stored in electronic format. This includes systems such as information management, SCADA, facility management, distribution control networks, and security systems—many of which are proprietary and contain or transmit intellectual data housing confidential business information. These systems span the entire spectrum of the enterprise's services and provide a major link in communicating both internally and externally. These systems are inseparable from Physical, Electronic, and Information Security, and need protection from loss of Confidentiality, Integrity, and/or Availability.

Exhibit 4.3 Corporate performance strategies

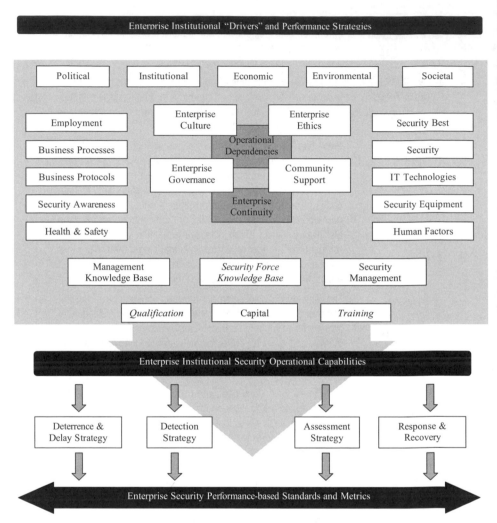

Exhibit 4.4 Enterprise security strategies

the established performance baseline. **Where no performance standard exists, there can be no meaningful evaluation of mission capability. Where no measurement standard exists, there can be no meaningful evaluation of mission effectiveness.** Without these criteria, the conduct of a security assessment is just an exercise in the waste of time, money, and resources, producing personal opinions rather than objective analysis and defendable professional judgments. Applied unified measurement criteria for determining criticality, prioritizing protection investments, and exercising preparedness provide creditability to the *S³E* **Security Assessment Methodology** and its conclusions and recommendations.

**Enterprise Security Program Performance-Based Standards
and Metrics to Measure the Performance Effectiveness of the
Overall Security Program**

The S^3E **Security Assessment Methodology** identifies and implements a
set of performance-based measurement standards to address risk exposure
and the effectiveness of capabilities are critical to establishing the creditability of the overall security-assessment process and its conclusions.
**These S^3E performance-based measurement evaluates the risk to sys-
tems and business operations based on the effectiveness of security
operations and security systems against specific identified malevolent
acts.** These standards of measurement establish the baseline from which
program enhancements can begin to take hold and materialize into actions.
   These standards are applied against:

- Assets that need protection
- Range and level of threats against asset vulnerabilities
- The effects of such threats on assets should they become actual losses
- The criticality of consequence loss on business services should a loss occur
- The likelihood that an actual loss may occur

   The relative values of each measurement standard offered have been
adopted by industry over time as best practices. They can, however, be
modified and assigned different values based on mutual agreement
between the security consultant and enterprise management, but caution
on making adjustments is in order. Experience dictates that whatever values are adjusted for whatever reason, the measurable result of the analysis
is not significantly affected. Furthermore, adding more levels of varying
performance standards only tends to cloud the decision-making process.
The current values as applied to each category of evaluation have a sufficient range of acceptability that any adjustments to the measurement criteria will not change the overall state of analysis. These standards and their
usage are explained below.

S^3E Performance Measurement Criteria

To adequately define the problem and achieve clear, distinct, and workable solutions, the S^3E **Security Assessment Methodology** employs these
measurement indicators:

- Peer-wise comparison analysis criteria
- Assigning protective ratings to assets

- Probability of occurrence [P_A] criteria
- Business criticality consequence factor [C]
- Probability of current program effectiveness [P_{E1}]
- Probability of proposed protective measures program effectiveness [P_{E2}]

S^3E PERFORMANCE MEASUREMENT INDICATORS

Performance measurement indicators are those sets of criteria accepted as being indispensable to the security-assessment process. They define the parameters of the performance measurement standard.

S^3E Peer-wise Comparison Criteria

A key challenge in identifying the importance of assets is to determine their usefulness to the enterprise in terms of their contribution to enterprise

Peer-wise comparison analysis criteria.	Also referred to as **Cardinal Rank Ordering** of assets, resources, and information.
Assigning protective ratings to assets.	Also referred to as categorizing and prioritizing the criticality of assets.
The **likelihood or probability** of defined threats becoming actual loss events. [P_A = Probability of Occurrence]	Also referred to as **Security Risk Characterization, Loss Event Probability, or Frequency.**
The **effectiveness of existing protective measures.** [P_{E1} − Probability of Effectiveness]	Also referred to as **System Effectiveness, Program Effectiveness, or Readiness.**
The **effectiveness of proposed protective measures.** [P_{E2} = Probability of Effectiveness]	Also referred to as **Risk Reduction or Mitigation.**
The **impact or effect** on the asset or operation if the loss occurs. [C = Consequence]	A sub element of **Security Risk Characterization.** Also referred to as **Loss Event Criticality.**

Exhibit 4.5 Performance measurement indicators

operations and business continuity and their loss consequence. Many factors are considered in reaching this determination, including:

- Are they primary or secondary assets?
- Do redundant systems exist?
- Are there backup assets?
- Can business operations be transferred to another location using other assets cost-effectively?

The peer wise-criteria shown in Exhibit 4.6 helps with this determination.

The S^3E measurement criteria are applied to each business criterion for every asset listed. The resultant totals first establish the cardinal-rank order of the importance of the asset. This is important because not all assets that have been defined as critical assets are equal with respect to their contribution to enterprise operations and business continuity and

Business Criteria	Rank Order of Assets						
	Asset 1	Asset 2	Asset 3	Asset 4	Asset 5	Asset 6	Asset 7
Production							
Governance							
Economic Security							
National Security							
Health & Safety							
Public Confidence							
Target Attractiveness							
Investment							
Repair/Replacement							
Total							

Legend:

Importance of one item relative to another item	Importance of the relative item to the initial item
Much greater than [.50]	Much lower than [.10]
Greater than [.40]	Lower than [.20]
The same as [.30]	The same as [.30]
Lower than [.20]	Greater than [.40]
Much lower than [.10]	Much greater than [.50]

Exhibit 4.6 Peer-wise comparison criteria

their loss consequence. The S^3E peer-wise comparison analysis introduces the systematic process of identifying the cardinal-rank order of critical assets. This rating system provides the necessary support rationale needed to defend the analyses and introduce prioritized program enhancements.

Assigning S^3E Protective Ratings to Assets

Protective ratings of **high, medium, low,** or **very low** establish the minimum performance-based security standard to be applied against designated assets while placing controls on expenditures. These S^3E protection standards are subsequently translated into design-criteria documents and specific policies, protocols, and processes to implement requirements. Exhibit 4.7 defines these standards.

Protection Rating	Corresponding Protective Standard
High	A high level of security is applied to areas containing a security interest or asset, the compromise or loss of which would have significant, immediate, and long-term effects on production, service capability, and public confidence in the enterprise to provide a safe and secure work and community environment.
	Unauthorized entry into these areas could result in significant damage or destruction of the areas or assets, serious business disruption to the enterprise, or the disclosure or compromise of proprietary and confidential information. Recommended security standards that apply include:
	• Boundary and area protective measures that afford the highest degree of deterrence, delay, detection and assessment of any unauthorized penetration.
	• High density boundary and area lighting to assist in the surveillance and assessment of the property during periods of reduced visibility.
	• Protective measures to assure the highest probability that only persons who actually have an appropriate right and need to enter the area are granted access.
	• Positive personnel identification techniques and methods to enter and depart the area including the use of verification and authentication protocols.
	• Positive techniques and methods to identify, control, and monitor commercial vehicles, deliveries, and materials entering and departing the area, including the use of verification and authentication protocols.

Exhibit 4.7 S^3E Levels of security standards

Protection Rating	Corresponding Protective Standard
 **High** 	• Positive techniques and methods to control employee, contractor and visitor vehicles including security, patrol, surveillance, and assessment of parking areas. • Facility access control capability to detect unauthorized items such as firearms, explosives, incendiary devices, and other contraband items. • Primary and secondary secure communications with multiple channels that are compatible to other emergency services and first responders. • Positive and direct system protective features to deter, delay, detect, and assess tampering with system components, functions, and processes. • Positive and direct duress notification techniques and assessment methods to protect security force members and other designated persons. • On-line redundancy and backup mechanisms to prevent catastrophic single-point failures, cascading subsystem and equipment failures, and dedicated infrastructure. • Primary and secondary Security Command, Control, and Communications Centers which display, annunciate, and archive transactions, activities, and system status. • Response capability to reach an area or asset to neutralize a threat is less than the designed penetration-delay time required for an adversary to cause damage or destruction of the asset. • Other protective measures as deemed necessary to meet regulatory compliance standards or deemed appropriate to remove, reduce, or control vulnerability against specified threats. The sum total of techniques, methods, and protocols to be employed provide a distinct layering of protective measures that offer the highest probability of defense against specified threats. **A 99.95 program effectiveness and system probability of success rating is the minimum acceptable measurement standard for these areas and assets.**
Medium	A medium level of security is applied to areas containing a security interest or asset, the compromise or loss of which would have a near-term effect on production and service capability and some degree of loss of public confidence in the enterprise to provide a safe and security work and community environment. These areas may be provided with selective special construction materials, barricades, technology, processes, and protocols to assure a sufficiently security environment. Unauthorized entry into these areas could result in significant damage or destruction of the areas or assets, serious business disruption to the enterprise, or the disclosure or compromise of proprietary and confidential information. Suggested security standards that apply include:

Exhibit 4.7 *(continued)*

Protection Rating	Corresponding Protective Standard
 **Medium**	• Boundary and area protective measures that afford a high degree of deterrence, delay, detection, and assessment of any unauthorized penetration. • Boundary and area security lighting to assist in the surveillance and assessment of the property during periods of reduced visibility. • Protective measures to assure a high degree of probability that only those persons who actually require entry and who have an appropriate right and need to enter the area are granted access. • Positive personnel identification techniques and methods to enter and depart the area. • Positive techniques and methods to identify, control, and monitor commercial vehicles, deliveries, and materials entering and departing the area. • Positive techniques and methods to control employee, contractor, and visitor vehicles including surveillance and assessment of parking areas. • Facility access control capability to detect unauthorized items such as firearms, explosives, incendiary devices, and other contraband items during high threat periods. • Primary and secondary secure security communications that are compatible to other emergency services and first responders. • Positive system protective features to deter, delay, and detect tampering with system components, functions, and processes. • Positive and direct duress notification techniques and methods to protect security force members and other designated persons. • On-line redundancy and backup mechanisms to prevent catastrophic single-point failures, cascading subsystem and equipment failures, and dedicated infrastructure. • Primary and secondary Security Command, Control and Communications Centers which display, annunciate, and archive transactions, activities, and system status. • Response capability to reach an area or asset to neutralize a threat is less than the designed penetration-delay time required for an adversary to cause damage or destruction of the asset. • Other protective measures as deemed necessary to meet regulatory compliance standards or deemed appropriate to remove, reduce, or control vulnerability against specified threats. The sum total of techniques, methods, and protocols to be employed provide a distinct layering of protective measures that offer a high probability of defense against specified threats. **A 95.90 program effectiveness and system probability of success rating is the minimum acceptable measurement standard for these areas and assets.**

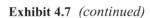

Exhibit 4.7 *(continued)*

Protection Rating	Corresponding Protective Standard
LOW	**A low level of security is applied to areas containing pilferable materials or items that have an attraction for an intruder in addition to monetary value.**

The area is provided with standard protection construction, use of electronic or mechanical aids, personnel movement controls, isolation, or a combination of these. Suggested security standards that apply include:

- Protective measures to detect any unauthorized attempt to remove vital or high-value equipment or material.

- Boundary and area protective measures that afford a reasonable degree of deterrence, delay, detection, and assessment of any unauthorized penetration.

- Boundary and area lighting to assist in the surveillance of the property during periods of reduced visibility.

- Protective measures to control the entry of persons who have an appropriate right and need to enter the area are granted access.

- Protective measures to control, and monitor commercial vehicles, deliveries, and materials entering and departing the area.

- Measures to control employee, contractor, and visitor vehicles including surveillance and assessment of parking areas.

- Response capability to investigate security incidents.

- Security communications compatible to other emergency services and first responders.

- Duress notification methods to protect security force members and other designated persons.

- Compensatory measures for catastrophic single-point failures, cascading subsystem and equipment failures.

- Security Command, Control, and Communications Centers, which display, annunciate, and archive transactions, activities, and system status.

- Other protective measures as deemed necessary to meet regulatory compliance standards or deemed appropriate to remove, reduce, or control vulnerability against specified threats.

The sum total of techniques, methods, and protocols to be employed provide a distinct layering of protective measures that offer a reasonable probability of defense against specified threats.

A 99.85 program effectiveness and system probability of success rating is the minimum acceptable measurement standard for these areas and assets.

Exhibit 4.7 *(continued)*

Protection Rating	Corresponding Protective Standard
VERY LOW	**A very low level of security is applied to an area containing pilferable materials or items that have an attraction for an intruder in addition to monetary value.**

These areas are provided with physical protection construction, use of electronic or mechanical aids, personnel movement controls, isolation or a combination of these. These areas may contain facilities, equipment, and materials necessary for the continual functioning of the activity but not necessarily a part of the immediate or near-term mission or operational capability. Suggested security standards that apply include:

- Protective measures to detect unauthorized attempt to remove materials.

- Boundary and area lighting to assist in the surveillance of the property at night.

- Personnel identification techniques to enter and depart the area.

- Methods to identify and control commercial vehicles, deliveries, and materials entering and departing the area.

- Response capability to investigate security incidents.

- Security communications compatible to other emergency services and first responders.

- Security Command, Control and Communications Centers, which display, annunciate, and archive transactions, activities, and system status.

- Other protective measures as deemed appropriate to remove, reduce, or control vulnerability against specified threats.

The sum total of techniques, methods, and protocols to be employed provide a distinct layering of protective measures that offer the highest probability of defense against specified threats.

A 85.80 program effectiveness and system probability of success rating is the minimum acceptable measurement standard for these areas and assets.

Exhibit 4.7 *(continued)*

S^3E Probability Measurement Criteria

S^3E Probability of Occurrence [P_A] Criteria or Risk Characterization

Probability of occurrence [P_A] measures the range of probabilities of an undesirable event taking place. It is deduced from historical data, actual events, and educated judgments based on situational circumstances and conditions and on the experience of the security assessment team. P_A

NARRATIVE DESIGNATION	NUMERICAL RATING	ID RATING	PERFORMANCE MEASUREMENT
OCCURRENCE HIGHLY PROBABLE	.85	H	Risk exposure is HIGH. The probability of an event occurring is much greater than the probability of the event not occurring.
OCCURRENCE MODERATELY PROBABLE	.65	M	Risk exposure is MODERATE. The probability of an event occurring is somewhat greater than the probability of the event not occurring.
OCCURRENCE LOW OR IMPROBABLE	.50	L	Risk is LOW. The probability of an event occurring is lower than the probability of the event not occurring.
OCCURRENCE PROBABILITY UNKNOWN	.20	I	Insufficient data available to evaluate circumstances and conditions.

Exhibit 4.8 S^3E Probability of occurrence $[P_A]$ criteria

is not a mathematical certainty. Rather, it addresses the likelihood that an attack, disruption of service, or disaster may occur in the future based on historical data of like events as well as other factors. P_A contains wide latitude for variation. Investigators, researchers, analysts, and scientists can assign different probabilities to an event based on different evaluations of the circumstances and their individual level of experience and area of expertise without jeopardizing the analysis. The advantage of using the technique of P_A is that absolute precision is neither important nor desired. **What is crucial is to be able to segregate all risks of probability from all others and to make similar distinctions between each of the other general classes.** Even competent professionals may debate what is highly probable and what may be moderately probable. To compensate for differing interpretations and inexactness, if a rating is in doubt after all available information has been gathered and evaluated and all discussion exhausted, the security consultant should then assign the higher of two probabilities to a particular threat.

S^3E *Business Criticality Measurement Criteria Ranking the Survivability of Enterprise Services*

A key challenge in identifying business criticality loss is the difficulty in estimating economic damage that could result from a terrorist attack or natural disaster. Such damage includes both the immediate loss of operations, equipment, and resources as well as any subsequent long-term economic

CRITICALITY RATING	NUMERICAL RATING	ID RATING	CONSEQUENCE FACTOR [C]
FATAL TO THE BUSINESS	.90	F	Loss would result in total recapitalization, abandonment, or long-term discontinuance of business operations.
OCCURRENCE VERY SERIOUS	.80	VS	Loss would require a major change in investment policy and have a major impact on the balance sheet assets.
OCCURRENCE MODERATELY SERIOUS	.50	MS	Loss would have a noticeable impact on earnings as reflected in the operating statement and would require attention from the senior executive management.
OCCURRENCE RELATIVELY UNIMPORTANT	.25	RU	Loss would be covered by normal contingency reserves.
OCCURRENCE SERIOUSNESS UNKNOWN	.10	SU	Insufficient data available to determine importance.

Exhibit 4.9 S^3E Business criticality consequence factor [C]

losses. The cascading effects often overshadow short-term repercussions over time, yet they are extremely difficult to estimate. Relatively short-term disruptions to critical operations can produce significant downstream economic effects [e.g., price changes, lost contracts and financing, and losses in insurability]. Predicting the extent of such effects accurately requires acute sensitivity to the myriad of interdependencies present in the modern industrial and financial markets. Certain aspects of criticality determination may also produce inadvertent consequences. Designating certain assets or facilities as "critical" in conjunction with enhancing their security may result in their becoming more difficult and expensive to insure, operate, and maintain.

S^3E *Probability of Program Effectiveness [P_E] Criteria*
Program effectiveness evaluates how well systems, functions, processes, protocols, and resources provide a reasonable degree of performance and capability against a defined threat or vulnerability based on a established performance-based measurement criteria. Like P_A, probability of program effectiveness [P_E] is not a mathematical certainty and contains wide latitude for variation. Investigators, researchers, analysts, and scientists could

assign different probabilities to effectiveness performance and capability based on different evaluations of the circumstances and their individual level of experience and area of expertise without jeopardizing the analysis. A significant factor associated with determining P_E involves the human element, the most subjective aspect of the overall formula. The advantage of using the technique of P_E is that absolute precision is neither important nor desired. **What is crucial is to be able to determine the degree of effectiveness of a "protection in-depth" security concept that separates one layer of protection from all others and to make similar distinctions from each of the other layers of protection.** Even competent professionals may debate the level of effectiveness for any specific measure. To compensate for differing interpretations and inexactness, if a rating is in doubt after all available information has been gathered and evaluated and all discussion exhausted, the security consultant should then assign the higher of two ratings to a particular capability. In customizing the measurement standard, we advance the use of the P_E criteria by dividing it into two segments, P_{E1} and P_{E2}:

- P_{E1} represents the inherent readiness status of in-place systems, functions, processes, protocols, and resources to provide reasonable deterrence, detection, assessment, and response capability against a defined threat. As previously mentioned, P_{E1} designates the preaction vulnerabilities identified during the evaluation of site conditions [a weakness that would exist absent any mitigating action].

- P_{E2} represents the inherent readiness status of proposed security enhancements to systems, functions, processes, protocols, and resources to reduce vulnerability and risk. Also as previously mentioned, P_{E2} is the

NARRATIVE DESCRIPTION	NUMERICAL RATING	ID RATING	PERFORMANCE MEASUREMENT
HIGH EFFECTIVENES	.85	HE	Positive evidence of specific capabilities to address specific protective measures.
MODERATE EFFECTIVENESS	.50	ME	Little or general evidence of protective measures, but no specific capabilities.
LOW EFFECTIVENESS	.10	LE	Little, ineffective, or no evidence of protective measures or no capability.

Exhibit 4.10 S^3E Probability of program effectiveness [P_E] criteria

residual or post-action vulnerability identified during the mitigation evaluation and selection process [that weakness that remains even after mitigating actions have taken place].

CONCLUSION

The S^3E assessment ensures that controls and expenditures are fully commensurate with the threat to which the enterprise is exposed.

Chapter 5

TASK 1—PROJECT STRATEGIC PLANNING: UNDERSTANDING SERVICE REQUIREMENTS

This chapter discusses planning and coordination of project activities including project organization, mobilization, and startup activities.

It identifies important missions and functions of the sites to be visited and describes enterprise operations. The examination focuses on services and customer bases to determine the enterprise's contribution to the community.

STRATEGIC SECURITY PLANNING

Strategic security planning defines requirements and sets project activities in motion. The chapter outlines a six-step approach, related tasks, and their respective interrelationships. For maximum productivity, efficiency, and effectiveness the tasks should be performed in the sequence presented:

- Subtask 1A—Project mobilization and startup activity
- Subtask 1B—Site investigation preplanning
- Subtask 1C—Plan, organize, coordinate project kickoff meeting
- Subtask 1D—Co-chair project kickoff meeting
- Subtask 1E—Review available project information and identify a vision
- Subtask 1F—Review technical project data and conduct workshop sessions

The following worksheets are suggested to assist in collecting and analyzing data:

- Worksheet 1—Contacts, key stakeholders, and persons interviewed
- Worksheet 2—Characteristics of an enterprise security strategy from concept to implementation
- Worksheet 3—Enterprise strategies and the extent to which they are addressed
- Worksheet 4—Enterprise security initiatives designed to enhance security performance effectiveness
- Worksheet 5—Review of available program technical data

Developing an effective project strategy requires the security consultant and the security-assessment team to have a comprehensive understanding of the enterprise culture and its strategic vision to pursue project goals and objectives in a systematic, organized, and aggressive manner. Having the security-assessment team establish communications and rapport with the enterprise early on in the project is extremely critical to overall success.

Task 1, project strategic planning: understanding service requirements, offers the security consultant and security-assessment-team members the opportunity to solidify the level of confidence established in the enterprise. One of the critical elements of this stage emphasizes the understanding of service requirements and the nurturing of the business relationship between the enterprise and the security-assessment team. Planning, organizing, and coordinating project activity are essential to executing a solid work plan. They establish the foundation for bringing together all the project stakeholders, outlining of roles and responsibilities, and laying the foundation for the sharing of information with all project participants. They also let the enterprise staff know what to expect during the process so they are better able to assist the security-assessment team in completing the task assignment.

COMPREHENSIVE WORK BREAKDOWN

Subtask 1A—Project Mobilization and Startup Activity

Quality security strategic planning includes a comprehensive work breakdown structure [WBS]. Every project has an official starting point—this task serves that purpose. Mobilization and startup activity are typically transparent to the enterprise and involve project-planning activity internal to the security-assessment team. The security consultant brings team members together to refine plans that were established during the proposal-development phase of the project. Typically this task should encompass the following:

- Assemble project team to establish project norms and governance parameters
- Introduce project organizational structure and reporting protocols
- Identify individual/group assignments through a work-breakdown structure
- Delineate individual/team expectations and goal-setting performance standards
- Define the initial work plan in schedule format
- Identify key deliverables and deliverable timelines
- Identify major milestones, hold points, and witness points
- Refine the work-breakdown structure and assignments
- Validate task time duration estimates for realistic application
- Obtain individual and team commitments
- Orchestrate the coordinated efforts of any teaming partners

A major output of this task is the formulation of the program implementation plan, commonly referred to as the PIP or the project schedule. The initial PIP is presented at the project kickoff meeting and serves as the roadmap to project progress and success. The PIP organizes the project work plan into small segments of work and outlines what is to be done, who will do it, and when the work is expected to start and finish. It represents a cost-effective allocation and assignment of resources to carry out the scope of work.

Subtask 1B—Investigation Preplanning

Once the security-assessment team is formed and roles and responsibilities are delineated, preplanning activity is formulated, and parameters are set for data collection, planning interviews, and developing questions. As a minimum, this task should encompass the following:

- Develop questionnaires
- Coordinate with the enterprise on the distribution of questionnaires to designated individuals
- Evaluate completed questionnaires returned by the enterprise staff
- Categorize/prioritize responses
- Develop and deliver a listing of enterprise data to be reviewed
- Develop and deliver a listing of enterprise management and staff candidates to be interviewed

Development of Questionnaires
Before the project kickoff meeting, the security-assessment team prepares a set of questionnaires for distribution and completion by designated enterprise staff. Whenever practical, the security-assessment team should review and evaluate the responses to these questionnaires before the project kickoff meeting. They provide valuable feedback to the team and help organize the review of documents, the interview process, and the field investigation effort. At times reviewing and evaluating questionnaires before the project kickoff meeting is not always possible. Staff members may be on travel status, vacation, or unavailable. Questionnaires are often returned at the meeting or during the interview phase. They are an important part of the process, however, and the security-assessment team should use patience and flexibility in filling any gaps at this point in project activity.

Subtask 1C—Plan, Organize, Coordinate Project Kickoff Meeting

Establishing Contact and Maintaining Open Communications
All security projects regardless of size and complexity require continuous effective communications between the security-assessment team and the enterprise. The basic steps to begin this open dialogue include:

- Develop and deliver detailed project kickoff meeting agenda to enterprise management
- Develop a listing of recommended attendees and their participation roles

- Provide enterprise management with a listing of support actions for the project kickoff meeting
- Develop listing of materials needed for the project kickoff meeting
- Establish jointly with the enterprise the vision and direction that will guide the project

Working Hand-in-Hand with the Enterprise Staff

Preparing for and coordinating the project kickoff meeting with enterprise management offers the security consultant and the security-assessment team the first formal opportunity to demonstrate excellence and competence in leadership, program management, and organization skills as well as presentation of expertise and strategic vision. The process is the beginning of team building between the enterprise staff and the security-assessment team and of accepting joint ownership of project success.

Developing the Project-Kickoff-Meeting Agenda

The project kickoff agenda is typically the first official notice to the enterprise's staff that the project is underway. The agenda is formalized by the security consultant and approved in writing by the enterprise. When published and distributed, the agenda becomes a guide to conducting and controlling the meeting and gives attendees a preview of topics to be discussed and who the participants are. To give the staff sufficient time to organize their schedules and prepare for the meeting, the agenda should be distributed at least ten working days prior to the meeting.

Subtask 1D—Co-Chair Project Kickoff Meeting

The project kickoff meeting brings all key stakeholders together in a formal face-to-face setting and sets the stage for project activity. The security-assessment teams meet with designated stakeholders to exchange introductions, establish project parameters and communications, discuss the overall security-assessment methodology, review the initial project schedule, and establish rapport. During this initial meeting, the enterprise management briefs the team on project expectations; presents an orientation to the general system configuration, processes, distribution, critical customers, and future growth plans; describes the general business culture of the organization; and answers questions. A well-organized project kickoff meeting should address the following topics:

- Enterprise management and security consultant to introduce respective members:

- ▸ Enterprise management to introduce staff with Homeland Security responsibility
- ▸ Enterprise to introduce state/local authorities with Homeland Security responsibility
- ▸ Security consultant to introduce security-assessment-team members present
- Enterprise to:
  - ▸ Describe business culture, mission, service area, customer base, and future growth plans
  - ▸ Describe security objectives and goals and overview issues and assets to be surveyed
  - ▸ Identify current initiatives that have strategic security implications and may impact the study
  - ▸ Discuss protocol of project norms and governance aspects
- Enterprise to present contact information as applicable for:
  - ▸ Steering committee, executive management, safety board, and security committee
  - ▸ A corporate single point of contact for direct interface and coordination of project activity
  - ▸ A site single point of contact for direct interface and coordination of each site to be studied
  - ▸ Individuals and agencies to receive deliverables and in what quantities
- Enterprise to provide protocols for access to:
  - ▸ Facilities as mutually agreed in the program implementation plan [PIP] or schedule
  - ▸ Site engineers and operations, maintenance, and security personnel for follow-up interviews
- Enterprise to discuss protocols for:
  - ▸ Protecting project sensitive information
  - ▸ Granting security-consultant and security-assessment-team members and teaming partners access to facilities and personnel, including permission to take photographs, calculations, and measures to perform contract obligations
  - ▸ Introduction to external dependency agencies announcing the security consultant, security-assessment-team members, and teaming partners as its security-consultant group authorized to contact and meet with the agencies with respect to security-related inquires

- Enterprise representatives or others in attendance to offer brief comments concerning vulnerability with regard to public health and welfare, risk management, operations continuity, security capabilities, and expectations, including expectations of the security-assessment team
- Enterprise to identify key personnel for security-assessment team to interview
- Security consultant to:
  - ▸ Introduce the assessment methodology to be used to execute project requirements
  - ▸ Discuss the systematic performance-based measurement criteria to be used to identify risk, consequences of loss, asset criticality, and program effectiveness
  - ▸ Identify key milestones and present initial PIP or schedule
- Security consultant to provide:
  - ▸ A project-team contact list
  - ▸ Schedule of proposed project workshops and future workshops and meetings
  - ▸ List of proposed internal interviews including follow-up meetings with each division chief
  - ▸ List of proposed external interviews with federal or state agency Homeland Security representatives
  - ▸ List of documents requested for review
  - ▸ Need to take photographs, calculations, and measures noting relevant security concerns
- Security consultant to define enterprise staff involvement for the various tasks and expectations
- Security consultant to discuss project procedures for safeguarding sensitive security information
- Enterprise staff and security consultant to jointly review and adjust requirements and costs:
  - ▸ Identify and resolve scheduling conflicts
  - ▸ Coordinate the integrated strategic implications of ongoing initiatives and site work with project activity
  - ▸ Identify security implications and interject security considerations into performing the work
  - ▸ Timely update and distribution of schedule
  - ▸ Identify and resolve project staffing conflicts
- Security consultant staff member to take attendance and minutes of proceedings

Subtask 1E—Review Available Project-Management Information and Identify a Vision of Security Requirements and the Road Map to Achieve Project Goals and Objectives

Following the project kickoff meeting, the security-assessment team reviews available data and the enterprise culture to baseline the evaluation process. The team analyzes the data for completeness and functionality and compares this information against the enterprise's goals and objectives for the security assessment in order to further solidify the project direction and the security organization's capability to perform its mission. Research tasks include but are not necessarily limited to the following:

- Review and become familiar with security-related reports, procedures, and operation plans for completeness, accuracy, and clarity of detail with respect to performing mission statement, including:
  - ► Late submissions of questionnaires
  - ► Previous security surveys, threat assessments, audits, and inspection reports
  - ► Enterprise annual business/financial reports and mission statement
  - ► Capital-improvement plan to identify ongoing or recent programs and projects that may be impacted or may have an impact on this study and proposed program enhancements
  - ► Fire and safety plans;
  - ► Disaster and/or emergency-preparedness plans to identify comprehensiveness of emergency-preparedness planning
  - ► Emergency-response and recovery procedures to identify event-driven planning and capabilities
  - ► Business-continuity plan to determine security integration and participation
  - ► Parking and lighting plans to identify completeness and flow-through
  - ► Security plan to identify comprehensiveness of security planning
  - ► Security procedures and post orders to identify arcane regulations or unnecessary and unworkable measures
  - ► Security force training records to identify types and frequency of training and certification;
  - ► Current security contract services to determine if expected services meet emerging enterprise requirements

- ► Security organization chart, staffing levels, and job descriptions to ensure the operating model meets enterprise's expectations and mandates for security performance
- ► Other security-related documents identified during the kickoff meeting or at other times
- Compare documents against previous security work undertaken by enterprise and identify standards, requirements, and considerations that may apply to this survey:
  - ► Compare information reviewed to enterprise security program goals and objectives
  - ► Identify and document information gaps
  - ► Summarize the status of information reviewed from all documents and/or persons within the final project report
  - ► Identify and report to enterprise management the nature and amount of research additionally required by the project team

This review of enterprise data, assessments, and studies is critical. It benchmarks the status of the existing security program and helps the security-assessment team determine the scope needed for the security project and their approach to developing an overall corporate security strategy when one has been requested by the enterprise. It is important to note that not all information requested for review may be available at this time and some data may have to be reviewed concurrent with the performance of other tasks, typically in concert with Task 2 [site characterization], Task 3 [threat characterization], and Task 4 [program effectiveness].

In the preceding chapter, six critical and inseparable corporate performance strategies were introduced. Two of these strategies, the entity mission and business operations, are initially examined here and further evaluated under Task 4 and Task 5 [program analysis]:

- Enterprise goals, objectives, and the security culture are looked at in terms of their contribution to the business world, standing within the community, and acceptance by employees
- Business operations, processes, techniques, and best practices come under review to expose vulnerabilities that impacts the security integrity of the enterprise

A major first step in reviewing the data is to define the enterprise's functional or operational requirements for providing services contrasted with a series of collective risks and vulnerabilities. Otherwise, no relationship exists between identified problems and recommended solutions, and the solutions in and of themselves become opinions rather than professional judgments. Likewise, in performing Task 3 we will see that the relationships between threats and their consequences, functional requirements, and recommended solutions are equally inseparable.

Subtask 1F—Conduct Workshop Sessions and Review Available Technical Project Information and Document Critical Information

Armed with the information reviewed under Task 1E, the team then holds a workshop or a series of workshops with enterprise representatives. The purpose of these sessions is to examine available site-layout drawings, to preliminarily identify critical security interfaces among system components, to baseline the operational system components and configuration, and to develop a prioritization plan for field investigation strategy. Tasks include:

- Review and become familiar with related construction plans, record drawings, diagrams, and sketches, including:
  - ▸ Master expansion plan
  - ▸ Facility drawings, water-system maps, hydraulic profiles
  - ▸ Site layout drawings and other plans
- Develop system tree to baseline system configuration and document:
  - ▸ Flow process, location and quantities of major system components, and facilities
  - ▸ Critical production and delivery operations, redundancy and backup systems
  - ▸ Supervisory Control and Data Acquisition [SCADA], System, distributed-control systems, and security systems
  - ▸ Critical technical and security interfaces among program/system components
- Identify internal and external protocols for dependency services and delivery of materials
- Identify operational constraints
- Identify status of proposed or new construction work
- Identify status of proposed or new security-upgrade work
- Discuss significant security problem areas
- Develop prioritized plan for field-investigation strategy
- Conduct open discussion with workshop representatives

- Summarize contact information and discussion contents within the final project report
- Assist enterprise in developing site-visitation schedule
- Assist enterprise in developing follow-up stakeholder interviews at site level
- Identify and report to enterprise the nature of research additionally required by the security-assessment team

Working-Group Sessions

For best results, workshop attendees should be comprised of enterprise representatives from the electrical, maintenance, operations, and security branches and be knowledgeable about the operations and the physical configuration of the facilities to be visited and assessed. The work sessions and the security-assessment team capture critical thoughts and factual data. At times it may be necessary to hold more than one session—for example, when representatives are not available or if the security consultant determines that it is best to conduct separate or sequential workshops with each discipline. One option is to stagger the timeline for attendance. For example, electrical and maintenance personnel could meet with the consultant at the same time, but operations personnel, who have different concerns, might best meet separately. The security division also has unique requirements, and so on. This is an efficient approach to use when focus on specific issues is desired. Based on the magnitude and complexity of the project, the security-assessment team can also hold multiple concurrent workshops, then regroup to compare and consolidate team observations and notes. Enterprise representatives should bring to these sessions a facility and system layout for each site to be visited that contains sufficient information to identify flow processes; location and quantities of major components such as intake sources, production facilities, storage facilities, distribution points, and redundancy and backup systems; SCADA network components; morphologies and communication pathways; network diagrams with connection details; networking address information; and other critical information that will help identify the characteristics of each facility or site.

Conducting Workshop Interviews

Two types of interviews are conducted during workshop sessions. The first is an open forum, but this approach only provides limited insight into key issues. Further individual interviews are almost always necessary to gain further knowledge of the technical, operational, and security issues. These workshops enable the security-assessment team to complete much of the field-interview process, thereby reducing time on site, including that needed for site-management interviews. Conducting site interviews with designated staff representatives rounds out the information-exchange process.

Candidate-interview representatives as applicable include:

- Plant superintendents or operations managers
- Electrical, maintenance, and facility managers
- Warehouse managers
- Administrative managers and purchasing chiefs or agents
- Business-continuity or emergency-preparedness-planning managers
- Security and safety managers
- Plant-control-room operators [on all shifts]
- Guard shift supervisors and security guards [on all shifts]
- Security monitoring-station operators [on all shifts]

In addition to the above, interviewing other employees is critical in obtaining a different perspective on the work environment than that of senior or middle management. However, some employees tend to have personal agendas and may not give answers based on objective facts. A seasoned consultant can detect this attitude and adjust accordingly. To ensure that no one person's perspective taints the findings, several employees from more than one work center should be sought out to corroborate information gathered by the team before the information is recognized as valid and included in the report.

Conducting Follow-on Site Interviews

Internal and external stakeholder interviews are critical to obtaining independent views from the staff and others and providing invaluable insight to the assessment team. An informal interview/discussion process is recommended to make personnel comfortable in relating individual perceptions, values, and experiences. This eliminates potential bias and pressure associated with group discussions. Predispositions include leadership influence, the dominance of individuals, and group pressure toward conformity. This approach enables the capture of a broad range of honest opinions, practical experiences, and the "actual" as opposed to "official" enterprise-culture mindset. Internal site interviews help the security-assessment team to understand the enterprise culture, gain insight into the strengths and weaknesses of existing practices, identify specific security functional needs, and address individual security concerns. External interviews provide invaluable information about the social order of the community, law-enforcement capabilities, relationships with the enterprise, and involvement with national, state, and local Homeland Security initiatives. Information collected during these interviews is used in the Task 2 [critical assessment], Task 3 [threat characterization], and Task 4 [program evaluation] analyses.

Developing Site-Visitation Schedules
At the completion of the workshop sessions, the assessment team works with the representatives in attendance to formulate the sequence of facility visits based on management's predetermined criteria. A facility-visitation schedule is developed including dates and times for selected site interviews with designated individuals. **Not setting a schedule—even one with flexibility—can lead to major time delays on the enterprise end and hold up the entire process.** When the enterprise approves the visitation schedule, the security-assessment team can then mobilize site investigations using the appropriate team configuration suited for the specific task. During this process responsibilities under Tasks 1 through 5 are executed on a site-by-site basis. As mutually agreed between enterprise management and the security consultant, the security-assessment team may remain in the field to complete all site surveys or return to its base of operations between each site or grouping of sites. There begins the formal analysis process, a discussion of preliminary strategies to mitigate observed vulnerabilities, and the formulation of status reports for presentation to enterprise executive management.

DOCUMENTING THE SECURITY-ASSESSMENT PROCESS

Much has been written about available software programs that can document the security-assessment analysis and generic checklists that can be customized to meet specific needs. Both instruments are noteworthy of discussion.

Using Proprietary Software Programs to Document Security-Assessment Results
There are many off-the-shelf software programs on the market today that proclaim to have the answer to security problem-solving. These programs range in capabilities from identifying and prioritizing assets to

quantifying the effects of recommended protective measures. Most programs are based on a series of drop-down menus and checklists to build a profile of a site or facility. Some programs calculate vulnerabilities in place and levels of acceptable risk, while others report residual vulnerabilities resulting from proposed protective measures. These programs are attractive because they emphasize advantages such as "save time, money, and resources," "special analytical skills not required," or "be your own security expert."

All of these programs have design constraints and limitations that may not meet the needs of the end user. They include a lock-step approach that prevents the user from navigating to other screens before "filling in all the blanks," even if such information is unknown, unavailable, or irrelevant. This prevents the user[s] from moving forward to screens particular to the task assignment. In other instances, steps cannot be retraced to update or change information until all screens in the lock-step sequence have been completed, at which time a prompt appears to validate and save all data entered. Some programs alert the user that information in certain screens is not correct or incomplete, forcing one to focus on "manufacturing a reply" to close the program. As such, many security practitioners shy away from using such programs. Nonuse also results from software programming that doesn't allow users to alter program parameters or report design criteria to meet their needs. Such programs may take a lot of time to maneuver, only to prove useless in the end.

Many enterprises and security consultants use these programs to conduct assessments without fully understanding their limitations or how they work. While these users may have created a "physical security program on paper," it is doubtful that such "cookie-cutter" replies are capable of producing accurate, realistic, effective, and efficient solutions. Equally misleading would be any cost data generated, as it would be such an extremely rough estimate that it may not be useful to the user. Missing variables including site-specific conditions, circumstances, and aspects such as dimensions, measurements, calculations, environments, and other considerations make such cost estimates suspect. Generic software programs cannot adequately and completely respond to such variables and conditions, and cost data and their respective cost-benefit analyses should best be left to the industry security construction and installation experts. A qualified security professional with extensive experience in security assessments may find such programs useful, but they would only be beneficial if they complement the security-assessment process, not replace human judgment. Used correctly and within their limitations, good software programs can aid the security-assessment process, but it is important to use a product that contains appropriate and relevant data and safeguards and one that can be modified for special applications such as hazardous materials and their effects, blast radius, and safety zones. Canned, closed-architecture programs don't have these capabilities. Rather, the manufacturer of such programs offers special costly software packages to perform these analyses. In many instances they have as many constraints and limitations as the more basic programs.

Using Checklists to Document Security-Assessment Results

Checklists are excellent tools to plan how to proceed through the site-characterization process, but they are of little value—no matter how excellently developed—in adequately documenting important information of subsequent critical value. While checklists can contain an enormous amount of valuable information, filling in the "Yes" or "No" columns with a simple checkmark does not inspire the insight necessary to describe site conditions in understandable terms for identifying weaknesses and vulnerabilities or depicting the broader range of problems that may require solutions. Often extensive research, further interviews, and detailed note taking are necessary to recall facts, circumstances, and conditions to clarify the placed checkmark, particularly when the "No" column is checked off. While developing checklists is useful for planning purposes and team assignments, using them to actually perform the assessment gives little insight and initiative to think outside the box. Moreover, a checklist usually applies to a single building, yet most sites have numerous structures with geographically separated operations. In this situation, a checklist for each building would be required, and their collective maintenance could be cumbersome and inhibit the field investigative process.

A VIABLE COST-EFFECTIVE TOOL TO DOCUMENT SECURITY ASSESSMENT RESULTS

Using self-developed worksheets to record essential security-assessment data has several distinct advantages over the use of proprietary software programs and checklists. Significant advantages include:

- Designing worksheets using Word, Excel, or Access gives the end user complete control over format, formula values, presentation, and the data baseline to fit unique reporting requirements.

- Worksheets are easy to complete and double as field notes—a task that has to be performed anyway during or after the completion of a checklist.
- The worksheet layout can be modified, revised, and adjusted to meet specific user needs. Information that is critical to enterprise management can be flagged to stand out. There is no need to rumble through pages of a checklist or software report to recall information.
- Worksheets can easily be inserted into the text of the security-assessment report as summary data, supporting the detailed narrative. Completed checklists may or may not be placed as an appendix to the report. More importantly, few readers of the report bother to go through hundreds of line items in the checklist to find a particular item of interest.
- Worksheets serve as excellent handout materials when briefing the enterprise staff on the status of the assessment and are easily imported into a Power Point presentation. Checklists do not serve either purpose well.
- Completing the worksheet saves valuable follow-up research time, resources, and money over the procurement of software programs or the development and use of checklists.

The worksheets presented on the following pages and in subsequent chapters represent one style of documenting the assessment process. The security-assessment team can realize maximum efficiencies by constructing these worksheets and tables as a Word document. The design of the table can be manipulated to meet specific project needs and the unique reporting requirements of the enterprise. This approach is particularly helpful in those instances where the enterprise needs to report specific information to a federal or state regulatory body or when seeking a federal grant to support project activity.

When the worksheets are filled in and identified as a particular work site, the data may require protection from unauthorized sources. For most government projects, the contract-classification guide instructs the user to protect data as either FOR OFFICIAL USE ONLY, CONFIDENTIAL, or SECRET. For the commercial nuclear industry, data is usually designated as SAFEGUARDS INFORMATION and protected under Nuclear Regulatory Commission [NRC] guidelines. For the commercial sector there is no uniform industry standard, except that the criteria established in 10 CFR 2.79[d] and the Patriot Act of 2002 can be applied to protect information from public disclosure. Typically, information is designated as COMPANY PRIVATE, PROPRIETARY, or SENSITIVE SECURITY INFORMATION.

Documenting Contacts, Key Stakeholders, and Persons Interviewed

Worksheet 1 documents all the individuals the security-assessment team has met to discuss project requirements. It serves as an official record of all contacts, key stakeholders, and personnel interviewed throughout the entire project and of assessment-team members. When completed, the listing can easily be inserted as an appendix to the final security-assessment report.

Documenting Enterprise Security Strategies

Worksheet 2 captures the enterprise core values, business culture, and business goals and objectives with respect to security operations. Collectively, the data-collection effort reflects the corporation image and mirrors its social values for the health and safety of the community. The effort also identifies the corporate standing within the community and its acceptance by employees for maintaining a safe and secure work environment.

Name & Position	Phone	Cell	FAX	Email

Exhibit 5.1 Example of Worksheet 1—Contacts, Key Stakeholders, and Persons Interviewed

Desirable Characteristic	Description	Example of Elements Significant of Reporting
Purpose, scope, and methodology	Addresses why the strategy was produced, the scope of its coverage, and the process by which it was developed.	• Statement of broad or narrow purpose • How it compares with other corporate strategies • Major mission areas or activities it covers • Principles of theories guiding its development • Impetus for strategy • Process that produced strategy • Definition of key terms
Problem definition	Addresses the particular problems and threats the strategy is directed towards.	• Discuss problems and their causes • Analysis of threats and vulnerabilities • Quality of data available and "unknown"
Goals, subordinate objectives, activities, and performance measures	Addresses what the strategy is trying to achieve, steps to achieve those results, as well as the priorities, milestones, and performance measures to gauge results.	• Overall results desired • Hierarchy of strategic goals and objectives • Specific activities to achieve results • Priorities, milestones, and performance measures • Process for monitoring and reporting on progress • Limitations on progress indicators
Resources, investments, and security management	Addresses what the strategy costs, the courses and types of resources and investments needed, and where resources and investments should be targeted based on balancing risk reductions with costs.	• Resources and investments • Types of resources required • Sources of resources • Economic principles • Resource allocation mechanisms • "Tool of governance" • Importance of fiscal discipline • Linkage to other resource documents • Security management principles
Organization roles, responsibilities	Addresses who is implementing the strategy, what their roles are compared to others, and mechanisms for them to coordinate their efforts.	• Roles and responsibilities of specific divisions, departments, or offices within the corporation, and external dependency agencies • Lead, support, and partner roles and responsibilities • Accountability and oversight framework • Potential changes to current organizational structure • Specific processes for coordination & collaboration • How conflicts have been resolved

Exhibit 5.2 Example of Worksheet 2—Characteristics of an Enterprise Security Strategy from Concept to Implementation

Desirable Characteristic	Description	Example of Elements Significant of Reporting
Integration and implementation	Addresses how the strategy relates to other strategic goals, objectives, and activities, and to subordinate levels of the corporation and plans to implement the strategy.	• Integration with other strategies [horizontal] • Integration with relevant documents from implementing organizations [vertical] • Details on specific strategies and plans • Implementation guidance • Details on strategies and plans for implementing enterprise architecture, facilities, and human capital

Exhibit 5.2 *(continued)*

Documenting the Extent to which Enterprise Strategies Have Been Implemented

Worksheet 3 summarizes the observations and findings that were documented in Worksheet 2. The management status of the security program can easily be summarized as ADDRESSED, NOT FULLY ADDRESSED, or NOT ADDRESSED. It quickly identifies significant strengths and weaknesses of the organization and provides a roadmap for further analysis leading to program enhancements. It serves as an excellent Power Point presentation slide.

Documenting Current or Proposed Security Initiatives

Worksheet 4 documents the current security initiatives of the corporation. It provides the security-assessment team valuable insight into plans to improve the security program. The status of each category identifies the degree to which these initiatives have been implemented. For Parts I and II of the worksheet, when the target date for the initiative has not been met, you should summarize the reasons for the delay. For Part II of the worksheet, summarize the reasons why the recommendation was not adopted.

Documenting the Status of Program Guidance

Worksheet 5 clearly summarizes the status of the document-review process. While the illustration shows only the title of sample documents, this task is better served by listing the table of contents of each document reviewed. This gives the evaluation criteria better meaning. Sections of a particular document that are rated SATISFACTORY only

Influence of Enterprise Security Strategies	Status of Strategies		Root Causes Leading to Not Fully Addressed and Not Addressed Categories
	Addressed	Not Fully Addressed	Not Addressed	
Purpose, Scope, and Methodology				
Problem Definition				
Goals, Objectives, Activities, Performance				
Measurements				
Resources Investment, Security Management				
Organization Roles, Responsibilities, Coordination				
Integration, Implementation				

Exhibit 5.3 Example of Worksheet 3—Corporate Strategies and the Extent to which They are Addressed

SECURITY INITIATIVE AND SOURCE	TARGET DATE FOR COMPLETION	TARGET DATE MET?	STATUS OF INITIATIVE AS OF [ENTER DATE]
Part I – Enterprise Generated Initiatives			
Establish a program to encourage innovation and excellence.			
Improve reporting and dissemination of security-related information.			
Part II – Recommendations Offered by Previous Studies			
Promote the development of performance standards and uniform implementation expectations.			
Establish procedures to govern the application of authority and jurisdiction and use of force.			
Part III – Recommendations Offered by Previous Studies and Not Adopted			
Create a multi-agency technical group to review facility security designs.			

Exhibit 5.4 Example of Worksheet 4—Enterprise Security Initiatives Designed to Enhance Security Performance Effectiveness

DOCUMENT	ACCEPTABLE	INADEQUATE	DEFICIENT	RECOMMENDED CORRECTIONS
Security Plan				
Security Procedures				
Post Orders				
Security Management Plan				
Security Training Plan				
Security Contract[s]				
Security-related Design Documents				
Capital Improvement Plan				
Security Emergency Response Plan				
Security Emergency Response Procedures				
Security Emergency Recovery Procedures				

Legend:

Acceptable Document addresses minimum requirements but selected portions of the document could be improved to provide for greater efficiencies and effectiveness. May not fully address corporate security goals and objectives.

Inadequate Sections of the document do not fully address significant security requirements. Presented guidance falls short of industry standards, best industry practices, and corporate security goals and objectives.

Deficient Document or portions thereof do not meet acceptable industry standards and best industry practices. Omissions, inconsistencies, or contradictions create confusion and misrepresentations with respect to mission capabilities. Information may indirectly contribute to program weaknesses and collateral vulnerabilities. Document the sections affected, needed to be rescinded and removed from publication.

Recommended Corrections Those corrections or enhancements needed to bring document up to acceptable industry standards, best industry practices, and corporate security goals and objectives.

Exhibit 5.5 Example of Worksheet 5—Review of Available Program and Technical Data

require a checkmark. For all other evaluation categories, a brief description of the shortfall will serve as a summary statement when the table is inserted into the final report and the review observations are expanded upon. The legend is a useful management tool for conducting a document-review process.

CONCLUSION

An effective project strategy permits the security consultant and the security-assessment team to pursue project goals and objectives in a systematic, organized, and positive manner. Regardless of the size or complexity of the project, establishing communications and rapport with the enterprise staff early on in the project develops a solid level of confidence for the delivery of services. Organizing project activity and issuing team assignments result in project effectiveness and cost efficiencies.

Meeting with the client, reviewing available project-management and technical information, and conducting group and individual interviews solidify project direction and clarify mission statements and business values. They also provide the platform necessary to organize the data-collection process, identify threats and vulnerabilities, measure and evaluate mission performance standards, and complete the analysis.

Developing solutions that are appropriate and cost-effective is often a complex and subjective process. The **S^3E Security Assessment Methodology** is a distinct approach that has carved out a systematic and integrated path by employing user-friendly strategies to identify an enterprise's strengths and weaknesses. As such, it offers the promise of reliable and dependable performance-based solutions across the entire spectrum of the enterprise and in any infrastructure environment.

Chapter 6

TASK 2—CRITICAL ASSESSMENT: UNDERSTANDING THE SERVICE ENVIRONMENT

This chapter characterizes the configuration of the site, its operations, and other environmental elements by examining conditions, circumstances, and situations relative to safeguarding public health and safety and to reduce the potential for disruption of services.

The critical assessment data-collection process focuses on what assets need protection to minimize the impact of undesirable consequences. It takes into account the impacts that could substantially disrupt the enterprise's ability to provide safe services and to reduce risks associated with the consequences of significant terrorist events, other criminal activity, and natural disasters.

The process of critical assessment and asset identification describe the work and threat environments. This chapter outlines a five-step process and its respective task interrelations. For maximum efficiency, the steps should be performed in the presented sequence:

- Subtask 2A—Enterprise characterization
- Subtask 2B—Data analysis
- Subtask 2C—Security characterization
- Subtask 2D—Capital-improvement characterization
- Subtask 2E—Engineering data

The following worksheets are suggested to assist in collecting and analyzing data:

- Worksheet 6— Facility characterization
- Worksheet 7—Defining critical operational criteria and business values
- Worksheet 8—Facility ranking based on enterprise operational criteria
- Worksheet 9—Time criteria
- Worksheet 10—Rank ordering assets
- Worksheet 11—Asset identification and physical security characteristics
- Worksheet 12—Security characteristics–strengths and weaknesses

DATA GATHERING

The security-assessment process involves the gathering of information about facilities, assets, operations, and resources across the entire spectrum of the enterprise. During this phase of the assessment the enterprise's staff is heavily involved, or should be involved, in sharing essential information with the security-assessment team. This provides the team with the benefit of the "corporate knowledge" of those employees who work at the site or various sites and are most familiar with the facilities, operations, programs, and protocols. Performing the security assessment without stakeholder participation loses critical insight from those people who best understand the enterprise's mission objectives and operations. **Unlike some existing security-assessment models, this approach looks at the total security profile of an enterprise.** How else does an enterprise decide what security enhancements are necessarily cost-effective? A comprehensive security assessment provides management with information to make informed decisions regarding which security initiatives to implement based not solely on the current threat exposure of the enterprise but also on projected operational needs, emerging threats, and cost.

Before real problems can be solved, they must be characterized in terms that are recognizable and accepted by the stakeholders. **The key to characterization is the ability to identify and define enterprise normal services, identify the level of service interruption the enterprise can sustain before its survival is threatened, identify the assets important to the enterprise, and identify what "response" and "recovery" should look like under the various threat scenarios developed. Enterprise buy-in of this characterization is indispensable to the successful outcome of the security assessment.**

Under this task, the security-assessment team describes and captures information, conditions, and circumstances important to the uninterrupted operations of the enterprise. This step of the data-collection process does not include passing judgment or analyzing the information with respect to vulnerability and threat assessment. The only judgments are of the relevance of the information with respect to the goals and objectives of the security assessment. Accurate and comprehensive documentation of observations, findings, and conclusions is critical to performing the varying analyses under Task 3 [threat assessment], Task 4 [evaluating program effectiveness], and Task 5 [program analyses].

PROTECTING AMERICA'S CRITICAL INFRASTRUCTURES

Homeland security is an enormous challenge—the infrastructures are a highly complex, heterogeneous, and interdependent mix of facilities, systems, and functions that are vulnerable to a wide variety of threats. Their sheer numbers, pervasiveness, and interconnected nature create an almost infinite array of high-payoff targets for terrorist exploitation. Given the immense size and scope of the potential target set, we cannot assume that we can completely protect all things at all times against all conceivable threats. As protective measures are developed for one particular type of target, terrorists no doubt will shift their destructive focus to targets they consider less protected and more likely to yield desired shock effects. To be effective, the characterization of a particular site or facility must be based on a thorough understanding of these complexities as the security team works with the enterprise to build a focused plan for action.

This understanding acknowledges that assets, systems, and functions that comprise an enterprise are not uniformly "critical" in nature. The first security assessment objective then is to identify and ensure the protection of those assets, systems, and functions deemed most "critical" in terms of production, governance, economic and national security, health and safety, and public confidence. This calls for a comprehensive prioritized assessment of facilities, systems, and functions. The second major objective is to assure the protection of those facilities, systems, and functions that face a specific, imminent threat. Finally, as the assessment team continues its analysis, it must remain cognizant that criticality varies as a function of time, risk, and market change. Assessing asset criticality is not a static analysis. It is continuous and evolving in nature.

The site characterization process describes a specific facility's contribution to overall business goals and impact against business loss. As a minimum, areas to be examined include:

- Primary and secondary mission and services
- Facility characteristics
- Asset and resource identification and criticality
- Physical geography
- Environmental attributes

PRIMARY AND SECONDARY MISSIONS AND SERVICES

Describing the particular mission and services of the site channels the assessment process to focus on what products and services are offered, including identification of the general and critical customer base. Where appropriate, distinguishing between primary and secondary services identifies internal and external dependencies that significantly impact the production of service capabilities of the enterprise that might require critical assessment consideration.

FACILITY CHARACTERISTICS

In performing the critical assessment the physical environment needs to be identified. The security-assessment team characterizes the facility in terms of its function and its construction as indicated.

Function

The facility population, throughput rates, hours of operation, and types of activities identify the facility's contribution to business goals and objectives. Such data as the site plan and facility schematic layout that identifies the physical configuration of activities and/or processes are crucial to the analysis.

Construction Category and Construction Type

Reviewing construction codes and construction design criteria helps the security-assessment team determine the facility's structural integrity against explosive effects, contamination, penetration delay times, and deterrence. Whether the facility is a permanent, semipermanent, temporary, shared with other facilities, or a new, altered, annexed, or converted structure helps to determine vulnerability. This analysis is particularly important in retrofit applications. The effectiveness of alternative solutions and cost-benefit analysis are of critical importance to enterprise executive management. More costly construction and installation options are often rejected in favor of enhanced procedural controls or increased security staffing. Alternatively, the continuing cost of manpower can be offset through a balanced mix of protective measures that include physical and electronic measures, protocols, and security awareness.

ASSET AND RESOURCE IDENTIFICATION AND CRITICALITY

The criticality of an asset is **four-dimensional**:

- The first dimension is the **importance of the asset** to the enterprise, its customer base, and the community.
  - ▸ What is its mission?
  - ▸ Is its production capacity indispensable to the enterprise?
  - ▸ What is the ease or difficulty surrounding its return to service or replacement if damaged or destroyed?
- The second dimension is the **asset's vulnerability**:
  - ▸ Is it a primary or secondary asset?
  - ▸ Does it have a redundant or backup capability?
  - ▸ Is it centrally located, dispersed, or remote and isolated?
- The third dimension is **adversary attractiveness**:
  - ▸ What is the adversary's perception of asset value?
  - ▸ Is it a soft or hardened target?
  - ▸ How does damaging or destroying the asset fit into the overall scheme of terrorism's objectives?

- The fourth dimension is **public reaction to endangering public safety and affecting community services:**
  - ‣ Will public confidence in the enterprise be tainted, seriously damaged, or lost altogether?
  - ‣ How will the stock market and financial sectors respond to the asset's loss and the enterprise's liability and reliability?

Assets include resources, facilities, equipment, material, and information of particular value to warrant some degree of protection. Resources are people. Assets may be either tangible or intangible. Examples of tangible assets include buildings, structures, land, equipment, material, and vehicles. Intangible assets include policies, protocols, and information in any form.

Asset value is based on many factors, including their relative importance to the enterprise's mission and function and the ease or difficulty with which they may be replaced. More direct means of measuring value include determining monetary costs, such as the potential publicity associated with the assets if compromised. Physical and cyber assets requiring characterization and documentation should include:

- Critical above-ground infrastructure, configuration, and boundaries
- Critical below-ground infrastructure, configuration and boundaries
- Supervisory control and data-acquisition [SCADA] systems
- Security networks and other management-information systems
- Assets, operations, processes, and protocols of importance to the enterprise
- Magnitude and duration of service disruption
- Chronic problems arising from any man-made and natural events
- Value of loss consequence including economic impact
- Impact on public and customer confidence to deliver services
- Other indications of impact consequences

Assets requiring protection should be identified as specifically as possible. Isolating assets within a facility or area will reduce the scope of protection required. If an entire facility or area requires protection, however, the scope of protection and its associated cost will be higher than that for an asset contained within a defined location. To the extent possible, assets should also be identified as primary or secondary.

Primary assets are potentially attractive targets for adversaries. They have a unique value to both the enterprise and an adversary because of:

- Visibility and prestige to the enterprise and the public
- Public perception of importance to the community and the enterprise
- Effect of compromise or loss on public safety
- Public confidence in the enterprise to provide a safe and secure work environment
- Business impact on enterprise's capability to sustain loss and continue operations
- Effects of publicity and media attention
- Impact of monetary loss to the enterprise, local community, and stock market
- Direct applicability of the asset to past, present, and future adversary goals

Secondary assets are those upon which the primary asset depends but which may not be directly related to it. For example, if a computer center is identified as a primary asset, the electrical-distribution system could qualify as a secondary asset. Destroying the electrical system could compromise the operation of the computer center as well as other assets relying on the same electrical system. Either the electrical system should be protected, or the computer system's protective features should compensate for the potential loss of the primary electrical system. A protective measure for a computer system could be a combination of any or all measures such as an uninterruptible power supply, emergency generator, or redundant computer-system operations.

Where a primary asset is supported by one or more secondary assets, the compromise of these secondary assets may significantly impact the compromise of the primary asset. The security-assessment team recognizes any primary-secondary asset relationships and ensures that the secondary assets are provided the necessary degree of protection. The value of a secondary asset is established by examining its importance to the mission of the primary asset. A secondary asset should not be considered more valuable or call for greater security than the primary asset it supports. When mission objectives, circumstances, and conditions warrant, providing secondary assets equal protection to the primary asset is generally justifiable.

Primary-Asset Considerations

In considering primary assets the security-assessment team takes into account the mission of the facility and the overall enterprise mission, tangible and intangible value, regulatory requirements, and enterprise directives.

Secondary-Asset Considerations

In considering secondary assets the security-assessment team looks at the operation of a primary asset that may require a secondary asset, such as utilities for continuous operations and other related support functions.

Criticality of Assets

Assets can be critical or noncritical. Noncritical assets are those whose loss, damage, or destruction has no significant effect on the enterprise, whose risk exposure is acceptable, and whose support requires only minimal protective measures. In this category are consumable products, office equipment, and furniture. Critical assets, on the other hand, have a significant effect on the enterprise if damaged or destroyed or if a particular process is significantly disrupted. In this category are assets that directly contribute to the production capability of delivered products and services. These include raw materials, manufacturing equipment, spare parts, energy and transportation systems, banking institutions, telecommunication and public-health services, and time-sensitive and time-dependent functions and processes. **The criticality of an asset therefore refers to its importance to the enterprise mission.** If facility or asset A has a greater value than facility or asset B, it stands to reason that protective measures for A should be greater than those for B. Criticality is a factor in establishing the level of protection the asset warrants to support critical business operations. It is important to identify critical assets for protection, but it is equally important to eliminate noncritical assets from consideration. Protective measures are expensive. Providing special protective measures for noncritical assets wastes funds that are already scarce. To stress a previous point made, in determining asset criticality, the security-assessment team must consider the asset's mission and whether or not its function is redundant; the enterprise's and adversary's perception of value; and the ease or difficulty with which an asset can be replaced or returned to service. An asset may be replaced by new construction, the modification of existing assets, or by an available alternative asset. Costs and the logistics of reconstruction or relocation and loss of revenues if the asset is lost are major factors of consideration.

PHYSICAL GEOGRAPHY AND ENVIRONMENTAL ATTRIBUTES

In a comprehensive security assessment the environment is viewed in a larger context than that surrounding a facility and asset. For instance geographic and environmental conditions influence facility siting and selection for new construction, and security equipment and technology are sensitive to weather and soil conditions, and geographic conditions indicate direction and modes of attack and which natural features can be used to the advantage of the security organization. Landscaping and site features influence location layout, equipment selection, and lines of protection. Landscaping features also impede or delay access and obscure or provide cover for aggressors or bombs.

Physical Geography

Identifying the physical geography not only complements the description of the facility's physical environment; it surfaces potential limitations and constraints that may impact on the results of the security assessment. These considerations include but are not limited to the following:

- Environmental boundaries such as terrain and major natural features
- Proximity of adjoining boundaries and land usage
- Facility legal boundaries, jurisdictional issues, and access routes
- Proximity of adjoining facilities, easements, and urban areas
- Nearby highways and commercial transportation routes [air, sea, truck, and rail]
- Political boundaries subject to federal, state, and local jurisdiction that provide law-enforcement and emergency-response services

Environmental Attributes and Physical Configuration

The presence of environmental attributes such as seismic activity [man-made, mechanical, or natural], radio-frequency and electromagnetic interference, and weather conditions contributes to subsequent potential considerations for the selection and siting of electronic security equipment. The physical configurations of barriers; lighting; heating, ventilating, and air-conditioning equipment; and other internal stimuli are also factors. Failure to identify and evaluate these and related factors may negate the validity of technology recommendations.

DOCUMENTING THE SITE CHARACTERIZATION PROCESS

Documenting the site-characterization process is critical to the data-collection effort: Because each site is unique in its security needs and vulnerabilities, documenting observations and findings is indispensable to establishing insight into the security needs specific to a particular location as well as into how those needs relate to the overall security plans of the enterprise. Capturing site conditions and circumstances can be accomplished using a series of scenario worksheets designed to systematically describe the business environment. By default, the security-assessment team is also able to define the threat environment during the data collection effort.

Documenting Facility Characterization

Worksheet 6 offers a comprehensive approach to documenting site characteristics and service capabilities in a user-friendly format. A completed sample of this worksheet is provided for illustrative purposes.

The information captured is useful to the security-assessment team in determining the vulnerabilities and target attractiveness of the facilities, their assets, and resources.

Documenting Critical Operational Criteria and Business Values

Worksheet 7 is an effective management tool for documenting the rank order of critical business values. It helps the security-assessment team determine which business considerations are most important to business-competitive advantages. For illustrative purposes Worksheet 7 identifies six business-performance measurement factors: **capacity, geographical extent, physical layout, critical customers,** and **quality**. Each criterion is defined in the illustration. Enterprises are encouraged to develop criteria of importance to their own business culture or accept the proposed values. The security-assessment team should work with the enterprise to refine these criteria as necessary.

In the above scenario, a peer-wise standard is used to measure the operational performance criteria based on the established definition of such criteria. A numeric rating is then applied to determine the value sum. The challenge in determining the importance of operational performance is making the clear and distinct definition of their purpose and value to business operations. In the above scenario, the physical layout of the plant and its ability to serve critical customers are more important criteria than the others. The establishment of these priorities is helpful to

Site Description:
Facility encompasses 15 acres, located in a commercial/residential area. It supports 85 operational locations. The surrounding terrain is very flat, sloping gently to the southwest. Adjacent land uses include industrial trucking and storage complexes and a salvage yard.

Population: 175 employees; 10 to 15 daily deliveries; 20 visitors

Population Throughputs	M	T	W	T	F	S	S
2nd Shift	275	275	275	275	275	59	59
3rd Shift	159	159	159	159	159	24	24
1st Shift	12	12	12	12	12	12	12

Key Services:
General Customer Base: [Self-explanatory]
Critical Customer Base: [Self-explanatory]

Facility Perimeter Boundary Configuration and Construction Type:
The facility is encircled by a 10-foot concrete wall topped with rolled razor concertina wire. Four entry points are secured by vehicle gates, including a Main Entrance, which is manned by a security officer 24/7.

Key Buildings:
Administration Building: The Administration Building is a two-story masonry structure that houses offices, a conference room, and a large auditorium. Major functions include a procurement division, quality assurance inspection office, and an auditing function. Several tenant organizations also occupy this facility. No formal access controls are employed to enter the building.

Warehouse: The Warehouse is a three-story steel-frame structure with metal siding. It is used for the storage and inspection of incoming materials and houses offices and training classrooms. Several tenant organizations also occupy this facility. During disasters or terrorist attacks, a designated area of this facility is used as a corporate crisis management center. No formal access controls are employed to enter the Warehouse complex.

Equipment Lay-down Area: Equipment is stored in an open area for rapid deployment as required. Except for area lighting no other security is provided in this area.

Fuel Station: The Fuel Station has diesel and unleaded gasoline pumps to support 75 vehicles assigned to the facility and to support fueling operations for company transit vehicles. Pump access is granted by an assigned gas card. No other security is provided.

Vehicle Maintenance Area: The Vehicle Maintenance Area performs necessary inspections and repairs of assigned vehicles, including company transit vehicles allocated use of this facility. The maintenance supervisor maintains a duplicate set of vehicle keys.

Vehicle Fleet Parking Area: 75 company light and heavy duty vehicles are parked in this area. Except for area lighting, no other security is provided in this area. Vehicle keys are controlled and issued by various supervisors.

Employee/Visitor Parking Areas: An adjacent area is used for employee and visitor parking. Except for area lighting no other security is provided in this area.

Site Emergency Generator: One generator is used to support all facility operations. Generator is tested on a quarterly basis.

HAZMAT Storage Areas: Two HAZMAT Storage Areas exist on the property. One is located in a secure area within the Vehicle Maintenance Area. The second storage area is located in an open area adjacent to the site emergency generator. Key to the HAZMAT storage containers are controlled by various supervisors.

Site Safety Characteristics and Facility Safety Features: Each HAZMAT container has a sump providing containment. Administration, Warehouse, and Vehicle Maintenance Areas have fire sprinkler systems, fire hoses, and fire extinguishers as appropriate to industry standards. The Fuel Station tanks are underground, double-walled, and have leak detection devices. There is a drum of absorbent at the Fuel Station that can be used to control spills.

Exhibit 6.1   Example of Worksheet 6—Facility Characterization

Criteria	Capacity	Geographic	Geographic Extent	Physical Layout	Critical Customers	Quality	Sum
Capacity: Ability to meet customer production demand		.5	.3	.3	.3	.3	.17
Geographic: Presence and service to general community	.1		.3	.4	.4	.4	.16
Geographical Extent: Ability to deliver to all service locations on time	.5	.1		.5	.4	.4	.14
Physical Layout: Importance of delivering quality services	.3	.4	.4		.4	.4	.19
Critical Customers: Delivering services 24/7 to identified critical customers base	.5	.2	.4	.3		.4	.18
Quality: Service rejection rate	.4	.1	.1	.3	.4		.13

Legend:
.5   Much greater than
.4   Greater than
.3   The same as
.2   Lower than
.1   Much lower than

Exhibit 6.2 Example of Worksheet 7—Defining Critical Operational Criteria and Business Values

the security-assessment team in identifying specific strengths and weaknesses in the security program to support these service needs. The collection of this data serves as a baseline to identify critical assets in relationship to business goals and operating priorities and provides input data for subsequent worksheets.

Documenting Facility Ranking Based on Enterprise Operational Criteria

When the enterprise has multidiscipline functions and geographically separated operations, Worksheet 8 is helpful for rank-ordering by peer-wise analysis each facility against accepted business criteria. To gather this information, the security-assessment team would have to complete Worksheet 7 for each separate plant or facility and then transfer the sum totals of each location to Worksheet 8 to complete the peer-wise analysis.

In the example on the facing page, Plants 1 and 5 are ranked at equal top importance, while Plants 4 and 6 are ranked equally at the second level of importance. The remaining plants follow in sequence of their importance. The information helps the security-assessment team to recommend a cardinal rank ordering of facilities, which in subsequent analysis will determine the level of security to be applied to each location.

Documenting Time-Sensitive Criteria

Every enterprise has some type of time-sensitive dependencies that guide operational performance parameters. Time-sensitive dependencies are often ignored during a security assessment, as these functions are traditionally viewed as site-management-planning responsibilities. However, these criteria often impact upon both normal and emergency security operations and should be identified by the security-assessment team. This worksheet captures **time-critical, time-sensitive,** and **time-dependent** crisis-management issues essential to the survival of the enterprise should it become the victim of a natural disaster or terrorist attack. Examples include the processing or transportation of hazardous materials and perishable goods, the breakage or damage of equipment or assembly lines, the disruption or loss of services, and work slowdowns or shutdowns. Subsequent emergency actions and the requisite transition recovery period to restore an acceptable level of operational capability directly

Operational Criteria of Specified Facilities	Facilities					
	Plant 1	Plant 2	Plant 3	Plant 4	Plant 5	Plant 6
Capacity: Ability to meet customer production demand	.5	.5	.3	.3	.3	.3
Geographic: Presence and service to general community	.1	.4	.3	.4	.4	.4
Geographical Extent: Ability to deliver to all service locations on time	.5	.1	.3	.5	.4	.4
Physical Layout: Importance of delivering quality services	.3	.4	.4	.3	.4	.4
Critical Customers: Delivering services 24/7 to identified critical customer base	.5	.2	.4	.3	.4	.4
Quality: Service rejection rate	.4	.1	.1	.3	.4	.2
Sum	.23	.17	.18	.21	.23	.21

Legend:
.5	Much greater than	.2	Lower than
.4	Greater than	.1	Much lower than
.3	The same as		

Exhibit 6.3 Example of Worksheet 8—Facility Ranking Based on Corporate Operational Criteria

TIME-CRITICAL 0–3 DAYS	TIME-SENSITIVE 4–7 DAYS	TIME-DEPENDENT 8+ DAYS
Loss of critical infrastructure	Loss of personnel	Government and industry relations
Loss of telecommunications, data, and other information systems	Treasury contingency cash plan	Corporate image
Degradation/loss of critical operations	Transition to recovery	Banking and finance
Degradation/loss of operational capability	Organization and resumption of critical business functions	Insurance and risk management plan
Loss of utilities	Infrastructure restoration	Work space outside primary facility

Exhibit 6.4 Example of Worksheet 9—Time Criteria

and indirectly impact on the ability of the enterprise's security organization to sustain its security mission. As such, it is essential that the security-assessment team work closely with the enterprise in identifying those criteria indispensable to the enterprise's business objectives in terms of their time-sensitive dependencies.

For example, in the above scenario the listed activities or events are categorized by their most immediate impact on the enterprise in terms of time criticality, time sensitivity, and time dependency. The losses identified in the respective columns indicate the length of satisfactory recovery periods before the enterprise reaches a critical performance failure point, loses monetary sustainability, or loses industry or community confidence needed to recover. This information is vital in identifying and assessing security services to support business operations as well as security priorities to response and recovery operations.

Identifying and Documenting Assets

A key challenge in prioritizing the importance of assets is a clear and distinct description of their mission statement, their importance to the enterprise, the asset's vulnerability, the attractiveness of the asset to an adversary, and public reaction to endangering public safety and affecting community services.

Business Value of Specified Asset at a Particular Location	Rank Ordering of Assets					
	Asset 1	Asset 2	Asset 3	Asset 4	Asset 5	Asset 6
Importance of asset production to the enterprise, its general customer base, its critical customer base, and the community.						
The vulnerability of the asset.						
Attractiveness of the asset to an adversary group or adversary.						
Public reaction to endangering public safety and affecting community services.						
Consequence of loss if asset is destroyed.						
Total						

Legend:

Importance of one item relative to another item	Importance of the relative item to the initial item
Much greater than [.50]	Much lower than [.10]
Greater than [.40]	Lower than [.20]
The same as [.30]	The same as [.30]
Lower than [.20]	Greater than [.40]
Much lower than [.10]	Much greater than [.50]

Exhibit 6.5 Example of Worksheet 10—Ran-Ordering Assets

A method for identifying and rank-ordering assets is illustrated in the Exhibit above.

Documenting the Physical Security Characteristics of Assets

Worksheet 11 is one way to identify major assets and describe their current security posture. The objective of this initial data-collection effort is to describe which security measures are in place for all assets. The analysis of this information is presented in Task 5 [program analyses].

Facility Assets	Site Access Controls	Perimeter Delay Barriers Facility Spacing	Facility Access Controls	Facility and Asset Barriers	Safety Alarms	IDS	CCTV	Communications	Area Lighting
Complex and Perimeter		10-foot wall w/wire				None	None		Industry Standard
Complex Entrances and Spacing Requirements	1 Main Gate Entry Controller 3 other entrances controlled by key	Main Gate spacing criteria inadequate							
Warehouse			Key Lock	Structure	Fire Alarm	None	None	Land Line Radio	Industry Standard
Equipment Parking Area			None	None		None	None		Industry Standard
Emergency Generator			None	None	None	None	None	Radio	Industry Standard
Crisis Management Center			Key Lock	Structure	Fire Alarm	None	None	Land Line Radio	Industry Standard
Administration Building			Key Control	Structure	Fire Alarm	None	None	Land Line Radio	Industry Standard
Fuel Island			Card Key Control	Bollards	Fuel Leak Spill Alarm	None	None	Radio	Industry Standard
HAZMAT Storage Areas			Key Control	Containers		None	None	Radio	Industry Standard
Guard Station	Manual		None	Structure		None	None	Land Line Radio	Industry Standard

Exhibit 6.6 Example of Worksheet 11—Asset Identification and Physical Security Characteristics

The information collected is useful to the security-assessment team in later evaluating the effectiveness of in-place protective measures. This analysis is presented in Task 4 [evaluating program effectiveness].

Recording the Observed Strengths, Vulnerabilities, and Adversary Attractiveness to Security Characteristics

Worksheet 12 is used to record relevant observations applicable to previous entries made on Worksheet 6 [facility characteristics] and Worksheet 10 [asset identification and physical security characteristics]. The transfer of this information from Worksheets 6 and 10 start the vulnerability-analysis process that is also carried over to Task 4 and Task 5. It is extremely valuable to jot down initial perceptions while in the field. This allows the security-assessment team the opportunity to walk the area,

OBSERVATIONS	STRENGTHS	VULNERABILITIES
Natural features that provide concealment or vantage points.		
Natural features that make access easy or difficult.		
Building density provides concealment or vantage points.		
Population density decreases visual detection.		
Open space that increases probability of visual detection.		
Distance relative to boundary, facility, and asset.		
Homogeneity of populace [conspicuousness of outsiders].		
Cultural attitude/acceptance of security compliance.		
In-house response capability.		
External local response capability.		
External state and federal response.		
Consistency of enforcement of security protocols.		
Restricting adversary escape.		
Expected results regardless of consequences of attack.		
Expected casualties and losses.		

Exhibit 6.7 Example of Worksheet 12—Security Characteristics Strengths and Weakness

retrace steps, and ask site personnel questions to fill in any voids or gaps that have been previously identified. Often such perceptions are helpful in steering the investigative process in a specific direction.

Strengths are physical features, security policies and protocols, operational processes, resources, and enterprise culture that address specific in-place protective measures that eliminate, reduce, or control vulnerability. A series of potential protective measures may exist to **deter, detect,** or **delay** entry, and **assess** and **respond** to an adversary action. These elements should be described as program strengths.

Vulnerability may be defined as the relative accessibility of an area or asset to specific risks; incomplete, out-of-date, or misleading protocols; or weak practices. For example, the outsider, working alone or with other outsiders, is confronted with a full range of protective measures to be bypassed. An insider may possess the ability to bypass one or more of these protective measures or assist the outsider by creating a diversion or by tampering with the performance of an operational or security system. Insider/outsider collusion threats require the full consideration of redundancy, diversity, and strict protocols and the resulting layering of security essential to maintaining the integrity of a successful level of program effectiveness. Poor processes and protocols produce ineffective controls and lead to inconsistent performance that gives birth to "vulnerability creep-in."

The collective data captured on these worksheets form the basis for analysis and problem solving presented in Task 4 and Task 5.

WORK BREAKDOWN

Quality site/facility characterization includes a comprehensive work breakdown structure.

Subtask 2A—Enterprise Characterization

Conduct in-depth site investigation to obtain and document specific information:

- Mission, service area, customer base, critical customer base
- Service capabilities, business culture
- Physical configuration, layout, and boundaries
- Hours of operation

- Population and population throughputs [including persons, vehicles, deliveries]
- Categories of tenants, as applicable
- Major structural, civil, mechanical, electrical infrastructure, primary equipment and subsystem equipment
- Identification and location of assets requiring protection, the surrounding environment, processes and time-sensitive deliveries, and associated security measures
- Identification and location of HAZMAT and high-value area delivery structures and conveyances, processes, handling, and associated security measures
- Communications
- Distribution control systems
- SCADA systems
- IT Networks
- Power Sources
- Redundant and backup systems
- Internal and external dependencies
- Land use and adjacent properties
- Other critical information

Subtask 2B—Data Analysis

Collect, review, and become familiar with related existing reports, procedures, operations, plans, construction and record drawings, and diagrams and sketches not previously available.

Subtask 2C—Security Characterization

Identify and describe in-place security measures as applicable:

- Physical barriers and security measures
- Security technologies and security operations
- Security advisory system and alert-notification system
- Information security
- Security awareness
- Personnel protection

- Security organization and support
- Security facilities and equipment
- Security qualifications and training
- Notification to enterprise management of any major system deficiencies noted during the data-collection process deemed to be of pressing importance

Subtask 2D—Capital-Improvement Characterization

Identify new construction that may have:

- Short-term security impact
- Long-term security impact

Subtask 2E—Engineering Data

Prepare as needed sketches to:

- Identify preliminary engineering calculations and measurements
- Identify preliminary technical interfaces
- Identify preliminary human factors engineering considerations
- Identify preliminary safety margins and system protective criteria
- Preliminarily identify location of security devices
- Preliminarily plot other categories of system components

In addition to documenting the status and condition of the site, taking photographs of noted relevant conditions provides valuable supporting data in developing conclusions and presenting facts, both in the final security-assessment report and when presenting findings to executive bodies.

CONCLUSION

The **site/facility characterization and asset identification process** describes the work and threat environments. It involves describing the site configuration, operations, and other elements by examining conditions, circumstances, and situations relative to safeguarding public health and safety and in reducing the potential for disruption of services. The

data-collection process focuses on which assets need protection to minimize the impact of undesirable consequences and takes into account the impacts that could substantially disrupt the ability of the enterprise to provide safe services and to reduce risks associated with the consequences of significant events.

Using worksheets to document findings keeps track of observations and decisions made during the data-collection process and helps the security-assessment team and enterprise management with the analysis process.

Chapter 7

TASK 3—IDENTIFY AND CHARACTERIZE THREATS TO THE SERVICE ENVIRONMENT

THE DESIGN-BASIS THREAT PROFILE

The threat analysis helps develop the design-basis threat profile that serves as the platform for protective measures, staffing requirements, and budget planning.

This chapter examines the possible threat, modes of operation, threat capabilities, threat level, likelihood of occurrence, and loss consequences. Focus is on threat characteristics and capabilities. Social-order demographics are examined to develop trend information. Based on the data collected and its significance, it provides the foundation for selection and implementation of actions to reduce the risk.

It outlines a three-step process and respective task interrelations. For maximum efficiency the steps should be performed in the presented sequence:

- Subtask 3A—Review available threat-related information
- Subtask 3B—Interface with external key players and document expectations
- Subtask 3C—Formulate initial threat analyses and preliminary design basis

The following suggested worksheets to assist in collecting and analyzing threat data are offered:

- Worksheet 13—Example of adversary characteristics by adversary profile
- Worksheet 14—Example of modes of adversary attack, weapons, and equipment
- Worksheet 15—Example of assets by adversary attractiveness
- Worksheet 16—Example of range and potential levels of malevolent acts and lesser threats
- Worksheet 17—Example of potential threats to assets by adversary attractiveness
- Worksheet 18—Example of malevolent acts and undesirable events by loss of consequence [C] and probability of occurrence $[P_A]$

Constant and increasing threats to national critical infrastructure assets are becoming more common and more complex to prevent and respond to. Networks of criminals and terrorists are now equal to or superior to our preventive and defensive techniques, methods, and processes. Therefore, to present a credible threat analysis, it is important to put the broader issue of targetability into context. **Before you can address security, you must first assess the threats, risks, and vulnerabilities that are present, including perceived threats, postulated threats, probable threats, and emerging threats.** The security-assessment process involves finding sources for credible comprehensive threat data that may be gathered from internal sources such as incident reports and security-system software, as well as external threat data such as crime statistics, industry standards, benchmarking data, and historical data about what has previously occurred in the enterprise and other similar organizations.

The **design-basis threat** formulates the platform for subsequent security decision-making by identifying:

- Postulated threats and other threat information
- Potential threats against the enterprise and adversary planning considerations
- Threat consequences to the enterprise
- Critical assets requiring protection
- The probability of occurrence of malevolent acts and lesser threats

Developing an effective security strategy requires a clear understanding of the threats of our time, the potential consequences they entail, and the best means to achieve the integrated security operational capabilities of **delay, deter, detect, assess,** and **respond** to defend against attacks. Threat and vulnerability analyses involve the interdependencies between business functions, the effectiveness of in-place measures, and roles and responsibilities, particularly those that blur functional boundaries.

The enterprise needs to know which threats can affect business operations and the consequences of those threats should an actual event occur. Only by understanding the range and level of threats can decision-makers make informed judgments to reduce or manage security risk more effectively. In many instances an enterprise expects this level of technical expertise to be provided by the security consultant; this is one of the reasons why the enterprise's threat statement is usually vague or even nonexistent. Another reason is economics: most enterprises don't maintain a current validated design-basis threat statement. Rather, they rely on obtaining generic threat information provided by professional organizations or obtain threat information specific to the enterprise directly from a security consultant.

The design-basis threat profile helps clearly define the ability of an enterprise to defend against a terrorist act and other criminal activities. It brings together, in a single reference, relevant and vital intelligence information so that security decision-makers can "connect the dots" with respect to the strengths and weaknesses of the enterprise in providing security. To be of significant value to the enterprise, the developed design-basis threat statement must be comprehensive and encompass several layers of analysis. The depth of each layer depends on the conditions and circumstances and the need for the enterprise to meet its business goals and objectives. Exhibit 7.1 illustrates the essential elements formulate a design-basis threat profile.

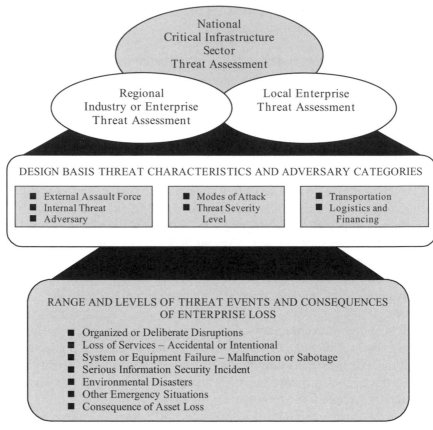

Exhibit 7.1 Composition of Design-Basis Threat Profile

A comprehensive threat profile begins with the summary of the threat affecting the national critical infrastructure sector to which the enterprise belongs. For example, an electric and gas company falls under the energy sector, and as such, a summary of the national energy threat would be appropriate. In this particular example, the national energy threat would be supplemented by a regional energy-threat assessment, followed by a local or site-specific threat summary. In addition, threats from environmental conditions or natural disasters relevant to the specific region or local site should be included.

The types of threats considered are acts that disrupt normal business operations, physical destruction of facilities and assets, contamination, cyber attack, theft of property, and vandalism. Consideration needs to be given also to each type of potential adversary to include disgruntled employees, disoriented persons, criminals, activities, and terrorists. The probable tactics each of these adversaries might use [e.g., deceit, stealth, force] also need to be addressed. As appropriate, six factors that should be considered in the development of a comprehensive design-basis threat profile include:

- Use of national-security assessments
- International, national, and local events
- Crime data and trends
- Industry incidents and trends
- Potential events and natural disasters
- Interviews with federal, state, and local authorities and professional organizations

Once the data-gathering process has been completed, the information is compiled and evaluated to determine a rank order of threats for each area, locale, or facility as appropriate.

The National Critical Infrastructure Sector Threat Assessment

The first layer—the **strategic analysis**—is a necessary element of threat analysis for Fortune 500 corporations and those enterprises that have multiple business locations throughout the nation or international interests. This national infrastructure sector analysis is particularly significant when larger terrorist organizations are exchanging intelligence and/or planning joint or independent strategic operations. Given the stated intention of organizations such as al Qaeda to destroy the economic prowess of the United States, this is very relevant to the business enterprises and the communities they service. Physical and cyber security risks that threaten the integrity of the national critical infrastructure include:

- Biological, chemical, and radiological contamination
- Physical destruction of systems, subsystems, and components
- Loss of services for fire protection, law enforcement, hospitals, and other critical customers
- Disruption of functions, activities, and operations
- Disruption of interdependent support systems
- Destruction of public confidence

In general, many enterprises do not consider a strategic threat assessment to be significant to their survivability. Many are unaware that a need exists to embrace a strategic security vision with respect to enterprise continuity.

The Regional Industry Enterprise Threat Assessment

The second layer—the **regional analysis**—is important for identifying the general threat environment of industries that may be geographically grouped or located together. As previously mentioned, most corporations are not specific targets of terrorist attacks. But when a particular corporation or industry—such as banks located within a financial district—are targeted, adjacent and neighboring enterprises must understand the effects of any cascading or collateral damage that may impact an enterprise's capability to sustain business operations and continuity. This analysis is relevant to both large corporations that need a national infrastructure threat assessment and to medium and small business enterprises that are grouped geographically with others in their industry, including competitors. Typical examples of such groupings include major industrial parks as well as the technology beltways around metropolitan areas such as Washington, DC; Houston and San Antonio, Texas; and the Silicon Valley in California, to name a few.

The Enterprise Threat Assessment

The third layer—**local analysis**—is the most prevalent threat analysis conducted by corporate America. This analysis focuses on the threat environment of a particular enterprise location and is more a tactical than a strategic assessment because only one facility or location is involved. When multiple enterprise locations require a threat assessment for the purposes of developing a strategic security plan, then an understanding of the national and regional threat assessments should be of interest to enterprise executive management in arriving at a realistic threat analysis.

Adversary Characteristics, Modes of Adversary Attack, Weapons, and Equipment

One cannot be successful in identifying security strengths and weaknesses and in evaluating the effectiveness of a security program without first knowing the adversary's organization and affiliates, tactics and weapons, and ideology. One must also respect their ability to harm us whenever and wherever they choose.

Adversary identification is based on intelligence data already or soon to be developed and validated. This information comes in a variety of forms and can be found in several places from government archives and private records, government regulatory bodies, security and law-enforcement agencies, private foundations, libraries, and the Internet. When available information is out of date or not applicable to an enterprise, it must be developed. The task of the security consultant is to "present" the information in an organized and systematic fashion that is user-friendly, clear, and oriented to decision-makers. To this point, the data should be assembled into these major categories:

- Adversary characteristics by adversary profile
- Modes of adversary attack and equipment
- Range and potential level of malevolent acts and lesser threats
- Potential threats and adversary attractiveness
- Malevolent acts and lesser threats by loss of consequence [C] and probability of occurrence [P_A]

Signally and collectively this grouping of information provides an invaluable resource that will lead to ultimate conclusions and recommendations that increase protective measures and enhance overall security-program effectiveness and efficiency.

DOCUMENTING THE DESIGN-BASIS THREAT

Documenting the design-basis threat is critical to subsequent strategic security planning and security program development and implementation. The first three elements of the design-basis threat statement [the general national critical infrastructure sector threat assessment, the regional industry threat assessment, and the local enterprise threat assessment] are best presented in narrative form with the use of illustrations, tables, and charts depicting recent and relevant data as appropriate and applicable to each observation. Not only do they support the analysis and conclusions presented, but they help those who understand concepts better in visual form grasp the arguments. The narrative should include specific facts. For example, a statement such as "recent threats and attacks against several utilities" does not have the sobering effect that a statement such as "Exhibit A illustrates the types and quantity of threats received and actual threat events that have occurred by location in the utility sector during the last 4 years."

Other data is best captured in summary-table format. A suggested approach is presented in the following sample worksheets. Explanations for preparing and using the data are provided for each as appropriate. The worksheets are presented in the sequence normally completed to complement the security-assessment-analysis process. However, for large, complex security projects that require a team of consultants, the data research and collection process may be performed concurrently and then pulled together by the security-assessment team during the final analysis process.

Identifying and Documenting Adversary Characteristics by Adversary Profile

To protect critical infrastructures requires a clear understanding of the intent and objectives of terrorism as well as the tactics and techniques its agents could employ against various targets. We must complement this understanding with a comprehensive assessment of the assets and resources to be protected, their vulnerabilities, and the challenges associated with eliminating or mitigating those vulnerabilities. Exhibit 7.2 [Worksheet 13] identifies a summary of adversary profile traits by group that would be of significant value during the decision-making process. The worksheet includes:

- Organization skills and recruitment
- Planning methods, tactics, logistics, and technical sophistication
- Motivation, willingness, and training

The sample profile is for illustrative purposes only. The profile column represents a sampling of organizational and culture traits that are applicable to each category of the adversarial groups identified. **The data collected to prepare this worksheet provides essential output source data for completing the various other worksheets used in the overall data collection and analysis process.** When properly completed, the information captured on this worksheet provides the basic criteria for "knowing the enemy."

Profile	Adversary Group					
	Terrorist Groups	**Extremist Protect Groups**	**Organized Criminal Groups**	**Miscel-laneous Criminal[s]**	**Disoriented Persons**	**Disgruntled Employees**
Organization	Well organized, hierarchical, bureaucratic, specialization, compartmental-ization practices.	No formal organization to well organized groups.	Efficient, hierarchical, bureaucratic; specific organization.	Little or no formal organization.	Little or no formal organi-zation. Anti-socials may belong to an organized criminal entity.	Little or no formal organization except organized strike violence.
Recruitment						Normally operates alone. Does not recruit others except sympathetic friends.
Financial						Use of personal funds as necessary.
International Concerns						None determined.
Planning						Little evidence of extensive or long-term planning.
Timing						Most acts timed to minimize risk of discovery.
Tactics						Bombing, sabotage, theft, intrusion, property damage, vandalism.
Collusion [inside]						Most often are insiders.
Motivation						Range of employment-related problems.
Dedication – Disclosure						Low
Willingness to Kill						Low

(continued on next page)

Exhibit 7.2 Example of Worksheet 13—Adversary Characteristics by Adversary Profile

Profile	Adversary Group					
	Terrorist Groups	Extremist Protect Groups	Organized Criminal Groups	Miscel-laneous Criminal[s]	Disoriented Persons	Disgruntled Employees
Willingness to Surrender						Low
Training Skills						Range is a function of individual's background, training, and experience.
Technical Sophistication						Range is a function of individual's background, training, and experience.
Group Size						Typically 1-2, 6 or more in strike related crime.
Weapons						Handguns, small arms, explosives, and incendiaries.
Equipment						Range of equipment is a function of individual's background, training, and experience.
Transportation						Privately owned vehicles, company vehicles, public transportation, on foot. *(continued on next page)*

Exhibit 7.2 Example of Worksheet 13—Adversary Characteristics by Adversary Profile *(continued)*

For each adversary group the security assessment team should define the appropriate profile using the following factors as a guideline for weighing capabilities:

ORGANIZATION	An adversary's structure ranges from the near-military organization of some terrorist groups to the informal organization. Define the one that applies.
RECRUITMENT	Consider how a group enlists its members, the quality of these members, and where it finds them. Consider mutual agreements with other adversaries to loan resources or conduct joint operations.
FINANCIAL	Consider how the adversary sustains itself and pays for weapons, tools, and explosives.
INTERNATIONAL CONCERNS	Consider aid in supplying, logistics, training, and safe havens.
PLANNING	Consider the ability to plan an act of aggression based on objectives, the tactics to be used, and the security posture of the asset.
TIMING	Consider the ability to determine the operational schedule and security of an operation and the vulnerability of the asset. An adversary attacks at the highest or lowest level of activity depending upon his objective.
TACTICS	Consider which tactics pose a threat to the asset and whether or not the adversary can employ those tactics.
COLLUSION [inside]	Consider an adversary's capability to develop collusion with an accomplice closely and legitimately associated with the identified asset and the usefulness of that collusion.
MOTIVATION	Consider the adversary's possible reasons for acting against the asset including politics, religion, nationalism, money, revenge, and irrational behavior.
DEDICATION – DISCLOSURE	Consider the adversary's resolve to carry out an act of aggression and to develop and maintain the capability necessary for that act.
WILLINGESS TO KILL	Consider the adversary's willingness to kill based on the asset, his objective, and the tactics anticipated to be used.
WILLINGESS TO SURRENDER	Consider the adversary's willingness to die based on his objective, the asset, and the tactics to be employed in the act of aggression.
TRAINING AND SKILLS	Consider whether or not an adversary has the necessary training and skills to successfully carry out an act of aggression. These include the ability to use weapons, tools, and explosives, and the skills needed to reach the asset.
TECHNICAL SOPHISTICATION	Consider whether or not an adversary possesses the level of tactical sophistication necessary to successfully compromise the asset. The manufacture and use of some weapons, tools, explosives, and the ability to defeat certain security systems require a degree of sophistication.
GROUP SIZE	Consider whether an individual or adversary group is needed to deploy the tactics necessary to successfully threaten the asset.
WEAPONS AND EQUIPMENT	Consider the adversary's possession of or ability to obtain the weapons, tools, and explosives necessary to successfully threaten the asset.
TRANSPORTATION	Consider the adversary's ability to transport to the site all of the necessary equipment to compromise the asset and to escape once the act of aggression has been carried out.

Exhibit 7.2 Example of Worksheet 13—Adversary Characteristics by Adversary Profile *(continued)*

Identifying and Documenting Modes of Adversary Attack and Equipment

Once we know who the enemy is, it is critical that we understand in which ways the enterprise can be placed in harm's way. Exhibit 7.3 [Worksheet 14] identifies the various adversary modes of attack by threat severity, weapons, and tools. It expands on the data presented in Worksheet 6 in the following profile areas:

- Tactics
- Collusion
- Technical sophistication
- Weapons
- Equipment

The security-assessment team will need to develop one worksheet for each asset and applicable adversary group, as illustrated below.

The data collected to prepare this worksheet establishes the platform for identifying the range and level of threats that will help to further develop security-planning criteria. When properly completed, the information captured on this worksheet gives us insight into "knowing enemy capabilities."

Documenting Assets by Adversary Attractiveness

In Task 2: (Critical Assessment) the security-assessment team identified and rank-ordered assets and documented their strengths and vulnerabilities for further analysis. Under the current task the security-assessment team examines these assets in terms of adversary attractiveness and considers adversary capabilities for each applicable group to determine the likelihood of aggression against the asset. Likelihood is rated as **high, moderate, low,** or **very low.** Only adversaries whose likelihood ratings are in the first three categories are carried forward to subsequent analysis. If the likelihood rating for a given adversary for a particular asset is determined to be **very low,** the security-assessment teams eliminate the adversary from further consideration. Exhibit 7.4 [Worksheet 15] illustrates an effective approach to performing this analysis.

Defining the Range and Potential Level of Malevolent Acts and Lesser Threats

Once we know the capabilities of the enemy, we must look at how they can do harm. Malevolent acts and other threats are any events that might damage or destroy property or cause injury or death. Such events have an

ASSET AND LOCATION: ADVERSARY: MODES OF ATTACK	THREAT SEVERITY	WEAPONS	TOOLS
Stationary or Moving Vehicle Bomb	Very High	200 lbs to 5,000 lbs TNT	Car, Van, Truck
	High		
	Moderate		
	Low		
Walk-in Bomber or Leave Behind	Very High		
	High		
	Moderate		
	Low		
Mail Bomb Delivery	Very High		
	High		
	Moderate		
	Low		
Airborne Contamination	Very High		
	High		
	Moderate		
	Low		
Standoff Weapons	Very High		
	High		
	Moderate		
	Low		
Forced Entry	Very High		
	High		
	Moderate		
	Low		
Covert Entry	Very High		
	High		
	Moderate		
	Low		
Insider Support	Very High		
	High		
	Moderate		
	Low		
Visual Surveillance	Very High		
	High		
	Moderate		
	Low		
Eavesdropping	Very High		
	High		
	Moderate		
	Low		

Legend:
Very High [VH]   High probability of using tactic, weapons, and tools successfully.
High [H]   Good probability of using tactic, weapons, and tools successfully.
Moderate [M]   Reasonable probability of using tactic, weapons, and tools successfully.
Low [L]   Little probability that tactic, weapons, and tools would be used successfully.

Exhibit 7.3 Example of Worksheet 14—Modes of Adversary Attack, Weapons, and Equipment

ASSET	ADVERSARIAL GROUPS					
	TERRORIST GROUP	EXTREME PROTECT GROUP	ORGANIZED CRIMINAL GROUP	MISC. CRIMINAL	DIS-ORIENTED PERSON	DIS-GRUNTLED EMPLOYEE
Complex and Perimeter	M	M	L	VL	L	VL
Complex Entrances	M	H	L	VL	L	VL
Warehouse	H	M	H	M	M	H
Equipment Parking Area	VL	L	H	H	M	H
Emergency Generator	H	M	M	L	M	H
Crisis Management Center	H	L	VL	L	M	H
Administration Building	L	H	VL	L	M	H
Fuel Island	H	H	VL	M	M	M
HAZMAT Storage Areas	H	H	VL	VL	M	L
Guard Station	H	H	L	VL	L	M
Human Resources	H	M	L	VL	M	H

Legend: H High   L   Low
          M Medium VL Very Low

Exhibit 7.4 Example of Worksheet 15—Assets by Adversary
Attractiveness

adverse impact on the enterprise's mission, physical assets, public relations, environment, and/or the health and safety of employees and the community. Through the years debate over categorizing malevolent acts and lesser threats has echoed throughout the security industry. Some professionals are of the opinion that such categorizing is not important to the overall task of identifying threat situations. This thinking is appropriate if the task of identifying threats has no further purpose. **The S^3E Security Assessment Methodology presented here, however, is based on the belief that threats need not receive equal attention, that threats pose a varying degree of loss consequences, and that a response to a threat condition need not always be a "charge up the hill."** To distinguish between the various types and degrees of threats, the security-assessment team should identify the specific range and potential level of malevolent acts and lesser threats that impact the enterprise. These include:

- Organized or deliberate disruptions—direct willful attacks against the enterprise
- System or equipment failure—malfunction or equipment breakage
- Serious information security incidents

- Industrial and environmental disasters
- Other emergency situations

As illustrated in Exhibit 7.5 [Worksheet 16], each threat range is populated with examples of its own unique potential level of malevolent acts and lesser threats. These call for a distinct set of conditions and circumstances that produce specific enterprise-security emergency-preparedness planning initiatives. In this sample, potential threats are listed in alphabetical order.

The listed threats are not intended to be all-inclusive. Rather, they are representative of potential terrorist events, criminal acts, and other emergencies that could result in consequence loss to an enterprise and deserve prudent and reasonable consideration for potential threat to business operations. The range and level of threats provide critical output information for determining adversary attractiveness, the next step in the analysis process.

Potential Levels of Threats and Adversary Attractiveness

Exhibit 7.6 [Worksheet 17] cross-references the identified potential level of threats against target attractiveness using the likelihood rating from high to very low as completed in Worksheet 15. Worksheet 17 illustrates one method of reporting the relationship of organized or deliberate disruptions against adversary attractiveness to a particular asset. The information is useful in determining which planning considerations must be addressed by each adversary group.

In Worksheet 17, the analysis indicates the range of potential threats and their attractiveness to adversaries by asset. **Developing this threat matrix helps decision-makers examine the strengths and weaknesses of the enterprise and determine program effectiveness with respect to security vulnerabilities and business continuance.**

THE ANALYSIS PROCESS

The last step in the threat-assessment process is the analysis of the information and formulating of threat conclusions. The security-assessment team uses information available through source intelligence and professional experience as a primary input to develop a threat-assessment statement unique to enterprise operations. All critical security intelligence information applicable to an enterprise's business interests is then pulled

Organized or Deliberate Disruptions – Direct Willful Attacks Against the Enterprise
Assault
Biological contamination
Bomb threat - planted
Bomb threat - vehicle
Bomb threat - walk-in
Chemical contamination
Communicating a threat
Cyber crime
Disclosure or theft of proprietary or sensitive information
Disruption or loss of courier or delivery services
Disruption or loss of recovery capability
Economic espionage
Extortion
Fraud, embezzlement, waste, and abuse
Hostage situation
Kidnapping
Radiological contamination
Robbery
Sabotage
Shrinkage
Tampering, damage, destruction of back-up power supply infrastructure
Tampering, damage, destruction of cable infrastructure
Tampering, damage, destruction of communications system infrastructure
Tampering, damage, destruction of drainage removal capability
Tampering, damage, destruction of energy system infrastructure
Tampering, damage, destruction of environmental control system infrastructure
Tampering, damage, destruction of fire system and safety system infrastructure
Tampering, damage, destruction of HVAC infrastructure
Tampering, damage, destruction of IT system, office system infrastructure
Tampering, damage, destruction of primary electrical power system
Tampering, damage, destruction of telephone system infrastructure
Tampering, damage, destruction of transportation system infrastructure
Theft
Trespassing
Vandalism
Workplace violence

(continued on next page)

Exhibit 7.5 Example of Worksheet 16—Range and Potential Level of Malevolent Acts and Lesser Threats

System or Equipment Failure – Malfunction or Equipment Breakage
Cable infrastructure
Communications systems
Electrical power supplies
Energy systems
Telephone systems
Environmental control systems
Financial systems
HVAC infrastructure
IT systems, office systems

Serious Information Security Incidents – Human Error
Data entry - employee error
Unintentional disclosure of proprietary or sensitive information
Loss, damage, destruction of records or data

Industrial and Environmental Disasters
Accidents - customers
Accidents – employees
Earthquake
Epidemic
Explosion
Fire

Industrial contamination and environmental hazards
Lightning
Neighborhood hazard
Production or business shutdown
Snow and ice storms
Water damage and rainstorms
Tornado and wind damage

Other Emergency Situations
Communicating a threat
Compromise of public confidence
Demonstrations and strikes
Inability of customers to conduct business in the normal way
Inability of key personnel to report to work
Liability issues
Local civil unrest
Loss of critical personnel
Long-term loss of external dependency systems
Negative publicity
Regional economic damage
Slowdown and interruption of courier/delivery services
Temporary closure of external dependency institutions
Temporary loss of external dependency systems

Exhibit 7.5 Example of Worksheet 16—Range and Potential Level of Malevolent Acts and Lesser Threats *(continued)*

RANGE OF POTENTIAL MALEVOLENT ACTS BY ASSET	ADVERSARY ATTRACTIVENESS					
	Terrorist Group	Extreme Protest Group	Organized Criminal Group	Miscellaneous Criminal	Disoriented Person	Disgruntled Employee
Organized or Deliberate Disruptions – Direct Attacks Against the Client						
Assault	H	H	L	M	M	H
Biological contamination	H	L	VL	VL	VL	VL
Bomb threat - planted	H	H	VL	VL	VL	M
Bomb threat - vehicle	H	H	VL	VL	VL	M
Bomb threat - walk-in	H	L	VL	VL	VL	VL
Chemical contamination	L	L	VL	VL	VL	VL
Communicating a threat	H	L	M	M	H	H
Cyber crime	L	H	M	L	L	H
Disclosure or theft of proprietary or sensitive information	M	M	L	M	M	H
Disruption, loss of courier or delivery services	M	M	L	L	L	VL
Disruption, loss of recovery capability	VL	M	H	L	L	VL
Economic espionage	H	L	H	L	L	M
Extortion	L	L	H	L	L	M
Fraud, embezzlement, waste, abuse	H	L	H	M	L	H
Hostage situation	H	M	M	L	L	M
Kidnapping	H	M	M	L	L	M
Radiological contamination	L	L	L	VL	VL	VL
Robbery	H	L	VL	M	M	L
Sabotage	VL	M	VL	L	M	M
Shrinkage	H	VL	VL	M	M	M
Tampering, damage, destruction of back-up power supply Infrastructure	H	H	M	M	L	M
Tampering, damage, destruction of cable infrastructure	H	H	M	M	L	M
Tampering, damage, destruction of communications system infrastructure	H	H	M	M	L	M
Tampering, damage, destruction of drainage removal capability	H	H	L	L	L	M
Tampering, damage, destruction of energy system infrastructure	H	H	M	VL	L	M
Tampering, damage, destruction of environmental control system infrastructure	H	H	M	L	L	VL
Tampering, damage, destruction of transportation system infrastructure	VL	M	M	L	VL	VL
Theft	VL	VL	H	M	M	H
Trespassing	VL	VL	VL	M	M	VL
Vandalism	VL	VL	VL	M	M	VL
Workplace violence	VL	VL	VL	VL	L	H
Tampering, damage, destruction of fire system and safety system infrastructure	M	H	M	L	L	M
Tampering, damage, destruction of HVAC infrastructure	M	M	M	L	L	M
Tampering, damage, destruction of IT system, office system infrastructure	H	H	M	L	L	H
Tampering, damage, destruction of primary electrical power system	H	H	M	L	L	M

Legend: H High L Low M Medium VL Very Low

Exhibit 7.6 Example of Worksheet 17—Potential Threats of Assets by Adversary Attractiveness

together. This serves as a blueprint for developing protective measures, determining security staffing needs, and security budget preparation. The threat conclusion involves those threats perceived to be significant by executive management, the senior staff, supervisors and employees in the field, law enforcement, and Homeland Security officials. Considering only one viewpoint produces a narrow, somewhat useless perspective. Therefore all viewpoints and data are equally valid, and must be analyzed, synthesized, and rank-ordered based on potential impact. Also evaluated are relevant incidents and losses of concern that have occurred in the past and current known threats, organizations, or groups. This includes crime statistical trends, national trends in similar industry operations, and postulated and emerging threats.

Identifying the Range and Potential Levels of Threat and Consequences of Enterprise Loss

The security-assessment team now prioritizes the range and potential levels of threats by rank-ordering the probability of occurrence [P_A] and the business impact resulting from a significant event. The previously reported threats are now illustrated in more detail. As some threats are less likely to occur than others and therefore may be less critical to the disruption of business operations, it is important to rank-order each threat to help decision-makers establish priorities for any corrective action. Some threats may occur only once in a period of years and others may occur on a more frequent basis—even daily or many times a day, particularly when a large corporation has multidiscipline operations that are geographically spread out throughout a region, across America, or in multiple locations abroad. Moreover, some threats are seasonal [e.g., natural disasters, vandalism, and internal theft] or situational [e.g., environmental hazards, white-collar crime, and terrorist attacks].

Once the data-gathering process is complete, the security-assessment team compiles this information in a user-friendly matrix format that clearly and distinctly reflects the rank order of threats by consequence of event occurrence, consequence loss factor, and probability of occurrence. Consideration is given to each type of potential adversary, including disgruntled employees, political activists, criminals, and terrorists. The probable tactics each of these adversaries might use are cataloged along with their capabilities, weapons, and expected number of participants. The information is useful when cross-referenced once more against prioritized critical assets.

Exhibit 7.7 [Worksheet 18] represents one method of reporting the threat analysis to the enterprise. It identifies the threat, distinguishes

MALEVOLENT ACTS AND UNDESIRABLE EVENTS	CONSEQUENCE OF EVENT OCCURRENCE	C	P_A
	Organized or Deliberate Disruptions		
Vehicle bomb threat	Can create uncertainty and fear and serve to destabilize the work environment. Can cause severe injury or death or destruction on a large scale, destroy power and communications lines, and disrupt services. Significant damage to a structure or building can occur, including total collapse of buildings and other adjacent structures. Loss of public confidence in the corporation to provide a safe and secure environment can also occur.	VS	H
	Loss of Center Services – Accidental or Intentional		
Loss, damage, destruction of electrical power	Can impact air conditioning, lights, telephones, communications, computers, and security equipment. Long-term outage can impact power operations.	MS	M
	System or Equipment Failure – Malfunction or Sabotage		
Loss, damage, destruction of IT systems, office systems	Can have consequences on the capability to maintain services where hardware failure, damage to cables, leaks, fires, air conditioning system failures, network failures, application system failures, and telecommunications equipment failures cannot be readily restarted and backup systems are not available. Loss of public confidence in the corporation to provide a safe and secure environment.	F	H
	Serious Information Security Incidents		
Loss, damage, destruction of records or data	Can be disruptive where poor backup and recovery procedures result in the need to re-input and recompile records. Can be embarrassing to organizational image where information is totally lost or not available. Loss of public confidence in the corporation to provide a safe and secure environment.	VS	M
	Environmental Disasters		
Industrial contamination & environmental hazards	Results from polluted air, polluted water, chemicals, radiation, asbestos, smoke, dampness and mildew, toxic waste, and oil pollution. Conditions can disrupt enterprise processes directly by mandated evacuation for long periods of time and, in addition, can cause sickness or death to employees and surrounding neighbors.	VS	L
	Other Emergency Situations		
Neighborhood hazard	Seepage of hazardous waste or escape of toxic gases or sewerage can cause injury or death to employees and community.	MS	L

(continued on next page)

Exhibit 7.7   Example of Worksheet 18—Malevolent Acts and Undesirable Events by Loss of Consequence [C] and Probability of Occurrence [P_A]

Legend:

Consequence Loss Factor [C]

CRITICALITY DESIGNATION	NUMERICAL RATING	ID RATING	CONSEQUENCE FACTOR [C]
FATAL TO ENTERPRISE MISSION	.90	F	Loss would result in total recapitalization, abandonment, or long-term discontinuance of business operations. Catastrophic loss, as in a WMD event.
VERY SERIOUS	.75	VS	Loss would interrupt service and require major effort and expenditure to repair.
MODERATELY SERIOUS	.50	MS	Loss would have noticeable but temporary impact on operations and would require attention from the senior executive management.
RELATIVELY UNIMPORTANT	.25	RU	Loss would be easily repaired and covered by normal contingency reserves.
SERIOUSNESS UNKNOWN	.10	SU	Insufficient data available to determine importance.

PROBABILITY OF OCCURRENCE [P_A] FACTOR

NARRATIVE DESIGNATION	NUMERICAL RATING	ID RATING	PERFORMANCE MEASUREMENT
OCCURRENCE HIGHLY PROBABLE	.85	H	Risk exposure is **HIGH**. The probability of an event occurring is much greater than the probability of the event not occurring.
OCCURRENCE MODERATELY PROBABLE	.65	M	Risk exposure is **MODERATE**. The probability of an event occurring is somewhat greater than the probability of the event not occurring.
OCCURRENCE LOW OR IMPROBABLE	.50	L	Risk is **LOW**. The probability of an event occurring is lower than the probability of the event not occurring.
OCCURRENCE PROBABILITY UNKNOWN	.20	I	Insufficient data available to evaluate circumstances and conditions.

DECISION MATRIX

Enterprise is exposed to **high or moderate threats and fatal or very serious consequences of loss.** Corrective action to resolve security vulnerabilities is critical to the business mission.

Enterprise is exposed to **moderate threats and very serious or moderate consequences of loss.** Action to resolve security vulnerabilities is essential to providing a safe and secure environment.

Enterprise is exposed to **moderate or low threats but the consequence of loss is relatively unimportant.** Action in some instances may be in the best interests of the corporation.

Exhibit 7.7 Example of Worksheet 18—Malevolent Acts and Undesirable Events by Loss of Consequence [C] and Probability of Occurrence [P_A] *(continued)*

between types of threats, and identifies the consequence of asset loss should a threat event occur. This collective information is then used to formulate strategic security protective measures and develop workable security emergency-preparedness planning.

Four analysis functions apply to the rank ordering of threats:

- The first focuses on defining the severity of consequence loss. This is best accomplished by graphically summarizing the expected level of loss, as illustrated in Worksheet 15 [potential threats by adversary attractiveness].
- The second pertains to the assignment of a consequence or business value to the acknowledged loss. Specifically, what are the short- and long-term impacts on the business operations should the described event occur, as illustrated in Worksheet 10?
- The third determines the probability of the event occurring, taking into consideration all factors as illustrated in Worksheet 15.
- The last analysis is the decision element facing enterprise executive management. This process assigns a priority value by applying the color-coded scheme below to each value, as illustrated in Worksheet 18.

The overall analysis is then brought into focus by rank-ordering the threats based on the collective analysis. The need to distinguish between the various types and degrees of threats and the establishment of a range of threats becomes quite clear. Each category or range of threats represents a distinct challenge to reducing vulnerability and risk, including collateral vulnerability and risk associated with external circumstances such as damage sustained to the enterprise by industrial contamination and environmental hazards from surrounding industries.

A COMPREHENSIVE WORK-BREAKDOWN STRUCTURE

SUBTASK 3A—Review Available Enterprise Threat-Related Information

Follow these procedures in reviewing information:

- Review of social-order status of surrounding community to include:
  - ► Previous incidents, losses that have occurred nationally or internationally to the enterprise

- ► Previous incidents, losses, or verifiable threats that have occurred against the enterprise
- ► Crime statistics obtained from the enterprise
- ► Neighborhood crime trends obtained from the law-enforcement agency of jurisdiction
- ► National reports developed by the FBI, state, or other agencies
- ► Crime statistics obtained through independent sources
- • Review of OSHA safety incidents
- • Review of human-resources turnover rate, discipline actions, and types of terminations
- • Identify and report to enterprise management the nature of research additionally required of the project team

SUBTASK 3B—Interface with External Key Players and Document Expectations

Here are the steps to take for this part of the task:

- • Conduct meetings and interviews with federal/state/local agencies such as the FBI and state, county and municipal law-enforcement agencies with respect to:
  - ► Identification of high-traffic drug areas
  - ► Nominal patrol coverage as applicable and their response times
  - ► Priority given to other business establishments and capabilities to support mission needs
  - ► Actual and postulated threats, crime, and threat trends.
- • Conduct meetings, interviews with health and fire-emergency response organizations with respect to:
  - ► Capability to support mission statement
  - ► Determining interface with enterprise in preparing and planning for disasters
  - ► Identifying response times and priority given to other business establishments
- • Conduct meetings and interviews with regulatory agencies or associations that govern enterprise business standards

- Conduct meetings and interviews with the area information sharing and analysis center with respect to potential threats and realistic risks that may potentially impact enterprise operations, including:
  - ▸ City representatives
  - ▸ Representatives of municipal, county, and state law enforcements
  - ▸ Representatives of state and national Homeland Security personnel
  - ▸ Representatives of the FBI
  - ▸ State, regional, and local emergency-management departments, as applicable
- Summarize the scope and content of all interviews within the final project report.

SUBTASK 3C—Formulate Initial Threat Analyses and Preliminary Design-Basis Threat

Here are the procedures for this stage:

- Use available national or agency reports or own analysis to:
  - ▸ Determine size and demographics of the community
  - ▸ Describe the local community to include sociopolitical, geographical, and economical status
  - ▸ Develop high-level process mapping
- Assess physical, biological, and chemical threats, internal and external, foreign and domestic
- Assess range and levels of threats and the consequences and effect on the enterprise's mission
- Identify categories of perpetrators, capabilities, and tactics
- Identify known organizations or individuals of concern in the general area if possible
- Evaluate publicly available sensitive information that might be useful to potential adversaries
- Match critical assets to the attractiveness of the design-basis threat by quantifying program or system vulnerabilities
- Rank-order the significance of threats based on probability of occurrence and the greatest potential for consequence loss [threat matrix]
- Formulate strategic threat-assessment statement for the region and local area

CONCLUSION

Understanding adversary capabilities, targets, means of attack, and techniques is vital to the security professional who works within the nation's critical infrastructure sectors. Without this knowledge, the design-basis threat statement and estimate of consequence loss cannot be adequately accomplished; assets and operations cannot properly be identified and prioritized; recommended protective measures certainly would not be prudent, reasonable, or cost-effective; and the presentation of a security assessment would be flawed in every manner from its conception through its conclusion.

Chapter 8

TASK 4—EVALUATE PROGRAM EFFECTIVENESS

Evaluating security program effectiveness identifies gaps in the program, helps prioritize solutions, and builds a platform for program development and implementation. This chapter examines and measures the effectiveness of in-place policies, processes, protocols, and protective measures that an adversary must defeat to carry out his or her mission. It outlines a 12-step approach, related tasks, and their respective interrelationships. The steps follow a logical sequence of analysis but do not necessarily have to be performed in the order given:

- Subtask 4A—Status of operating-system features
- Subtask 4B—Status of SCADA and distributed-control systems
- Subtask 4C—Status of IT network system
- Subtask 4D—Status of facility security operations
- Subtask 4E—Status of electronic security systems

- Subtask 4F—Status of security operations methods and techniques
- Subtask 4G—Status of information security programs
- Subtask 4H—Status of personnel-protection program and human-resources policy
- Subtask 4I—Status of practical ability to detect, assess, and respond to incidents
- Subtask 4J—Status of security organization structure and management
- Subtask 4K—Status of security emergency planning and execution capability
- Subtask 4L—Status of security training

The following suggested worksheets are offered to assist in collecting and analyzing program effectiveness:

- Worksheet 19—Recording status of current enterprise institutional "drivers" and Performance Strategies
- Worksheet 20—Recording status of current physical security effectiveness
- Worksheet 21—Recording test and exercise results by organizing sector

EVALUATING PROGRAM EFFECTIVENESS AND ACCOUNTABILITY

Publicly traded corporations whose shareholders demand that management protect their assets ultimately share the same diverse concerns and issues as private enterprises. Among them are:

- **Duty to care**—The inherent responsibility to provide a safe and secure work environment by exercising reasonable and prudent initiatives across the entire spectrum of the corporation
- **Management**—For those corporations that rely on national and international services, global partnerships, on-site expatriate employees, and position-jockeying executives, security issues can be critical to their success or failure
- **Protection of shareholder interest**—The commitment and initiative to meet business goals, market targets, and profit levels as well as providing employees with a positive work environment

- **Exposure to litigation**—The possible increase of legal action being taken against the corporation
- **Business continuity**—The ability to keep the company up and running even during high threat conditions and in particular in the aftermath of a catastrophic attack or disaster

Identifying Program Shortfalls

Program effectiveness is the measurement of the enterprise's capability to perform its security mission. The process involves analyzing all the data previously collected, reviewing new data, conducting additional interviews, and performing tests and exercises. Under this task existing policies, processes, protocols, and protective measures are analyzed to determine the present effectiveness of the security organization and its dependency partners in preventing terrorist attacks or undesired events and their consequences from occurring. It identifies both the general strengths and weaknesses of the enterprise and those specific to the security organization as well as barriers to performance. The program-evaluation process focuses on providing answers to the following questions:

- What are the objectives and strategies of the overall security program and the mission of in-place security systems and other protective measures?
- Which facilities, systems, functions, and resources need to be protected?
- What is the threat against multifunctional activities and resources?
- Which physical, cyber, and procedural protective measures must the insider/outsider adversary defeat to successfully penetrate the protected area, carry out the mission, and effect an escape?
- How well do the enterprise institutional "drivers" and performance strategies contribute to the effectiveness of the overall security program?
- How well has the enterprise integrated the institutional security operational capabilities into its routine?
- What are the restrictions, limitations, and constraints of protection?

The very nature of protecting an enterprise's infrastructure is a dynamic and evolving process that requires attention to detail and focus—which when applied and integrated into best practices will lead to improved awareness, preparedness, prevention, response, and recovery from acts of terrorism, criminal activities, and natural disasters.

Profiting from the Lessons Learned by Others

While many enterprises profess to implement, operate, and maintain integrated security programs and systems, the results of evaluating thousands of such programs reveal otherwise. This assessment is not an indictment against industry managers. Rather, it is an observation that commercial enterprises simply are not qualified and lack the in-house resources and infrastructure to adequately address or cope with the terrorist threat. Attempts to develop protective measures at some locations only address symbolic cosmetic solutions at best. Reporting and coordinating an integrated response in many locations are also haphazard. Security measures employed by many enterprises fall short of providing the capability needed to detect an incident during the development stages. Examples of relevant observations include:

- More often than not many security programs are comprised of several activities and functions strung together in a rather haphazard fashion with little coherence at the highest level:
  - ► Employees do not buy into the security. This results from a poor security-awareness program, no method of communication to alert employees of an emergency, and no means of telling employees which actions to take in an emergency.
  - ► Enterprise business objectives and goals and OSHA requirements are often not a part of security planning, and security plans and procedures are not kept up to date.
  - ► No designated security leadership is present, and management actions are not integrated. In many locations a "stove-pipe" mindset exists that fosters poor or no security planning.
  - ► Security contracts and agreements throughout the industry are weak. Contract guard forces lack tactical-response and search-and-rescue expertise and adequate training to meet the demands of the security mission.
  - ► Most security programs lack standard metrics to evaluate their effectiveness.
  - ► Poor security emergency planning and lack of organization readiness capability are rampant.

- ► Lack of integrating communications and data networks into security assessments is common.
- Electronic security systems are also comprised of several systems and parts that are not integrated:
  - ► Where electronic security system operations are involved, little or no attention is given to human-factors engineering.
  - ► Conflicting security and auxiliary responsibilities distract control-room or equipment operators from effectively performing their primary security duties of monitoring system activity.
  - ► Distractions coupled with inadequate or no system integration creates the inability to adequately detect and assess displayed scenes on a console monitor in a timely manner.
  - ► Many enterprises employ cameras that are neither integrated into the access-control or intrusion-detection systems nor viewed at a monitoring station.
  - ► At many locations these cameras are only used as recording devices to review and analyze incidents on a demand basis after the fact.
  - ► Some enterprises do not have an integrated intrusion-detection system that centrally reports alarms to a central security monitoring station. The various reporting configurations observed include security alarms reporting to various operational control centers including facility-management control centers, private alarm-monitoring companies, or the local police station.

Under this reporting configuration it was observed that many security organizations were reliant on third parties to receive and evaluate alarm conditions without providing the security organization with a security activity report. In other instances, third-party alarm-monitoring agencies were responding to an alarm without notifying the enterprise security organization of their actions or the severity of the alarm incident. One example in point was a recent breach of security at a Fortune 500 company that involved the dispatch to the scene of local law-enforcement agencies and the FBI as well as other special first responders and several of the corporation's senior managers. The incident was detected late on a Friday evening, and the event was broadcast on a national TV network the next day. The enterprise security organization was not made aware of the incident until Tuesday morning. Under this scenario the security organization was unaware of the status of a critical facility it was responsible to protect and had no centralized control, command, and communications over the events that occurred. In the cited example control-room operators, shift supervisors, and other managers, perhaps unfamiliar with established security reporting protocols or lacking such clear and distinct security protocols, made a decision to call for outside help and ignored their own security organization.

EXPOSING VULNERABILITY

Vulnerability "Creep-In"

Over the years I have experienced a distinct phenomenon that permeates the industry, which I call "vulnerability creep-in." Vulnerability can often creep into an enterprise undetected. It usually takes a long time to mature into an identifiable problem, making it difficult for the enterprise staff to recognize the weakness and its subsequent impact on the security integrity of the enterprise. Several factors contribute to vulnerability creep-in. The following examples illustrate its potential widespread occurrence.

Many complex and diverse enterprises practice a management culture in which each division, directorate, or department is almost a separate entity within the corporation with significant autonomy to conduct its business affairs. Security is an inherent responsibility of selected staff members—none of whom are security professionals or have any significant background in the application of security principles, techniques, methods, or technology. Therefore planning, coordinating, developing, and implementing security efforts at best are cumbersome, ineffective, and costly. Under this approach:

- Operational business goals and objectives, security priorities, and the application of resources often take a separate path.
- The significance of threats is usually minimized, while management actions to implement protective measures are often based on incomplete information and misguided analysis.
- Security and security-related roles and responsibilities often become blurred and may be duplicated or overlooked—creating cascading contradictory results.

This management style is often referred to as the "stovepipe" mentality. It fosters vulnerability creep-in and allows weaknesses in security to reside undetected until such time as an actual security incident occurs, at which point it becomes too late to react except for the exercise of lessons-earned.

A second factor that leads to vulnerability creep-in is change in management, high employee turnover, restructuring of the enterprise, change in the business culture and its goals, changes in processes and protocols, the construction of new facilities, the modification of existing facilities, or the transfer of systems and functions to new locations.

A third is the inexperience of enterprise executive management and the staff in security matters. This fosters a narrow and perhaps misguided application of security principles, techniques, methods, protocols, technology, and resources. This in turn creates a corporate culture that establishes unintentional barriers to the formulation of a corporate-wide strategic security mission. This situation can implant gaps into the security program.

A fourth is those instances in which the enterprise does not have a corporate security director or single security professional assigned the responsibility for planning, coordinating, developing, and implementing an integrated, comprehensive security program to combat terrorism.

Another involves a weak security director who lacks the clout to influence decision-makers, one who is unable to educate executive management and the staff to change their approach to security or one who has not been fully delegated authority to carry out the security charter.

Detecting Vulnerability is a Challenge

Certainly, one of the goals of the security assessment is to identify any such gaps in the security program, tear down barriers, and remove obstacles where they exist. Successive analyses performed under Task 5 [program analysis] determine the range of security enhancements that may be effectively employed to remove, reduce, or control vulnerability.

The vulnerability analysis breaks out each facility, system, and function and—given the adversary characteristics previously established in the design-basis threat statement and the criticality of the asset to the enterprise—proceeds to identify weaknesses in the site's capability to **deter, detect, delay,** and **respond** to threat conditions. Vulnerability is then determined by the dynamic relationship between asset and threat. The range of assets and their criticality [refer to Chapter 6 and Worksheets 6 through 12] and the range and levels of threats and probability of assumptions [refer to Chapter 7 and Worksheets 13 through 18] are key elements in this determination process.

Analytical Skills, Breadth of Experience, and Strategic Vision

Since vulnerability is viewed from the perspective of the evaluator, the analysis task is not one to be assigned to the novice or one who has a narrow viewpoint of what constitutes security. Hence, a security assessment requires multiple participants. The security consultant and security-assessment-team members must be well versed in all aspects of security, have a diverse security and business background, be experienced in multi-disciplines [and be an expert in one or more fields], and have a broad base of perspectives on and understanding of the enterprise including business, technical, and programmatic issues.

MEASURING PROGRAM EFFECTIVENESS

Program effectiveness measures the enterprise's capability to **deter, delay, detect, assess,** and **respond** to terrorist attacks and other security-related incidents. **Program effectiveness evaluates and performs a thorough analysis of the physical and operational environments; constraints; policies, processes, and protocols in all aspects of the security program; and enterprise resources assigned security roles and responsibilities.**

It also involves the planning, developing, coordinating, and performing of security tests and exercises and the analysis of test and exercise results. Specifically, program effectiveness does the following:

- Continues the examination of data processes and protocols previously started under Task 2 [critical assessment] and not fully completed
- Identifies the enterprise's current capability to mitigate and counter threat occurrences
- Identifies the absence of measures that are needed to mitigate and counter threat occurrences
- Identifies residual vulnerabilities remaining when new protective measures are integrated with existing protective measures

In Chapter 4 six critical and inseparable corporate performance strategies were introduced, two of which were briefly discussed in Chapter 5. The evaluation process measures the effectiveness of the enterprise's security strategies. These include the review of program data not previously examined, the institutional drivers and business performance standards, institutional security operational capabilities, and security performance-based standards and metrics.

Reviewing Inexamined Processes, Policies, Protocols, and Data

The security-assessment team continues its performance of reviewing processes, policies, and procedures not adequately addressed under Task 1 and adds the results of their findings to Worksheet 5. This analysis allows team members to understand the demands of operational timelines and operational protocols.

Next the security-assessment team analyzes the current security program, such as access points, perimeter fencing, approaches, and lighting, at each site to determine the present effectiveness of protecting critical assets, the probability of undesired events, and the consequences against those threats previously identified.

Enterprise Institutional "Drivers" and Business Performance Standards

Exhibit 8.1 illustrates how the security-assessment process blends these strategies into a measurement of the effectiveness of a comprehensive security program.

These standards include the enterprise mission goals, objectives, and culture that represent its contribution to the business world, its standing within the community, and its acceptance by employees. The processes, techniques, and practices used by the enterprise to deliver services and products in quality fashion, timely pursuit, and at reasonable market value reflect its image and mirror its social value within the community.

To the extent that these institutional drivers and performance standards complement the overall security program—or conversely pose obstacles that may hinder program application—they are evaluated to measure the degree to which they contribute to establishing and maintaining a safe and secure work environment and sustaining the continuance of business operations.

Enterprise Institutional Security Operational Capabilities

In previous chapters much reference has been made to the six indispensable security operational capabilities that must permeate all aspects of the enterprise. These entail the timely integration and interdependency of **deterrence, delay, detection, assessment, response,** and **recovery.**

Deterrence

Deterrence is mostly an intangible strategy. It includes administrative processes, protocols, and practices as well as physical and technological enhancements to defend against attacks. When the adversary [including

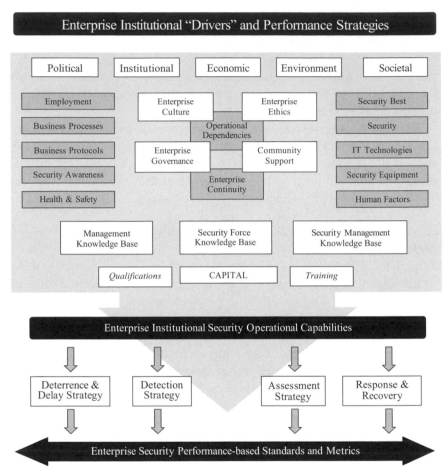

Exhibit 8.1 Enterprise Security Strategies

the insider] perceives an unacceptable risk associated with causing the damage or destruction of a critical asset, the effectiveness of deterrence impacts the adversary's mode of attack, tactics, capability, and necessary delay-penetration time. The exception to the general rule is a suicide assault. Deterrence aids the other processes and provides employees a level of confidence that a safe and secure work environment exists.

Delay

Delay encompasses those techniques employed to slow down the adversary [including the insider] or prevent him or her from reaching a vulnerability point or asset until a response force capable of defeating the adversary arrives at the scene. It measures function differently depending on the mode of attack and tactics used by an adversary. The delay-penetration time required by the adversary to reach an objective is dictated by the configuration of the area, the types and quantity of physical barriers to be crossed, the distance the adversary has to travel, and the types of tools, weapons, and equipment used. Delay techniques aid the **detection, assessment, and response** processes, giving the security organization time to react in a timely manner. Ideally, the capability of a timed response should not exceed the delay-penetration time needed by the adversary to complete his or her assignment once detected. In a corporate environment this capability seldom exists, and this inherent program weakness must be compensated with other protective measures.

Detection
Detection is performed through direct human observation including protocols and supervision or assisted by security technology including intrusion detection sensors, video surveillance equipment, and procedural checks and balances, such as the "buddy system," "two-person rule" or verification and authentication systems. It monitors for potential breakdowns in the application of protective mechanisms, processes, and protocols that could result from security breaches, criminal behavior, or compromise of system integrity. It includes those activities undertaken to detect illegal, unlawful, or unauthorized actions as far away as possible from the critical assets requiring protection. Detection aids the assessment and response processes.

Assessment
Assessment of security breaches, criminal behavior, and general activity may either be human or procedural and assisted by technology such as video systems, display and annunciation panels, and computer management-information summaries. The annunciation of remotely dispersed intrusion detection and surveillance devices creates a need for the security organization to be aware of the validity, severity, and nature of the event that triggers the alarm in near real time. This may require the strategic placement of area patrols, fixed posts, and CCTV cameras installed to be activated in conjunction with a sensor in alarm status. A balanced mix of protective measures enhances the safety and response effectiveness of first responders and reduces the need for expensive use of patrols and fixed posts. Security lighting and advanced imaging technology also enhance surveillance and assessment capabilities. Beyond the technical capability to visually annunciate an alarm or provide the

video surveillance of areas, the responsibility rests solely with the human element to assess security conditions and the supervisory element to direct an immediate security response. Assessment is critical to determining an adequate response.

Response

Response is the weak link in all security programs and often the most misunderstood and ignored of all the security operational capabilities. The existence and capability of the response is operationally dependent on human evolvement, both in decision-making and execution, to respond to detected breaches of security to thwart attacks before serious damage can be done or to respond to developing events to control the threat situation. The previous factors are meaningless unless the security organization has the capability to control, contain, or neutralize the threat. First responders may be a proprietary security force; a contract guard service; designated local, state, or federal law enforcement, or a combination of several separate but integral agencies including the Department of Defense and elements of the Department of Homeland Security at both the state and federal levels.

Security technology plays a vital support role when systems are programmed to respond to specific stimuli. For example, where a harmful contaminant is released within a research laboratory, the environmental-monitoring system may be programmed to react when the agent is detected. Such a response could include sealing all the portals and utility openings, releasing neutralizing agents into the affected zone, or sanitizing the area through a secondary and protective ventilation system, followed by a human response to remove personnel from the affected area when it is safe to enter. As such, the overall technical solution to a security response must assist and never hinder the security organization when executing a human response to an incident.

Recovery

Recovery is often a capability that is overlooked in both strategic security planning and in the execution of a response. It is generally an issue only addressed after the response is complete. Security recovery, however, must receive equal management attention, as it offers the security organization the ability to operate well enough to meet obligations at an acceptable level, to protect the life and safety of employees, and to protect property in any given emergency while transitioning from an emergency-response posture back to normal security operations.

Enterprise Security Performance-Based Standards and Metrics

In addition to the performance-based standards and metrics introduced in Chapter 4 **the dimension of timely reaction and the interdependency of capability efficiencies directly apply to the measurement of security operational capabilities.**

Timeliness and Interdependencies of Security Capabilities

Security is a reactionary business in both strategy planning and execution. The processes of detection, assessment, response, and recovery are time-sensitive and time-dependent functions for the security organization. They vary from one situation and set of circumstances to another and under varying conditions. In most instances, however—particularly where technology is employed—the ability of the security organization to detect, assess, and respond to a given event is measured not in hours or minutes but in near real time that spans a dimension of 3 to 5 seconds. Therefore, the principle of timeliness is crucial to effective security operations as indicated below.

Principle of Timely Delay

The principle of timely delay introduces a series of strategic obstacles and nuisances into the path of an adversary that otherwise increases the time it takes the adversary to reach a protected facility, system, or function, thereby raising the risk of exposing the adversary and permitting time for an adequate response to develop and take effect.

Principle of Timely Detection

The principle of timely detection requires that the intrusion of a protected asset[s] be detected as far from the facility, system, or function as is possible and soon enough to prevent the adversary from completing his or her mission.

An effective detection capability must have a proper balance of three basic elements: detection, delay, and response. A security program's detection capability is only as good as the capabilities offered by these elements. For example, an aggressor may spend days or weeks planning to defeat protective barriers, processes, and protocols. If the delay mechanisms employed are too short, an aggressor can trip one or more lines of sensor detection, open or bypass one or more portals to gain access to an area, or circumvent several protocols to complete his or her assignment and leave the area before the response force arrives. Or an aggressor can set off a series of alarms in the system and still take the time needed to complete the mission, knowing the response to be inadequate. To be effective, detection and response must occur before the aggressor has time to complete his or her mission.

Principle of Timely Assessment

The principle of timely assessment permits security management to observe and assess the situation and associated conditions and provide the necessary guidance or direction to the response force to control, contain, or neutralize the threat.

An effective assessment capability has two elements: **event identification** and **event tracking.** Assessment must be able to distinguish between normal acceptable activity and behaviors and any deviation from these norms. The human element can accomplish this through a motivated and dedicated workforce, security-awareness training, and the reporting of unusual activity or behavior. When such observations are complemented by video technology, the application of cameras must provide for scene activation of a detected event to clearly distinguish the severity of a threat including, if possible, the identification of the perpetrators. It must also maintain the capability to track the sequence of events and the movements of the perpetrator[s] from the start of the engagement through the conclusion of the incident.

Principle of Timely Response

The principle of timely response requires that the response be flexible, gradual, and sufficient to prevent, control, contain, or neutralize the event that justified the call to react.

Response involves those immediate actions that prevent loss of life or property and provide emergency assistance to others. For security it involves the ability to provide a timely, gradual, and flexible response to one or more event-driven threat scenarios simultaneously. The response may be to neutralize or contain a specific threat, establish and secure the perimeter of an area to preserve its integrity and control access; assist in the evacuation of an area or building; guide and assist other first responders to the scene as appropriate, and report conditions, circumstances, and situations to management.

Principle of Timely Recovery

For security, **the principle of timely recovery** involves the coordinated efforts to ensure the smooth and rapid transition from an emergency-response posture back to normalcy or improved security operations and standards without jeopardizing the integrity of the overall security program.

Recovery involves the short- and long-term activities that transition the security organization from its current state of emergency operations back to its normal operating posture. It may include such activities as assisting in returning the affected area to normal business operations;

reestablishing the security of the area; adjusting the threat-alert notification system; gradually withdrawing the security force from the area and adjusting work priorities; aligning the staff and rescheduling normal security operations, inventory, and inspection; and the repair or replacement of security assets.

DEVELOPING AND CONDUCTING VULNERABILITY TESTS AND EXERCISES

Conducting tests is indispensable to determining the applied capability of an enterprise to **deter, detect, assess,** and **respond**.

Vulnerability testing determines the likelihood that an adversary or adversarial group can successfully attack a particular facility, system, function, or resource and details the process, protocols, and techniques to achieve this objective. The results serve several purposes in terms of mitigating the vulnerabilities from such attacks. Vulnerability testing or exercises allow management to:

- Assess plans and procedures to determine feasibility under actual conditions
- Assess whether personnel understand their emergency-response functions and duties
- Identify areas for improvement
- Enhance coordination, communication, and proficiency among response staff
- Enhance the ability of management and the staff to respond to emergencies

Experience gained and errors committed during testing and exercises can provide valuable insights and lessons learned to factor into the planning process. Solutions are then prioritized based on the highest P_A and C for each facility and asset.

IDENTIFYING AND DOCUMENTING PROGRAM EFFECTIVENESS

In Task 2, the security-assessment team identified and documented site conditions, facilities, systems, and functions and their respective security features. In addition, they started the process of characterizing the

OBJECTIVES AND SCOPE OF TESTS AND EXERCISES	Exercises using simulation provides the opportunity to test skills and knowledge to identify strengths and weaknesses; learn new skills; practice decision-making, techniques and communications; and determine gaps in existing plans and procedures. The scope of any exercise will be to carry out the approved scheduled activity in a comprehensive and exhaustive manner.
PLANNING EXERCISES	The security assessment team works with the enterprise to determine the types of exercises to be performed, the scope of involvement, and exercise limitations. Exercises are structured and organized in a way in which the results can measure performance in the following areas:

- Mobilization of personnel and equipment in response to an emergency
- Evaluation of response time
- Inspection of equipment performance and demonstrated knowledge of its usage
- Interaction and coordination within defined groups
- Response tactics and decision-making
- Capability to implement extended work schedule
- Timeline to provide expanded security coverage and duration of coverage
- Integration of dependency groups both internal and external
- Crisis Management organization and leadership

SCENARIO DEVELOPMENT	Developing scenarios requires an extensive knowledge of adversary modes of attack, their capabilities, and facility activities. They involve a great deal of research, planning and coordination, strict controls, and approval at all levels of management. The security assessment team develops a realistic set of scenarios to simulate as far as possible disruptive situations that will effectively task responders while not placing personnel at risk of injury. In developing test and exercise scenarios, the focus is on vulnerability and vulnerable access points where penetration may be accomplished and how. Emphasis is placed on the physical characteristics of vulnerable points, accessibility to the site and protected area, and physical and procedural obstacles an adversary would likely have to overcome to reach a critical facility, system, or function.
CONDUCTING EXERCISES	The security assessment team develops a schedule of planned exercises at each site to be visited, along with a copy of the approved exercises to be conducted. These exercises are scheduled after the analysis process. The actual control of exercises follows existing enterprise protocol for implementing emergency and response plans with the security assessment team serving as exercise evaluators.

Exhibit 8.2 Program Exercise and Test Development Model

strengths and weaknesses of the overall security program. In this chapter, the analysis is taken to the next step. The security-assessment team now applies a distinct performance measurement value to each facility, system, and function and their respective security features to evaluate the effectiveness of in-place policies, processes, protocols, and protective measures. The analysis is expressed as a probability of program effectiveness [P_{E1}].

With respect to measuring enterprise institutional "drivers" and performance standards, the P_{E1} principle can quantify the degree to which each contributes to the success of the overall security program.

With respect to the operational and physical security characteristics of deterrence, delay, detection, assessment, and response, the P_{E1} principle can quantify program effectiveness by determining the cumulative P_{E1} up to the point that there is sufficient delay [time remaining for the adversary to complete the undesired actions and create an undesired consequence] to allow adequate response to a given event. An adequate response is defined as getting a sufficient number of properly equipped and trained response-force personnel in place in time to prevent a given undesired consequence.

Measuring and Recording the Status of Enterprise Institutional "Drivers" and Performance Strategies

Worksheet 19 categorizes how each of the enterprise performance strategies and standards contribute to the security program. A rating of **high effectiveness [H_E], moderate effectiveness [M_E],** or **low effectiveness [L_E]** is applied to each program element for each specific operational capability. The rating defines how well the enterprise-integration process has taken hold.

The evaluation of enterprise SCADA, distributed control systems, and the IT network uses a top-down systems approach, integrating the security-assessment team's knowledge and experience in the myriad of techniques being used to compromise information systems. The objective is to identify and quantify the weakest link in these systems, be it physical, technical, or human. This necessitates investigating every aspect of the processing, storage, and transfer of information.

The evaluation of installed electronic security systems includes automated-detection devices, access-control devices, vehicle gates, surveillance equipment, and monitoring capabilities. The security-assessment team examines these subsystems to determine their current performance level, whether they meet current enterprise expectations of performance, and whether their remaining life expectancy, supportability, and capability to support growth match the solutions.

Enterprise Institutional "Drivers" and Performance Strategies	Enterprise Institutional Security Operational Capabilities					
	Deter	Delay	Detect	Assess	Respond	Recover
Operating System Features	.10	.50	.50	.50	.50	.90
SCADA and Distributed Control Systems	.10	.50	.50	.50	.50	.50
IT Network Systems	.50	.50	.50	.50	.50	.50
Facility Security	.50	.50	.50	.50	.50	.50
Electronic Security Systems	.90	.90	.90	.90	.90	.90
Security Operations, Security Methods, Techniques	.50	.50	5.0	.90	.50	.90
Information Security Program	.90	.90	.90	.90	.90	.90
Personnel Protection and Human Resources	.10	.10	.50	.50	.50	.50
Practical Ability to React	.50	.50	.50	.50	.50	.50
Emergency Planning	.50	.50	.50	.50	.50	.50
Training	.10	.10	.10	.50	.50	.50

ENTERPRISE SECURITY PERFORMANCE-BASED STANDARDS AND METRICS

NARRATIVE DESCRIPTION	NUMERICAL RATING	ID RATING	PERFORMANCE MEASUREMENT
HIGH EFFECTIVENESS	.85 to .99	H_E	Positive evidence of specific capabilities to address specific protective measures
MODERATE EFFECTIVENESS	.50 to .84	M_E	General evidence of protective measures, but no specific capabilities
LOW EFFECTIVENESS	.01 to .49	L_E	Little, ineffective, or no evidence of protective measures or capability

Exhibit 8.3   Example of Worksheet 19—Recording Status of Enterprise Institutional "Drivers"and Performance Strategies Exercise

In the sample exhibited above, the established criteria applied represent a program performance effectiveness rating of 0.55 out of a possible rating of 0.99, placing the effectiveness rating in the mid-range on the evaluation scale. While the rating reflects evidence of general protective measures, the goal of providing security enhancements would be to raise this effectiveness level by providing specific capabilities to address the lack of specific protective measures.

Measuring and Recording the Status of Current Physical Security Effectiveness

Worksheet 20 was previously used in Chapter 6 to identify assets and document their security features. In this task it is expanded to incorporate the assignment of a performance-measurement effectiveness rating of **high**

[H], moderate [M], or **low [L]** based on the adequacy of in-place protective measures as well as the asset's defined level and range of threats. Other factors that contribute to the assigned effectiveness rating include the status of site conditions and circumstances, adequacy of processes and protocols, and security awareness. Applying the various shades of grey depicted in the legend of the exhibit helps management to quickly assess the overall status of the program.

The evaluation of physical security features focuses on the physical layout and configuration of facilities, systems, and functions and the surrounding areas to help the security-assessment team recognize the safety and security elements supporting and preserving community needs. The analysis also examines security operational capability. In the above example, applying the established criteria given this site represents an overall effectiveness rating of 0.37 out of a possible rating of 0.99, placing it at the low end of the evaluation scale. For this particular site the rating is influenced by deficiencies in control processes, ineffective entry

Critical Facility Assets	Site Access Controls	Perimeter Barriers	Facility Access Controls	Facility Barriers	Safety Alarms	IDS	CC TV	Com	Area Lighting
Perimeter		10-foot wall w/wire 8-foot wall				No	No		Yes
Effectiveness Rating		H_E				L_E	L_E		M_E
Warehouse			Key Lock	Structure	Yes	No	No	Yes	Yes
Effectiveness Rating			M_E	H_E	H_E	L_E	L_E	M_E	M_E
Equipment Parking Area			No	No		No	No		Yes
Effectiveness Rating			L_E	L_E		L_E	L_E		M_E
Emergency Generator			No	No	No	No	No	Yes	Yes
Effectiveness Rating			L_E	L_E	L_E	L_E	L_E	M_E	M_E
Emergency Management Center			Key Lock	Structure	No	No	No	Yes	Yes
Effectiveness Rating			M_E	M_E	L_E	L_E	L_E	M_E	M_E
Administration Building			Key Control	Structure	Yes	No	No	Yes	Yes
Effectiveness Rating			M_E	H_E	H_E	L_E	L_E	M_E	M_E
Fuel Island			Card Key Control	Bollards	Yes	No	No	Yes	Yes
Effectiveness Rating			L_E	L_E	H_E	L_E	L_E	M_E	M_E
HAZMAT Storage Areas			Key Control	Casing		No	No	Yes	Yes
Effectiveness Rating			M_E	M_E		L_E	L_E	M_E	M_E
Guard House	Manual		No	Structure		No	No	Yes	Yes
Effectiveness Rating	L_E		L_E	H_E		L_E	L_E	M_E	M_E

PROBABILITY OF EFFECTIVENESS [P_E] CRITERIA

NARRATIVE DESCRIPTION	NUMERICAL RATING	ID RATING	PERFORMANCE MEASUREMENT
HIGH EFFECTIVENESS	.85 to .99	H_E	Positive evidence of specific capabilities to address specific protective measures
MODERATE EFFECTIVENESS	.50 to .84	M_E	General evidence of protective measures, but no specific capabilities
LOW EFFECTIVENESS	.01 to .49	L_E	Little, ineffective, or no evidence of protective measures or capability

Exhibit 8.4 Example of Worksheet 20—Recording Status of Current Physical Security Effectiveness [P_{E1}]

AREA	WHAT WENT WELL?	WHAT WENT POORLY?	LESSONS TO BE LEARNED
MANAGEMENT	Decisions were communicated effectively to all levels within the organization.	Dependency on technology left decision-makers little option regarding manual override of systems.	Manual override of technology is necessary to assure backup operation of systems.
PLANNING			
CONTROL OF OPERATIONS			
TIMELINESS OF DETECTION			
TIMELINESS OF ASSESSMENT			
TIMELINESS OF RESPONSE			
ON-SCENE COMMAND, CONTROL AND COMMUNICATIONS			
TACTICS			
ACTIONS AT THE SCENE			
RECOVERY			
SECURING THE SCENE			
TIMELINESS OF RECOVERY			
LOGISTICS			
ADMINISTRATION			
INFRASTRUCTURE			
EXTERNAL RELATIONS			

Exhibit 8.5 Example of Worksheet 21—Recording Exercise Evaluation by Organizing Sector

procedures, and a lack of capability to adequately detect, assess, and respond to security incidents in a timely manner. The conditions identified in the above sample place the enterprise at an extremely high risk.

Documenting Test and Exercise Results

Worksheet 21 provides one approach to recording test and exercise results. It is also a good management tool to use for completing an after-action report when responding to actual security incidents.

COMPREHENSIVE WORK-BREAKDOWN STRUCTURE

SUBTASK 4A—Status of Operating System Features

Here are the procedures to follow for this task:

- Assess the configuration and technical features of the prime operating systems with respect to potential threats and vulnerabilities, including:
  - ▸ Effectiveness of current production and/or delivery services, configuration, and distribution and delivery technologies
  - ▸ Effectiveness of current processes and business practices
  - ▸ Identification of hazards from process chemicals and other acutely hazardous materials
  - ▸ Identification of acutely hazardous materials from adjacent establishments and facilities
  - ▸ Tracking mechanisms to account for process chemicals and other acutely hazardous materials received, stored, and used at the facility

SUBTASK 4B—Status of SCADA and Distributed-Control Systems

These are the steps to follow for this assessment:

- Application, capability, performance level, expectations, single points of failure
- Supportability and capability for growth, remaining life expectancy
- Report of SCADA and distributed-control-system technology application effectiveness; identify strengths and weaknesses, including:
  - ▸ Subsystem components affected by recommendations, which may require upgrading and/or replacement
  - ▸ Subsystem components affected by recommendations but that may not require upgrade and/or replacement
  - ▸ Subsystem components that may no longer be supportable or whose useful life expectancy has been reached
  - ▸ Performance level of subsystem components and the existing maintenance program
  - ▸ Other related elements, which may be affected by business processes and practices
  - ▸ An acquisition strategy and milestone timetable for executing actions

SUBTASK 4C—Status of IT Network Systems

Following are the stages for this task:

- Application and capability
- Performance level and expectations
- Single points of failure
- Screening of network traffic for viruses and attacks
- Virus protection for computers
- Security architecture implemented for external communications
- Access via modem to enterprise's wide-area network
- Vulnerability/penetration evaluations or test of enterprise networks
- Modems attached to end-user desktop systems on a security local-area network
- Supportability and capability for growth
- Remaining life expectancy
- Report of IT technology application effectiveness; identify strengths and weaknesses, including:
  - ► Subsystem components affected by recommendations, which may require upgrading and/or replacement
  - ► Subsystem components affected by recommendations but that may not require upgrade and/or replacement
  - ► Subsystem components that may no longer be supportable or their useful life expectancy has been reached
  - ► Performance level of subsystem components and current maintenance program
  - ► New subsystem components not currently part of the existing security system that may need to be added to support upgrade effort
  - ► Other related elements, which may be affected by security interface issues
  - ► An acquisition strategy and milestone timetable for executing actions
  - ► Evaluation of constraints of existing system for expandability, modification, and upgrade

SUBTASK 4D—Status of Facility Security Features

Follow these steps for this stage:

- Assess the configuration, layout, site conditions, and processes that contribute to the effectiveness of the security program or, conversely, lead to program gaps and weaknesses, including the following for physical security measures:

- ► Approaches and other property boundary controls
- ► Configuration and condition of perimeter physical barriers such as fencing, gates, or walls
- ► Security posts and patrols
- ► Personnel, vehicle and material access controls, inspections, and lighting
- ► Vehicle barrier systems, parking controls, and lighting
- ► Control and tracking of company vehicles
- ► Elevator and utility-area controls including key control program
- ► Special high-value security needs
- ► Buildings and/or areas that may require an explosive-threat analysis
- ► Report of new or revised physical protective measures where greater [or lesser] security needs address excessive or unwarranted expenditures

- • Evaluate site environmental and operational conditions that impact security operations
- • Evaluate security posture, current security upgrades, and projected program changes

SUBTASK 4E—Status of Electronic Security Systems

These are the steps to follow for this stage:

- • Where electronic security systems are installed, assess and test security-system performance levels and identify status of effectiveness:
  - ► System configuration and performance
  - ► Automated identification system and personnel access controls
  - ► Personnel portal controls, use of identification technology
  - ► Two-person rule or other administrative controls
  - ► Screening stations, use of metal-detection equipment and x-ray equipment
  - ► Personnel and package searching procedures
  - ► Automated identification system for vehicle portal controls
  - ► Use of identification technology
  - ► Vehicle clearance and searching procedures

- For external and internal intrusion detection systems:
  - ▶ Fence-mounted, buried, freestanding, and boundary-penetration sensors
  - ▶ Motion and proximity detection sensors
- For external and internal CCTV surveillance and assessment systems:
  - ▶ Integrated visual assessment with alarm annunciation, camera assignment and controls
  - ▶ Event-capture and digital-recording-playback capability
- For communications:
  - ▶ Dedicated system, multiple-channel capability
  - ▶ Secure network, primary and backup systems
- For command, control, transmission data, and display systems:
  - ▶ Protected and supervised means of transmission
  - ▶ Dual-loop redundant auto-reconfigurable media
  - ▶ Alarm-reporting protocols, user-friendly graphical display and interface
  - ▶ Redundant computer system, UPS and backup power supplies
  - ▶ Integrated CCTV functionality
  - ▶ Primary and secondary monitoring stations
- For management reporting systems:
  - ▶ System status reports, event reports, customized programmable and on-demand reports
- For security-system maintenance:
  - ▶ System maintenance strategy plan
  - ▶ Maintenance response times
  - ▶ System maintenance schedule and type of maintenance performed
  - ▶ Status of parts inventory and replacement
  - ▶ Warranty program
- Report of security technology application effectiveness; identify strengths and weakness including:
  - ▶ Subsystem components affected by recommendations, which may require upgrading and/or replacement
  - ▶ Subsystem components affected by recommendations but that may not require upgrade and/or replacement
  - ▶ Subsystem components that may no longer be supportable or whose useful life expectance has been reached

- ► Performance level of subsystem components and the existing maintenance program
- ► New subsystem components not currently part of the existing security system that may need to be added to support upgrade effort
- ► Other related system elements, which may be affected by security interface issues
- ► An acquisition strategy and a phased project milestone timetable for executing actions
- ► Evaluate compatibility constraints of existing security system for expandability, modification, and upgrade

SUBTASK 4F—Status of Security Operation Methods and Techniques

Here are the procedures to follow:

- • Evaluate effectiveness of security plan and security procedures to determine whether they live up to the needs of the enterprise or if it's time to rethink enterprise strategy
- • Research and identify regulatory requirements and industry guidelines
- • Evaluate security policies, procedures, and methods to determine their accuracy, completeness, and integration of industry best practices to detect, assess, and respond to undesirable events
- • Report on new or revised operational measures where greater [or lesser] security needs address excessive or unwarranted expenditures

SUBTASK 4G—Status of Information Security Program

For this step, follow these procedures to evaluate the effectiveness of measures to protect enterprise sensitive information:

- • Evaluate the receipt, handling, and distribution of information
- • Evaluate the physical security measures to protect the information
- • Evaluate practices for identifying authorized personnel to receive information
- • Evaluate disposition of information when no longer needed
- • Evaluate reporting practices for loss or unauthorized release of information

- Assess and measure the degree to which personnel understand their functions and duties for safeguarding sensitive information
- Report on areas for improvement

SUBTASK 4H—Status of Personnel Protection Program and Human-Resources Policy

Here are the steps to follow for this stage:

- Policies on background checks for potential employees before hiring
- Policies on periodic criminal checks for existing employees
- Policies for maintaining a drug-free work environment
- Policies for controlling and managing workplace violence
- Plan for management to react effectively if employees refuse to come to work during an incident
- Plan to transport personnel to and from work if roads and streets are closed due to police order or physically blocked as a result of an incident
- Plan to mitigate the concern employees may have for their families' well-being during a disaster or incident
- Management discussion of security issues, emergency-response plan, and disaster plan with union representatives
- Assess and measure the degree to which personnel understand their functions and duties for safeguarding sensitive information
- Report on areas for improvement

SUBTASK 4I—Status of Practical Ability to Detect, Assess, Respond to Incidents

These are the steps to follow for vulnerability testing:

- Define the adversary plan, sequence of interruptions, and path analysis to determine points of vulnerability
- Develop test objectives, define scope, and plan tests and exercises including:
  - ▸ Mobilization of personnel and equipment in response to an emergency
  - ▸ Evaluation of response time
  - ▸ Inspection of equipment performance and demonstrated knowledge of its usage

- ▸ Interaction and coordination within defined groups
- ▸ Response tactics and decision-making
- ▸ Capability to implement extended work schedule
- ▸ Timeline to provide expanded security coverage and duration of coverage
- ▸ Integration of dependency groups both internal and external
- ▸ Crisis-management or business-continuity organization and leadership
- ▸ Develop and coordinate scenarios for approval
- ▸ Conduct actual tests and exercises
- ▸ Critique results and present written findings for improvement

SUBTASK 4J—Status of Security-Organization Structure and Management

Follow these steps for this evaluation stage:

- Evaluate the security organization's ability to achieve its mission statement and mandates:
  - ▸ Evaluate management expectations and voluntary commitments
  - ▸ Evaluate effectiveness of security contracts and contract management
  - ▸ Evaluate the effectiveness of existing program security strategy and management initiatives to develop a framework for implementing program enhancements
  - ▸ Evaluate the current organizational structure of the security organization including qualifications, capabilities, expectations, and staffing to determine its level of program effectiveness
  - ▸ Examine the level of training of the security organization and measure its effectiveness
  - ▸ Assess current standards of performance to detect, assess, and respond and evaluate remaining vulnerability
  - ▸ Evaluate security-organization management, command and control and public relations
  - ▸ Assess current site and state/local security and operational capabilities to detect, deter, assess, and respond to identified or postulated threats for each threat level including WMD and conventional explosive attacks

- Report on improvements to:
  - ▸ Detect, deter, and respond to potential WMD/bombing attacks at and around designated locations
  - ▸ Enhance future planning needs, organizational and governance issues, technical issues, and equipment procurement to reduce risk

SUBTASK 4K—Status of Emergency Planning and Execution Capability

For this stage, follow these steps:

- Assess enterprise readiness emergency preparation planning, including conditions and response actions that state/local authorities and site staff would take under each threat level:
  - ▸ Defined/established alert levels
  - ▸ Defined/established emergency levels
  - ▸ Roles and responsibilities
  - ▸ Recall procedure and resource capability
  - ▸ Organization structure, command and control, and reporting relationships
  - ▸ Coordination with external/internal agencies
  - ▸ Trained spokesperson as point of contact for public notification and/or media communications
- Assess corporate and site emergency-response capabilities and tactics including conditions and response actions that state/local authorities and site staff would take under each threat level
- Assess corporate and site-recovery operations, including conditions and response actions that state/local authorities and site staff would take under each threat level:
  - ▸ Whether customer base is informed about plans to mitigate the effect of service interruptions
  - ▸ Whether customer base is informed about how to cope with service interruptions
  - ▸ Whether customer base is aware of what activities they should report [and who to call] if they witness something unusual with a company vehicle or actions of an employee, contractor, or other suspicious individuals
  - ▸ Report on areas for improvement

SUBTASK 4L—Status of Training

Here are the steps to take to assess and measure the degree to which personnel understand their emergency-response functions and duties:

- Employee training to properly handle a threat that is received in person or by phone, email, U.S. mail, or other delivery service
- Employees' understanding of procedures to follow should an incident occur
- Management understanding of whom to contact to report a threat or emergency
- Procedures for determining when and how to evacuate a building
- Employee training in security awareness and protective measures
- Employee training in emergency preparedness in accordance with the enterprise's plan
- Employee training to detect symptoms of a chemical, biological, or radiological attack
- First-aid training for employees

CONCLUSION

The flexibility of the ***S³E Security Assessment Methodology*** allows portions of Task 4 [evaluating program effectiveness] to be performed in concert with Task 3 [identifying threats to the service environment] to integrate the evaluation of site conditions, operations, in-place security technologies, practices, and dependencies.

The status of site physical conditions, security operations and practices, dependencies, and in-place technologies is also fed back into Task 2 [critical assessment] to update the threat analysis.

The effectiveness ratings establish the platform for designing solutions and formulating prioritized actions, which are discussed in Chapter 9 [program analyses]. In that chapter, solutions are then based on the highest P_A and C for each facility, system, and function evaluated. The results of that analysis are expressed as P_{E2}.

Chapter 9

TASK 5—PROGRAM ANALYSES

PROGRAM ANALYSES OFFER ENTERPRISE DECISION-MAKERS COST-EFFECTIVE CHOICES

This task synthesizes strategies, priorities, options, and alternatives to bring together the right balance of people, information, facilities, oper¬ations, processes, and systems to deliver a cost-effective solution.

Emphasis is placed on assessing the status and level of protec¬tion provided against desired protective measures to reasonably counter the threat. Operational, technical and capital improvement plans are examined. This chapter outlines a three-step approach, related tasks, and their respective interrelationships. The steps follow a logical sequence of analysis and should be performed in the order presented:

- Subtask 5A—Finalize and refine the design-basis threat profile
- Subtask 5B—Assess vulnerability

- Subtask 5C—Determine and finalize rank order for protection
- Subtask 5D—Develop workable solutions

The five steps of program analysis are:

- Data gathering and characterization
- Comparison and value
- Program effectiveness and measurement
- Calculation and prioritization
- Capability and results

Under this task, previously developed threat, vulnerability and asset-identification worksheets are cross-referenced and refined. The process formulates the foundation for establishing enterprise security strategy as well as the development of an implementation plan to execute recommendations including identifying reasonable timelines to complete actions. Lastly, a preliminary cost estimate is developed. Inputs from this task are fed into Task 6—the project site deliverables—discussed in Chapter 10.

Program analysis is not a separate activity but rather the structured synthesis of the data collected under Tasks 1 through 4. **The purpose of this analysis is to discover the causes, effects, and severity of vulnerabilities and threats.** The process forces organized, purposeful thought to eliminate, reduce, or contain the effects of undesirable events and their potential consequences. It involves the process of planning, identifying, qualifying, quantifying, and selecting workable solutions for resolving vulnerabilities and associated risks. In performing this analysis the security-assessment team draws on methods, techniques, and tools available in decision-analysis disciplines to determine recommended course of actions. These include, as appropriate:

- Threat analysis
- Vulnerability analysis
- Consequence analysis
- Utility analysis
- Time-motion analysis
- Work-breakdown-structure analysis

- Tradeoff analysis
- Cost-benefit analysis
- Cost-economy analysis

Program analysis uses the elements of strategic planning to find workable solutions to identified problems. It blends enterprise goals and objectives, performance strategies, and organizational capabilities into an overall process that measures enterprise performance. The analysis strategy focuses on solutions that embrace an integrated top-down, bottom-up review of operations, processes, information, people, and technologies, such as security computer systems and communications and data-transmission networks. Exhibit 9.1 illustrates the program-analysis approach.

The ultimate goal is to establish the critical security operational capabilities of **deter, delay, detect, assess, respond,** and **recover** with a high assurance of success.

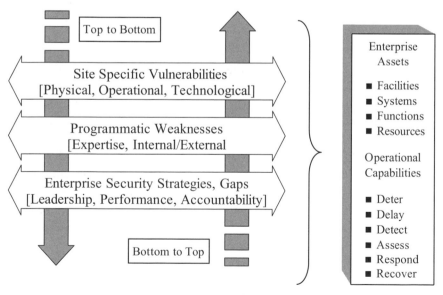

Exhibit 9.1 Program-Analysis Model

SYNTHESIZING STRATEGIES AND PRIORITIZING OPTIONS AND ALTERNATIVES

Under Task 5, the security-assessment team synthesizes the information collected during Tasks 1 through 4 to formulate the functional requirements and the conceptual need to address solutions based on the likely threats, resultant vulnerabilities, and potential consequences. Proper analysis allows for the comparison of proposed upgrade alternatives— before any money is expended—to determine which solutions deliver the most efficient balance of risk, cost, and impact on enterprise activity. This involves validating the strategic and programmatic data gathered in Worksheets 2 through 5 (see Chapter 5).

The security-assessment team also uses information available through numerous independent sources and their professional knowledge as inputs into the development of solutions and brings together the analysis of site conditions, operational efficiencies and effectiveness, and technology capabilities and performance standards to reduce vulnerability.

FACILITY, SYSTEM, AND FUNCTION CHARACTERIZATION, ASSET IDENTIFICATION, AND THE RANK ORDERING OF ASSETS

Physical and cyber infrastructure are interconnected. The devices that control physical systems including electrical distribution, dams, water systems, communications, transportation, and other important infrastructures are increasingly connected to a variety of IT networks and management-information systems. Thus, the consequences of an attack on a cyber infrastructure have dramatic cascading effects across the enterprise.

The criticality of assets is not a static assignment that fits all circumstances, conditions, and situations. The security-assessment team rank-orders assets based on the criticality of the area requiring protection, the degree of threat against the area, the special motivation and skill level of the adversary group attracted to the asset, the vulnerability of the asset, and its consequence of loss to the enterprise.

The security consultant, the security-assessment team, and the enterprise staff work hand-in-hand to reduce vulnerabilities to terrorism, prevent and prepare for terrorist attacks, minimize public-health impacts and infrastructure damage, and enhance recovery from any attacks or disasters that may occur. This involves validating the characterization of facility, system, and functional information and their physical security features using the following worksheets:

- Worksheet 6—Facility characterization
- Worksheet 7—Defining critical operational criteria and business values
- Worksheet 8—Facility ranking based on corporate operational criteria
- Worksheet 9—Time criteria
- Worksheet 10—Rank-ordering of assets
- Worksheet 11—Asset identification and physical security characteristics
- Worksheet 12—Security characteristics strengths and weaknesses

Once this step is completed, the assessment team can begin to assess their relative value in terms of the level of protection measures to be applied.

REFINING THE DESIGN-BASIS THREAT PROFILE

When first developed and collected under Task 3 the data on the threat environment only represented a preliminary judgment. This initial determination helped the security-assessment team to identify the "state of condition" at the time of the security survey. **Security intelligence information, however, is never a stagnant phenomenon; it requires continuing attention and modification.** Under this task, not only are the malevolent acts and undesirable events previously identified reexamined, but the potential loss of consequence [C] to facilities, systems, and functions as well as the probability of occurrence [P_A] are given a thorough second review. Refining the design-basis threat involves validating the threat profile data collected on the following worksheets:

- Worksheet 13—Adversary characteristics by adversary profile
- Worksheet 14—Modes of adversary attack, weapons, and equipment
- Worksheet 15—Assets by adversary attractiveness
- Worksheet 16—Range and potential level of malevolent acts and lesser threats
- Worksheet 17—Potential threats to assets by adversary attractiveness
- Worksheet 18—Malevolent acts and undesirable events by loss of consequence and probability of occurrence

The security-assessment team uses this analysis as the final baseline to determine solutions and appropriate security strategies to achieve an effective balance of protective measures. During this process, the completed information on Worksheets 13 through 18 is validated to determine their continued significance to the decision-making process. Adjustments to any of the previous criteria developed in Worksheets 13 through 18 are made at this time and the rationale for significant change in judgments recorded.

Validating and measuring vulnerability and risk guides decision-making. Security vulnerability involves three-dimensional analysis: assessing criticality, asset vulnerability, and the likelihood of threat based on target attractiveness. This decision-making model is illustrated in Exhibit 9.2.

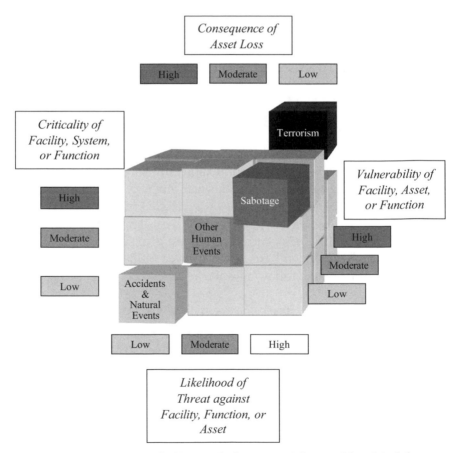

Exhibit 9.2 Risk Shifting and Threat Decision-making Model

VALIDATING PROGRAM EFFECTIVENESS AND SECURITY STRATEGIES

It should be clear at the beginning of this analysis that a threat to a facility, system, function, or resource does not warrant implementing new protective measures if existing measures sufficiently minimize the threat, particularly when changes to processes, protocols, and procedures heighten the degree of overall protection to the area or asset. In fact, it is conceivable that changes in management operational directives could result in a reduction of physical security measures for a particular asset. The same principle applies when assets are grouped together inside an area that has adequate protection measures in an outer zone.

DEVELOPING WORKABLE SOLUTIONS

One method of tracking and recording problem solving is the use of Worksheets 22 and 23. Their use helps remove ambiguity in the program and introduces clear, distinct, and positive security strategies and initiatives that give rise to the acceptance of specific positive protective measures that address specific operational capabilities. This may not always be possible. If not, then the variance between what exists and what is to be is "residual vulnerability" that must be documented and acknowledged by the enterprise as "acceptable risk."

Exhibit 9.3 displays conceptually the analysis process for increasing program strategic effectiveness based on the data shown in the following worksheets:

- Worksheet 2—Characteristics for an Enterprise Security Strategy from Conception to Implementation
- Worksheet 3—Enterprise Security Strategies and the Extent to which They are Addressed
- Worksheet 4—Enterprise Security Initiatives Designed to Enhance Security Performance Effectiveness
- Worksheet 5—Review of Available Program and Technical Data
- Worksheet 6—Facility Characterization
- Worksheet 7—Defining Critical Operational Criteria and Business Values
- Worksheet 8—Facility Ranking Based on Corporate Operational Criteria

Enterprise Institutional "Drivers" and Performance Strategies	Enterprise Institutional Security Operational Capabilities					
	Deter	Delay	Detect	Assess	Respond	Recover
Operating System Features	.50	.50	.50	.50	.50	.90
SCADA and Distributed Control Systems	.50	.50	.50	.50	.50	.50
IT Network Systems	.50	.50	.50	.50	.50	.50
Facility Security	.50	.50	.50	.50	.50	.50
Electronic Security Systems	.90	.90	.90	.90	.90	.90
Security Operations, Security Methods, Techniques	.50	.50	.50	.90	.50	.90
Information Security Program	.90	.90	.90	.90	.90	.90
Personnel Protection and Human Resources	.60	.60	.50	.50	.50	.50
Practical Ability to React	.50	.50	.50	.50	.50	.50
Emergency Planning	.85	.85	.85	.85	.85	.85
Training	.85	.85	.85	.85	.85	.85

ENTERPRISE SECURITY PERFORMANCE-BASED STANDARDS AND METRICS

NARRATIVE DESCRIPTION	NUMERICAL RATING	ID RATING	PERFORMANCE MEASUREMENT
HIGH EFFECTIVENESS	.85 to .99	H_E	Positive evidence of specific capabilities to address specific protective measures
MODERATE EFFECTIVENESS	.50 to .84	M_E	General evidence of protective measures, but no specific capabilities
LOW EFFECTIVENESS	.01 to .49	L_E	Little, ineffective, or no evidence of protective measures or capability

Exhibit 9.3 Example of Worksheet 22— Recording the Effectiveness of Enterprise Performance Strategies [P_{E2}]

- Worksheet 9—Time Criteria
- Worksheet 10—Rank Ordering of Assets
- Worksheet 13—Adversary Characteristics by Adversary Profile
- Worksheet 14—Modes of Adversary Attack, Weapons, and Equipment
- Worksheet 15—Assets by Adversary Attractiveness
- Worksheet 16—Range and Potential Level of Malevolent Acts and Lesser Threats
- Worksheet 17—Potential Threats of Assets by Adversary Attractiveness
- Worksheet 18—Malevolent Acts and Undesirable Events by Loss of Consequence and Probability of Occurrence
- Worksheet 19—Recording Status of Enterprise Institutional "Drivers" and Performance Strategies

In the sample exhibit, the programmatic and strategic enhancements applied show an increase in security effectiveness from the previous rating of 0.55 shown in Worksheet 19 to a rating of 0.66, which represents a reduction in vulnerability to the enterprise.

Exhibit 9.5 displays the recommended physical security enhancements based on the program weaknesses identified in the following worksheets:

- Worksheet 13—Adversary Characteristics by Adversary Profile
- Worksheet 14—Modes of Adversary Attack, Weapons, and Equipment
- Worksheet 15—Assets by Adversary Attractiveness
- Worksheet 16—Range and Potential Level of Malevolent Acts and Lesser Threats
- Worksheet 17—Potential Threats by Assets of Adversary Attractiveness

Critical Facility Assets	Site Access Controls [4]	Outer Barriers [4]	Facility Access Controls [1] [4]	Facility Barriers [4]	Safety Alarm [4]	IDS [2] [4]	CCTV [3] [4]	Com [4]	Area Lighting [4]
Perimeter		10-foot wall w/wire 8-foot wall				No	Yes		Yes
Effectiveness Rating		H_E				L_E	H_E		M_E
Warehouse			Card Reader	Structure	Yes	Yes	Yes	Yes	Yes
Effectiveness Rating			H_E	H_E	H_E	H_E	H_E	M_E	M_E
Equipment Parking Area			No	No		No	Yes		Yes
Effectiveness Rating			L_E	L_E		L_E	H_E		M_E
Emergency Generator			Card Reader	No	No	No	Yes	Yes	Yes
Effectiveness Rating			H_E	L_E	L_E	L_E	H_E	M_E	M_E
Emergency Management Center			Key Lock	Structure	No	Yes	Yes	Yes	Yes
Effectiveness Rating			M_E	M_E	L_E	H_E	H_E	M_E	M_E
Administration Building			Card Reader	Structure	Yes	Yes	Yes	Yes	Yes
Effectiveness Rating			H_E	H_E	H_E	H_E	H_E	M_E	M_E
Fuel Island			Card Key Control	Bollards	Yes	No	Yes	Yes	Yes
Effectiveness Rating			L_E	L_E	H_E	L_E	H_E	M_E	M_E
HAZMAT Storage Areas			Key Control	Casing		No	Yes	Yes	Yes
Effectiveness Rating			M_E	M_E		L_E	H_E	M_E	M_E
Guard House	Card Reader		Key Lock	Structure		No	Yes	Yes	Yes
Effectiveness Rating	H_E		M_E	H_E		L_E	H_E	M_E	M_E

PROBABILITY OF EFFECTIVENESS [P_E] CRITERIA

NARRATIVE DESCRIPTION	NUMERICAL RATING	ID RATING	PERFORMANCE MEASUREMENT
HIGH EFFECTIVENESS	.90	H_E	Positive evidence of specific strategic initiatives and protective measures that address specific operational capabilities
MODERATE EFFECTIVENESS	.50	M_E	General evidence of operational capabilities and protective measures, but no specific capabilities
LOW EFFECTIVENESS	.10	L_E	Little, ineffective, or no evidence of operational capabilities and protective measures

Exhibit 9.4 Example of Worksheet 23—Security Effectiveness of Recommended Protective Measures [P_{E2}]

- Worksheet 18—Malevolent Acts and Undesirable Events by Loss of Consequence and Probability of Occurrence
- Worksheet 20—Recording Status of Current Physical Security Effectiveness
- Worksheet 21—Recording Test and Exercise Results by Organizing Sector

The implementation of proposed protective measures represents a significant reduction in vulnerability, raising the security effectiveness level higher on the scale than the previous 0.37 recorded on Worksheet 20 to 0.60.

The rating represents a significant increase in security effectiveness. The implementation of integrated security enhancements will offer a level of acceptable risk that will assist in deterring an attack against the site:

- An automated access-control system reporting to the guard house and the security monitoring station (SMS) will significantly increase control and supervision of assets
- An integrated-intrusion detection system reporting to the guard house and the SMS will significantly increase monitoring-system status
- An integrated CCTV system reporting to the guard house and the SMS will significantly increase the assessment of alarms
- Frequent monitoring of all areas will strengthen detection and assessment capability

Based on the P_{E2} analysis, the security-assessment team introduces cost-effective security enhancements to increase the P_E to an acceptable risk level.

COMPREHENSIVE WORK-BREAKDOWN STRUCTURE

SUBTASK 5A—Finalize and Refine Design-Basis Threat Profile

Following are the elements of this stage:

- System technical features, including distributed-control systems and SCADA systems
- Physical security measures
- Electronic security systems
- IT networks

- Security operation methods and techniques
- Information security
- Personnel protection and human-resources practices
- Emergency planning and execution capability
- Practical ability to detect, assess, and respond to incidents
- Security organization structure and management
- Training

SUBTASK 5B—Assess Vulnerability

These are the steps to take for this task:

- Finalize likelihood of malevolent occurrence
- Finalize defined consequences of loss as determined by identified ranges and levels of threats or postulated threats
- Finalize analysis of current risks, selection of specific risk-reduction actions

SUBTASK 5C— Finalize Rank Order for Protection

Follow these steps for this task:

- Finalize rank ordering of critical operations, processes, and assets in relative importance to the mission, undesirable events, and consequence of loss
- Identify and rank-order other issues for possible consideration by executive management
- Finalize undesirable consequences that can affect rank order

SUBTASK 5D—Develop Workable Solutions

These are the steps for the last task:

- Determine acceptable PE_2 ratings
- Identify workable options
- Perform cost-economic analysis
- Select workable prioritized options
- Identify residual vulnerabilities and their consequences

As appropriate, develop alternative approaches or short-term compensatory measures.

CONCLUSION

The thoroughness with which program analysis is performed determines the quality and effectiveness of selected workable solutions. There is no "cookie-cutter" solution to address every security concern. With unlimited resources and funding, developing solutions would be less of a challenge, but there are limited and tough choices that have to be made.

Program analysis helps the security-assessment team to understand threats and danger signals that may indicate that vulnerability exists, determine the seriousness of the weakness including consequences, and prioritize corrective actions as appropriate. It serves as a focal point for the security consultant to discuss with the enterprise staff and executive management, both to sensitize them to issues that will require future management support as well to obtain a consensus for direction in the early phases of the analysis process prior to the submittal of the security-assessment report.

Chapter 10

REPORTING SECURITY ASSESSMENT RESULTS

The singular objective of reporting is to deliver information and ideas to management to permit decision-makers to select the most creditable and cost-effective solutions to identified problems.

Recipients of information include but are not limited to the following:

- Executive management
- Staff and site management
- Enterprise governing bodies
- Community governing bodies and agencies

Reporting focuses on achieving project goals and objectives. No other task is more critical to executive management than that of reporting clear, distinct, factual, and comprehensive results that provides decision-makers with the tools to make exacting choices. At this stage executive management steps in to accept the report findings and implement its recommendations. To achieve this goal, it is important that the preparation of the security-assessment report and other deliverables receive a level of professional quality-assurance review commensurate to their complexity, use, and potential consequences when not followed due to inconsistency, omission of facts, and lack of clarity and organization.

This chapter outlines a four-step process for effective reporting:

- Subtask 6A—Develop enterprise security strategies that outline program upgrades to mitigate risk and cost estimates to implement program enhancements
- Subtask 6B—Present findings and recommendations to executive management
- Subtask 6C—Present findings and recommendations to governing authorities
- Subtask 6D—Project-management reports and data management

REPORTING SECURITY-ASSESSMENT OBSERVATIONS, FINDINGS, AND RECOMMENDATIONS

The singular objective of reporting assessment results is to deliver information and ideas to executive management and staff to permit decision-makers to select the most creditable and cost-effective solutions to identified problems. An effective and comprehensive security-assessment report distinctly identifies unusual difficulties, deficiencies, or questionable conditions of interest to enterprise executive management such as:

- Site conditions or processes that contribute to the effectiveness of the security program or, conversely, lead to program weaknesses deserving management attention
- Vulnerabilities that are identified and categorized as high, moderate, or low against a series of potential threats on specific assets based on professional judgment and experience, available statistics and historical data, and intelligence information developed during the assessment process
- The prioritization of near-term and long-term recommendations, presenting proposed new or revised operational and protective measures where greater [or lesser] security needs address excessive or unwarranted expenditures
- Improvements, enhancements, future planning needs, organizational and governance issues, technical issues, and equipment procurement that impact assessment results

- Improvements to detect, deter, prevent, and defend against potential weapons of mass destruction and bombing attacks at and surrounding each site as applicable
- Recommendations that encompass enhancements and improvements in policy, protocol, physical security upgrades, training, and exercises consisting of one or more well-designed, responsive, unbiased, cost-effective solutions to bring about changes that fit the enterprise culture and complement business objectives
- With respect to security, SCADA, controlled-distribution systems, and IT network technology recommendations, whether presently installed and used or planned for future use, identify:
  - Subsystem components affected by recommendations that may require upgrading and/or replacement
  - Subsystem components affected by recommendations but that may not require upgrade and/or replacement
  - Subsystem components that may no longer be supportable or that have reached their useful life expectancy
  - Performance level of subsystem components and the existing maintenance program
  - New subsystem components not currently part of the existing security system that may need to be added to support upgrade efforts
  - Other related system elements that may be affected by security interface compatibility issues
- With respect to distributed control systems, SCADA systems, IT networks, communications, and security technology recommendations, recommend a system integration and maintenance program for each system that:
  - Brings systems online with minimal operational disruption
  - Maintains systems with minimal downtime
  - Includes a staff plan to operate and maintain systems and associates' skills
  - Identifies system components to be maintained by user, contractors, or suppliers
  - Establishes parts inventory and service agreements
  - Incorporates a cost estimate for each maintenance program
- A comprehensive discussion of the program security strategy and management initiatives necessary to develop a framework for the implementation including:

- ▸ Recommendations on how enterprise efforts lead to enhanced protection of critical infrastructure assets
- ▸ An acquisition strategy and a phased project milestone timetable for executing actions
- ▸ An order-of-magnitude cost estimate

The report should be carefully crafted to avoid potential enterprise or security-consultant liability. For example, a well-crafted report should never say that the enterprise "must" undertake a specified course of action. A better approach would be to emphasize both the risk of not taking action and the benefit of doing so. Moreover, security-assessment reports should only deal with facts and professional judgments, not opinions or speculations.

Exhibit 10.1 illustrates an example of a security-assessment-report outline.

Each and every recommendation in the security-assessment report should be addressed by management. Under no circumstances should a recommendation be rejected. To do so exposes the enterprise to potential liability for not taking reasonable and prudent action to provide for both a safe and secure work environment and public safety. A more prudent course of action is to divide recommendations into two categories: accept and defer. Recommendations placed in the "accept" category commit the enterprise to allocating the necessary budget and resources to implement actions on an acceptable timetable.

Deferring recommendations rather than rejecting them offers the enterprise significant flexibility across a wide spectrum of alternatives and avoids a bad practice that leaves the enterprise with no options. The threat environment is not static but dynamic and ever-changing. A threat or vulnerability not considered viable today may become a critical concern in the weeks, months, and years ahead. Therefore, the door should not be closed on these potential events, conditions, and circumstances. A viable solution for an enterprise should take the following into consideration:

- It might be determined that a particular enterprise facility at one location can implement a recommendation, while another facility may not reasonably have the capability. In the later instance, the enterprise should record this deviation to security at the site in question and develop a system of comprehensive measures to enforce until such time it becomes reasonable and prudent to move forward with the recommendation.

HIGHLIGHTS AND EXECUTIVE SUMMARY

Why This Study Was Conducted

Findings

Analysis

Recommendations

Status of Current Security Program Effectiveness

Security Effectiveness of Recommended Protective Measures

Summary of Design Basis Threat Statement

Summary of Program Implementation Plan

Conclusions

SECTION 1–PROJECT INTRODUCTION AND PROTOCOLS

1.1	GENERAL BACKGROUND INFORMATION
1.2	PURPOSE OF THE SECURITY ASSESSMENT STUDY
1.3	SCOPE OF TASK ASSIGNMENT
1.4	LIMITATIONS TO THE SCOPE OF ASSIGNMENT
1.5	SECURITY ASSESSMENT METHODOLOGY
1.6	RESULT EXPECTATIONS
1.7	PERFORMANCE PERIOD
1.7.1	Chronology of Steps Completed
1.7.2	Contacts, Staff Acknowledgments, Identification of Key Stakeholders and Persons Interviewed
1.7.3	External Experts Interviewed
1.8	SAFEGUARDING THIS REPORT
1.8.1	Protection, Designation and Markings
1.8.2	Distribution and Handling
1.8.3	Storage

SECTION 2–OBSERVATIONS AND RECOMMENDATIONS

2.1	BACKGROUND
2.2	SUMMARY OF OBSERVATIONS AND RECOMMENDATIONS

OBSERVATION 1

OBSERVATION 2

OBSERVATION 3

OBSERVATION 4

OBSERVATION 5

OBSERVATION 6

OBSERVATION 7

OBSERVATION 8

OBSERVATION 9

OBSERVATION 10

OBSERVATION 11

OBSERVATION 12   *(continued on next page)*

Exhibit 10.1 Example of Security-Assessment-Report Outline

(continued on next page)

Exhibit 10.1 *(continued)*

8	Assessment of Threats Against Critical Assets
9	Effectiveness of Current Protective Measures [P_{E1}]
10	Effectiveness of Mitigation Selection Protective Measures [P_{E2}]
11	Prioritized Recommendations
12	Security Enhancement Milestone Schedule and Non-Binding Budget Estimate
13	Integrated Security Platform Strategy to Establish Security Standards, and Metrics
14	Integrated Process for Implementing Security Strategies
15	Six Indispensable Security Operational Capabilities
A1	Safety and Security Characteristics
B1	Design Basis Threat Integrated Process
B2	Preliminary Potential Threat, Adversary Planning Considerations, and PA
B3	Preliminary Malevolent Acts, Consequence of Loss [C], and PA
C1	Security Assessment Methodology
C2	Six Pillars Essential to Enterprise Security Operations
C3	Performance Measurement Indicators
C4	Probability of Occurrence [PA] Criteria
C5	Probability of Effectiveness [PE] Criteria
C6	Pair-wise Comparison Criteria
C7	Enterprise Criticality Consequence Factor [C]

APPENDICES

A	Identification and Characterization of Facility
B	Preliminary Design Basis Threat Statement
C	Contacts, Staff Acknowledgements, Key Stakeholders, and Persons Interviewed
D	Security Assessment Methodology
E	Acronyms and Abbreviations
F	Glossary of Terms
G	References

Exhibit 10.1 *(continued)*

- In another instance, reducing vulnerability may require that realistic measures be exercised by outside neighbors. For example, the threat of an aircraft flying into a critical facility may require the joint effort of the enterprise and the airport authority. The enterprise may only be able to partially reduce the vulnerability, but hardening the facility may be beyond the financial ability to protect from this type of attack. In this instance it may be incumbent on the airport authority to contribute to the cost of hardening or put into place measures that reduce the threat [e.g., by changing the flight patterns if possible].

PRESENTING THE ROUGH-ORDER-OF-MAGNITUDE COST ESTIMATE

In this section the security consultant identifies a spending methodology appropriate to the milestone schedule, including any estimate constraints. The cost estimate offers executive management a baseline to develop a budget for taking the project to the next level. Cost estimates generally contain a labor and material line-item descriptive accounting for each recommendation presented. Where optional solutions are offered within a particular recommendation, the cost estimate—where appropriate—

Cost Estimate Element	Labor	Materials	Total
STRATEGIC INITIATIVES			
Develop enterprise security awareness program			
Develop security seminar for enterprise management			
Establish an enterprise-wide security advisory board			
Expand screening program to other critical positions			
SUBTOTAL			
PROGRAMMATIC INITIATIVES			
Develop enterprise security plan			
Develop enterprise functional system specification			
Update security emergency response procedures			
SUBTOTAL			
PHYSICAL SECURITY UPGRADES			
Design and Engineering			
Equipment Procurement and Installation			
System Testing and Turnover			
System Training			
Maintenance and Logistics Support			
Warranty			
Insurance, Freight, Other			
SUBTOTAL			
CONTINGENCY FACTOR [12% TO 18%]			
TOTAL COST ESTIMATE TO IMPLEMENT PROGRAM ENHANCEMENTS			

Exhibit 10.2 Example of Rough-Order-of-Magnitude Cost Estimate

should also include a cost-benefit economy analysis. Some enterprises request the security consultant to use their preferred cost model, but most enterprises leave the cost-format-reporting structure up to the security consultant. One method of reporting security-assessment cost data is shown in Exhibit 10.2.

The above model is not intended to be all inclusive but only to illustrate an example of cost elements that may require reporting.

A QUALITY SECURITY-ASSESSMENT-REPORT MODEL

The sensitivity and complexity of a security-assessment report requires that a strict standard of quality be employed in preparing and in controlling the dissemination of information. Exhibit 10.3 illustrates a proven method that clearly illustrates the data-collection, research, and analysis effort behind the development and quality-assurance review of a security-assessment report and other technical program documentation as well:

- The security consultant performs quality reviews and holds working-group sessions with the researchers, analysts, and authors to validate the data-collection process and use of supporting rationale. The report organization and the initial writing outline are reviewed and then submitted to enterprise management for review. This approach affords enterprise executive management an early opportunity to understand the makeup and organization of the report or other deliverable and gives the enterprise the opportunity to participate in the process. Many enterprises elect not to participate or get involved at this stage, and that is their choice. That being the case, the security consultant should submit to enterprise management the report outline for information purposes only.
- Executive management of the security firm [if not the security consultant] should participate in the quality-review process at the preliminary and final submittal stages. The draft is then presented to enterprise management for its review and comments.
- Enterprise review comments are evaluated by the security consultant and the security-assessment team and those that are appropriate are incorporated into the final deliverable when deemed necessary. The security consultant firm's executive management should conduct a final review of the report before it is released to the enterprise for acceptance.

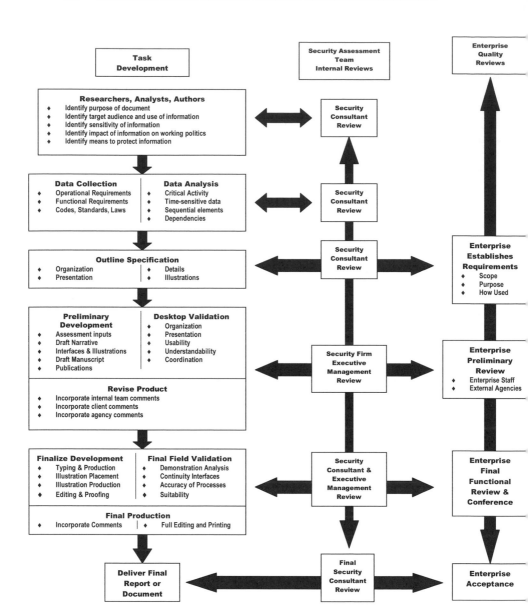

Exhibit 10.3 S^3E Documentation Development Model

QUALITY REPORTING

The quality-assurance effort ensures that the security-assessment methodology was followed; that the analysis identified assets, vulnerabilities, and threats; and that cost-effective protective measures were addressed. The quality-assurance review validates assessment results, lending credence to the overall assessment process. To achieve the most beneficial results, the review should be conducted by a person with ample experience in assessment methodologies who did not directly participate in the planning and conduct of the security assessment.

Internal to the security-consultant firm that performed the security assessment, the quality-assurance review should initially be conducted before the delivery of the draft report to the enterprise. The quality-assurance review is concluded after the security consultant meets with the enterprise staff to discuss the report, accept review comments, and incorporate valid comments into the delivery of the final security-assessment report.

Should the enterprise choose to select an unbiased third-party security firm experienced in conducting independent quality-assurance reviews, such a process should take place before any expenditures are authorized for implementing recommendations to avoid misdirecting resources. When requesting such an external review, the enterprise may have the independent firm conduct a review of the draft or the final report.

Conducting an Internal Review of the Draft Security-Assessment Report

Conducting an external review of the draft report is the prevalent practice within the industry and the one that I recommend. This approach has several distinct advantages. It gives the enterprise the greatest flexibility and affords the opportunity to evaluate both staff comments and independent review comments. Moreover, it permits the enterprise to bring together all interested parties, management; staff; the security consultant; and the independent reviewing firm, around the conference table in a constructive dialogue to exchange views and resolve outstanding issues. Review comments are then prepared by enterprise management and given to the security consultant to validate and incorporate into the appropriate areas of the report and prepare it for final delivery.

Conducting an External Review of the Final Security-Assessment Report

This alternative does not produce the expected results and may have dire cascading effects:

- The enterprise staff has already expended time and resources to pass judgment on the draft security-assessment report and met with the security consultant to address the findings and recommendations and approve the approach taken [subject to incorporating staff comments into the final report]. Such a follow-up review may be counterproductive and a waste of resources.

- Such a review typically limits any exchange between third-party reviewers and enterprise executive management, shutting out the staff and the security consultant. This approach produces minimal benefits, may not be cost-effective to either the enterprise or the security consultant, and may strain the business relationship between them.

- Perhaps most importantly, conducting an external review at this point in the process may be contrary to best industry practices and often tends to alienate both the enterprise staff and the security consultant. Since management has previously accepted the staff's position and that viewpoint has already been transmitted to the security consultant to incorporate into the final report, such an intrusion may be seen as offensive, a breach of trust, and a lack of confidence in performance. This approach may also be viewed by industry peers as an unethical practice—a stigma the enterprise can do without.

Incorporating Enterprise Staff Review Comments into the Final Security-Assessment Report

The purpose of issuing a draft of the security-assessment report is to offer interested stakeholders the opportunity to participate in the security-assessment process by meeting with the security consultant to share information with respect to reporting observations, findings, conclusions, and recommendations. This is the final opportunity for stakeholders to collect their thoughts and generate comments for clarification, bring to light potential errors and omission, or express opposing views. These review comments are an essential element of the overall quality-assurance effort, and the process takes the security assessment to the contractual state of "substantial completion."

The security consultant has a professional and ethical obligation to act on the review comments presented by either accepting or rejecting them. All review comments should be in an appendix to the final report and those actions accepted by the security consultant summarized and incorporated into the report. Comments that the security consultant disputes should be appropriately explained with defendable rationale and specific implication or impact.

Many security consultants fail to adopt the concept of placing stakeholder review comments as well as rebuttals into the final report. Many tend to address these issues in the transmittal letter that accompanies the final report. Transmittal letters, however, have a tendency of becoming separated from the report, and many enterprises do not include them when distributing the report to the staff and other stakeholders.

In addition, if a threat event occurs at the enterprise after the delivery of the final security-assessment report, without a doubt the report is likely to be scrutinized by investigating authorities and government agencies for those enterprises that are regulated. The lack of complete and accurate documentation can be dooming. Neither the enterprise nor the security consultant should be placed in a position of having to rely on memory. Often under inquiry it is human nature to forget, misinterpret, or fail to recount verbal discussions held months or even years earlier. People may retire, be promoted into other areas, or move onto other jobs—leaving a void in corporate knowledge. Including review comments and rebuttal statements as a separate appendix of the final report establishes an audit trail and closes the entire security-assessment process. It provides a historical path for actions taken by stakeholders and the security consultant for future review and analysis. Lastly, the approach limits the liability of both the enterprise and the security consultant. For these collective reasons including review comments and rebuttal statements into the final report is a wise choice.

OTHER ESSENTIAL REPORTING

The security-assessment team uses status reports to identify to enterprise management progress, potential problem areas, and anticipated activity for the next reporting period. Reports summarize activity concluded during the reporting period. Progress reports can be weekly, biweekly, monthly, or quarterly, depending on the enterprise's reporting needs. Typically, most progress reports are submitted on a monthly basis. Exhibit 10.4 illustrated a sample format for reporting progress.

Work Started This Reporting Period
Work Remaining To Be Completed
What's Going Well and Why
What's Not Going Well and Why
Suggestions/Issues/Action
Budget Status
    Total Contract Value
    Paid to Date
    Invoice This Reporting Period
    Remaining Contact Balance

Enclosures:
    Stakeholders, Key Points of Contacts, Persons Interviewed
    Sites or Agencies Visited
    Summary of Meetings and/or Conferences Attended, and List of Attendees
    Project Schedule
    Project Document Control Register, if applicable

Exhibit 10.4   Example of Progress Report Outline

The progress report should also include a performance-based program-implementation plan [PIP] that establishes a roadmap to progress. The PIP should list every major project activity to be performed. It should identify past, actual, and projected activity, including projected and actual task start and completion dates, critical milestone/paths, and percentage of the task completed thus far. For large, complex projects the PIP may also contain staff loading and detailed cost data. It is generally first initiated at the start of the project, but it is a living document. The PIP is a stand-alone document usually attached to the progress status report. A good tool for presenting the PIP is Microsoft Project Management Scheduler. This software program has the capability to measure and identify progress and costs for defined activities. For example, when used effectively the PIP can:

- Track schedule, critical-path elements, and resource allocations
- Outline the entire assessment process
- Quickly identify areas that are not being addressed, are ahead of schedule, and fall short of expectations
- Establish the basis of payment.

The enterprise and the security consultant can track progress through updates as well as through progress meetings.

PRESENTING SECURITY-ASSESSMENT RESULTS TO EXECUTIVE MANAGEMENT AND GOVERNING BODIES

Oral Presentations to Executive Management

Presenting security-assessment observations, findings, conclusions, and recommendations to enterprise management is as important as the report itself. It provides the security consultant and assessment team the opportunity to summarize in person key issues that warrant management attention, including the presentation of priority actions, ensuing challenges, and a roadmap to implementing recommendations.

Presenting this information also permits the security consultant and team and management and staff members to engage in information sharing, to ask questions, to plan a course of action for the next step, and to assign responsibility for any follow-on actions to be taken.

Oral Presentations to Governing Bodies

Presentations given to authoritative bodies such as councils, commissions, boards, and special committees take on a slightly different agenda and need to recognize the special interests of such bodies. These governing bodies are usually composed of elected or appointed officials. When they meet, their agenda typically is open to the general public and the media and includes a wide range of community topics. Presenting any results of a security assessment to a public forum is a distinct challenge, and such a presentation should only be given with the full content and review of the affected enterprise to assure that enterprise-sensitive security information is not inadvertently disclosed. A recommended forum for presenting such a public briefing is to have enterprise executive management give the presentation and only use the security consultant to help field any specific questions that may be asked by a board representative. Under no circumstance would it be appropriate for a security consultant to engage in a discussion with the media or public regarding the specifics of any security assessment. Because of the potential of such an exchange occurring, most enterprises elect not to have their security consultants attend such gatherings or participate in presenting such a briefing, except in those instances where such a presentation is incorporated within the contractual statement of work. Under this condition, presentations given to such bodies are typically done in closed sessions with both enterprise management and the security consultant participating in the process as appropriate.

COMPREHENSIVE WORK-BREAKDOWN STRUCTURE

SUBTASK 6A—Develop Enterprise Security Strategies that Outline Program Upgrades to Mitigate Risk and Cost Estimates to Implement Program Enhancements

Here are the steps to follow:

- Assessment methodology
- Identify existing strengths and weaknesses of operations and installed distributed control systems, SCADA systems, IT networks, communications, and security technology equipment
- Identify risks and vulnerabilities categorized as high, moderate, or low against a series of threats for each threat level
- Strategies for countering high-risk scenarios, which identify requisite deterrence, prevention, detection, assessment, and response methods for each level of threat
- Strategies that offer the most technically operational, and procedurally economically viable mitigation alternatives
- Strategies associated with facility reconstruction or demolition, redundant infrastructure, new technology, spare-parts enhancements, and reduced off-line maintenance
- Designs and diagrams as appropriate to highlight recommendations set forth
- Development of an implementation plan for executing recommendations that identifies:
  - ▸ Prioritization of near-term and long-term risk reduction recommendations
  - ▸ Vulnerability and criticality of threat
  - ▸ Consequence of loss
  - ▸ Presenting proposed new or revised operational and protective measures where greater [or lesser] security needs address excessive or unwarranted expenditures

- ▸ Conditions under which state/local authorities and site staff anticipate having to complete tasks
- ▸ Standard to which the task must be performed
- ▸ Current local capabilities regarding the identified tasks
- ▸ Desired capabilities regarding the identified tasks
- Milestone schedule for both short-term and long-term actions
- Order-of-magnitude cost estimate

SUBTASK 6B—Present Report to Executive Management

Follow these procedures for this step:

- Prepare report and perform internal team review of draft to draw on reviewers' expertise
- Submit draft deliverable for enterprise review and comment
- Submit draft deliverable to designated external community agencies for review and comment
- Meet with enterprise representatives to discuss strategies and recommendations, treating differing views as opportunities for developing alternative strategies
- Incorporate review comments, publish, and distribute final report.

SUBTASK 6C—Make Presentations of Findings and Recommendations to Governing Authorities

Here are the steps to follow:

- Prepare briefing and perform internal team review to draw on reviewers' expertise
- Finalize briefing and publish materials for subsequent distribution
- Present findings as appropriate to:
- Steering committee
- Executive management
- Security and safety-board committees
- City governing council
- Governing board of commissions

SUBTASK 6D—Project-Management Reports and Data Management

Here are the steps to follow:

- Prepare and maintain a document register, which identifies all documents and information exchanged between the enterprise and the security consultant
- Identify and mark deliverables to enterprise with:
- Contract or purchase-order number
- Task assignment number, if applicable
- Date of deliverable
- Number of copies distributed
- Title of deliverable
- Protection level required
- Furnish identification and qualification of key personnel
- Submit changes to key personnel assignments for approval
- Obtain approval to pursue joint participation agreements with teaming partners
- Prepare joint participation agreements with others for performance of portions of basic agreement
- Prepare and deliver a project quality-assurance plan, if applicable
- Develop and provide project safeguards procedures, if applicable
- Review and approve subcontractor quality-assurance program if applicable
- Notification of spending for task assignment exceeding $ _____, if applicable
- Submit monthly outreach program reports, as applicable
- Obtain insurance certificates and submit proof of insurance and/or extension within __ days of contract award
- Submit notice of insurance cancellation 30 days prior to effective date
- Submit invoices for payment of services no later than the _____ day of the month for services performed the previous month
- Prepare and deliver monthly progress reports:
  - ▸ Work accomplished during reporting period
  - ▸ Work left to be done
  - ▸ Work to be done in the coming months and the estimated completion dates

- ► Task number, task coordinator, task title, start and completion date, if applicable
- ► Authorized expenditure and total of dollars received to date
- ► Total dollars received under agreement to date
- Prepare and distribute notice of progress meetings with agenda at least ___ days before meeting
- Prepare and distribute minutes of progress meetings not later than ___ days after completion of meetings
- Prepare and deliver project schedule updates as major changes occur
- Prepare and deliver a project closeout report

CONCLUSION

The security-assessment report must be presented in clear and distinct terms and organized in a logical sequence. The narrative must be direct, factual, and accurate and complemented when necessary by illustrations that graphically display main themes. Worksheets that were used to present data in a user-friendly format document field observations and findings and help develop the analysis and conclusions.

The report must also present practical, defendable recommendations that provide the pertinent information to assist enterprise executive management in making sound business decisions. Prioritized recommendations identify the best solutions based on the integration of facilities, operations, people and existing or planned programs. The security-assessment report should inform executive management, the staff, and the budget committee exactly what enhancements are necessary and why.

Lastly, the results of the security assessment need to be protected. If their contents are disclosed, the enterprise could suffer increased vulnerabilities.

Part III

TAILORING THE S^3E SECURITY METHODOLOGY TO SPECIFIC CRITICAL INFRASTRUCTURE SECTORS

Following the terrorist attacks of September 11, 2001, the administration developed and published seven national strategies that relate, in part or in whole, to combating terrorism and enhancing homeland security. These included the:

- National Security Strategy of the United States of America, September 2002
- National Strategy for Homeland Security, July 2002
- National Strategy for Combating Terrorism, February 2003
- National Strategy to Combat Weapons of Mass Destruction, December 2002
- National Strategy for the Physical Protection of Critical Infrastructures and Key Assets, February 2003

- National Strategy to Secure Cyber Space, February 2003
- National Money Laundering Strategy, July 2002

Part III provides the missing link for the serious security practitioner by integrating the national strategies into the security-assessment process. This is accomplished by tailoring the general S^3E **Security Assessment Methodology** previously introduced in Chapter 4 to specifically address these national critical infrastructures:

- Water sector
- Energy sector
- Transportation sector
- Chemical and hazardous-materials sector
- Agriculture and food sector
- Banking and finance sector
- Telecommunications sector

Each tailored security assessment presents its own set of unique challenges. These issues are highlighted to alert those conducting security assessments of potential constraints and limitations they may face. While the range and level of risk may differ from one critical infrastructure sector to another, from one enterprise to another, and between industry and government, **the execution of the S^3E Security Assessment Model fundamentally is the same irrespective of the sector, asset, or environment being examined.** Only the investigative methods, techniques, and tools used might vary to fit unique requirements. The actual security-assessment process, however, remains the same.

Recognizing these complexities, examining the elements that make up the hierarchy of the security program, focusing on the interrelationships and impact on the enterprise, and developing workable solutions provides the valuable insight management needs to make critical decisions for implementation planning.

Part III is dedicated to showing how to adopt these variances into the overall security assessment without violating the integrity of any part of the process. Chapters 11 through 17 introduce the S^3E **Security Assessment Methodology** to each sector. Although overlap may appear to be present in the discussion, the goal of each chapter is to stand alone, without having to refer excessively to other sections of the book, yet complement those sections while permitting the reader to focus on a particular sector of interest. To our knowledge, this approach under a single cover sets a new assessment standard. It offers the diversified consultant, researcher, practitioner of multidiscipline responsibilities, and the academic a quick, reliable, and practical reference to use in the office, on the road, or in the classroom.

At the local level, the result of security assessments should be viewed as an integral part of the cycle for homeland-security issues. Such assessments allow local and national planners to project the consequences of possible terrorist attacks against other sector facilities or facilities in different sectors of the economy or government. These projections allow local and national authorities to strengthen defenses against noted vulnerabilities and different threats.

The S^3E **Security Assessment Methodology** model is depicted for each sector identified above in Chapters 11 through 17. For ease of comparative analysis to the generic model first introduced in Chapter 4, each tailored exhibit introduced in Part II has highlighted in grey shaded areas those specific elements that are tailored for application to each particular sector.

Chapter 11

THE WATER SECTOR

This chapter outlines the:

- Water Sector's Contribution to America's Economy
- Water Sector's Attractiveness to Terrorist and Criminal Elements
- Water Sector Vulnerabilities
- Tailoring the S3E Security Methodology
- Water Sector Challenges Facing the Security Assessment Team
- Applying the Security Assessment Methodology to the Water Sector
- Historical Overview of Selected Water Incidents
- Preparing the Water Security Assessment Report
- U.S. Government Water Initiatives

CRITICAL TO NATIONAL INTERESTS

The nation's water sector is critical to national interests from both a public health and an economic standpoint. It consists of two basic yet vital components: fresh water supply and wastewater collection and treatment. Water-system infrastructures are diverse, complex, and widely distributed, ranging from systems that serve a few to those that serve millions. The customer base includes the general public, businesses, agriculture, and critical services such as fire suppression and the medical industry. An attack—or even a credible threat—on water infrastructure could seriously jeopardize the public health and economic vitality of a community.

Drinking-water systems vary by large measure, as do the means of securing them. Many rely on groundwater as their primary water source, but most systems, particularly larger ones, rely on surface water such as lakes, rivers, and streams.

Public Law 107-188 pushed water system security to the top of the agenda of many United States utilities and resulted in the mobilization of effort and resources almost unprecedented in the water industry. The Bioterrorism Act amends the Safe Drinking Water Act by adding Section 1433. Section 1433[a] requires that certain community water systems conduct vulnerability assessments, certify to the Environmental Protection Agency [EPA] that these assessments were conducted, and to submit a copy of the assessment report to the EPA. Section 1433[b] requires that certain community water systems prepare or revise their emergency response plans and certify to the EPA that the plans have been completed. The Act betters the nation's ability to prevent, prepare for, and respond to bioterrorism.

The water infrastructure is comprised of many components that must work together as an integrated whole in order to function properly. Thus, **the general design and configuration of a typical water-supply system has vulnerabilities to attack by default that must be addressed.** It is spread throughout the community—in many cases for hundreds or even thousands of square miles, covering acres of often vulnerable real estate too large and too costly to protect against intrusion, with little definition in terms of boundaries, fences, and control points.

On the supply side, the primary focus of critical infrastructure protection efforts is the nation's 170,000 public water systems. As of 2006, public water supplies serve a total of 273 million residential and commercial customers, although the vast majority of water systems serve fewer than

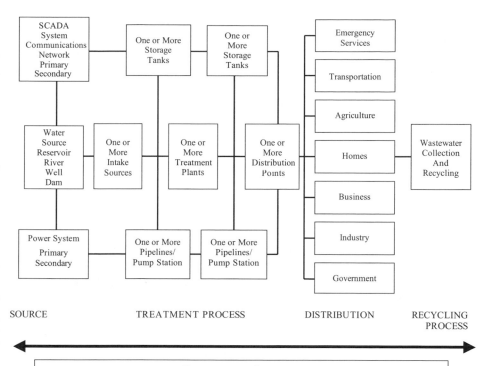

SOURCE　　　　　　TREATMENT PROCESS　　　　　DISTRIBUTION　　　RECYCLING PROCESS

CONSEQUENCE SUMMARY

- A loss of pumps or valves anywhere in the system would cause loss of treated water.
- A break or misuse of control units anywhere in the system may cause loss of treated water.
- Loss of power anywhere in the system may cause loss of treated water.
- Contamination at certain points in the system may cause unacceptable consequences.
- Loss of communications may cause delays in detecting, assessing, and responding to events.
- Any attack would cause economic impact and widespread apprehension.

Exhibit 11.1 Typical Water-Utility Configuration

10,000 people each. Public utilities account for 40 percent of the total water usage in the United States and depend on reservoirs, dams, wells, and aquifers as well as treatment facilities, pumping stations, aqueducts, and transmission pipelines. Distribution systems of public drinking-water supplies include the pipes and other conveyances that connect treatment plants to consumer taps. They span almost one billion miles in the United States and include an estimated 154,000 finished water facilities. These distribution systems constitute a significant challenge from both operational and public-health standpoints. Furthermore, they represent the vast majority of the physical infrastructure for water supplies, such that their repair and replacement represent an enormous financial liability. The wastewater industry's emphasis is on the 19,500 municipal sanitary sewer systems, including an estimated 800,000 miles of sewer lines. Wastewater utilities collect and treat sewage and process water from domestic, commercial, and industrial sources. The wastewater sector also includes storm-water systems that collect and sometimes treat storm-water runoff. [1]

An Attractive Target

Of all the national critical assets, only the water-utility infrastructure is absolutely crucial to human, plant, and animal survival and proves to be one of the most challenging matches for the security professional. Municipal drinking-water supplies and raw water resources have been a target and political tool of terrorists, armies, and governments since the beginning of time. Not only does water have the power to annihilate everything in its path, but the destruction of its resources or contamination of its supply can bring an enemy to its knees in a very short period of time. Accordingly, the criticality of water to life means that providing for water needs and safeguarding its availability will never be free of politics, tension, or war. Terrorists use water resources and systems both as targets and tools of violence or coercion. Armies have done the same during military action. Governments use them for political advantage as a major source of contention and dispute in the context of economic and social development.

Since 9/11 threats to America's water system have occurred nationwide, with similar incidents occurring in Indonesia, Japan, England, France, Italy, and other parts of the world. For security reasons, some of these incidents are summarized without detail below:

- A large amount of trespassing and photography of utility property
- A higher than average unexplained cutting of utility fences and damage to storage tanks
- Drawings found in the U.S. showing major cities to poison
- Documents found in U.S. cities that show specific locations to poison city water system
- Discovery of explosives and water contaminants on utility property.

Since the start of the war on terror, plots uncovered to attack water supplies and reports of incidents confirm that the American water system and the systems of other Western cultures remain a target of interest. A historical overview of selected terrorist attacks, criminal incidents, industry mishaps, and government actions within the water sector is displayed in Appendix A.

Water-Sector Vulnerabilities

Water vulnerability and damage consequence remain a pressing concern across the nation. Tremendous efforts have been made by the U.S. government and the water industry to complete the initial round of security assessments and develop strategic sector initiatives, but much work remains to be done. The road from risk assessment to implementation of security improvements is rocky, uphill, and slow. The industry as a whole is only in the first phase of a general national response.

Water systems have long been recognized as potentially vulnerable to terrorism of various types. Those concerns were greatly amplified after 9/11 when intelligence gathered over time indicated that both the capability and the desire of al Qaeda to attack U.S. interests here and abroad continue. This includes the U.S. drinking water-supply using such tactics as the disruption of water delivery through a physical attack on the water supply or infrastructure and the introduction of chemical or biological agents into water-distribution systems and post-treatment facilities. With respect to biochemical contamination of the drinking-water supply, al Qaeda has shown interest in cyanide, botulinum toxin, *Salmonella typhi* [the causative agent of typhoid fever], and *Bacillus anthracis* [the causative agent of anthrax]. The documents in which these agents were identified indicated that al Qaeda was developing plans to produce or acquire these agents. In addition, al Qaeda discussed plans to hyperchlorinate treated water as another means of disrupting the drinking-water supply. Of particular importance are the documents captured in Afghanistan, Iraq, and other countries that reveal diagrams and plans of American water facilities, dams, and nuclear-power plants. Training manuals discovered in Afghanistan detail how terrorists could support attacks on drinking-water systems in the United States. With respect to physical attacks, al Qaeda discussed targeting critical components of the water infrastructure such as the primary and backup high-service pumps.

TAILORING THE S^3E SECURITY ASSESSMENT METHODOLOGY FOR THE WATER SECTOR

Apply the investigative principles outlined in Part I. The security-assessment team can use the tailored revision of the S^3E **Security Assessment**

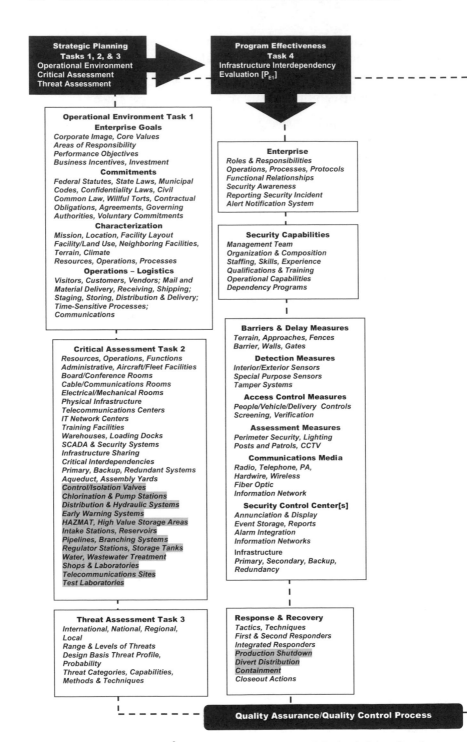

Exhibit 11.2 The S^3E Security-Assessment Methodology for the Water Sector

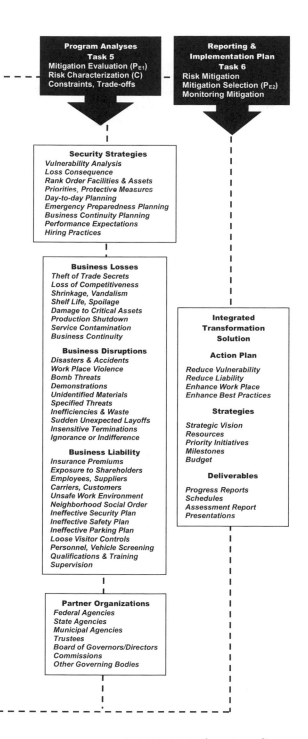

Program Analyses
Task 5
Mitigation Evaluation (P_{E1})
Risk Characterization (C)
Constraints, Trade-offs

Reporting &
Implementation Plan
Task 6
Risk Mitigation
Mitigation Selection (P_{E2})
Monitoring Mitigation

Security Strategies
Vulnerability Analysis
Loss Consequence
Rank Order Facilities & Assets
Priorities, Protective Measures
Day-to-day Planning
Emergency Preparedness Planning
Business Continuity Planning
Performance Expectations
Hiring Practices

Business Losses
Theft of Trade Secrets
Loss of Competitiveness
Shrinkage, Vandalism
Shelf Life, Spoilage
Damage to Critical Assets
Production Shutdown
Service Contamination
Business Continuity

Business Disruptions
Disasters & Accidents
Work Place Violence
Bomb Threats
Demonstrations
Unidentified Materials
Specified Threats
Inefficiencies & Waste
Sudden Unexpected Layoffs
Insensitive Terminations
Ignorance or Indifference

Business Liability
Insurance Premiums
Exposure to Shareholders
Employees, Suppliers
Carriers, Customers
Unsafe Work Environment
Neighborhood Social Order
Ineffective Security Plan
Ineffective Safety Plan
Ineffective Parking Plan
Loose Visitor Controls
Personnel, Vehicle Screening
Qualifications & Training
Supervision

Integrated
Transformation
Solution

Action Plan
Reduce Vulnerability
Reduce Liability
Enhance Work Place
Enhance Best Practices

Strategies
Strategic Vision
Resources
Priority Initiatives
Milestones
Budget

Deliverables
Progress Reports
Schedules
Assessment Report
Presentations

Partner Organizations
Federal Agencies
State Agencies
Municipal Agencies
Trustees
Board of Governors/Directors
Commissions
Other Governing Bodies

Exhibit 11.2 *(continued)*

Methodology model shown below to conduct security assessments within the water sector.

The exhibit on the previous page highlights in the grey shaded areas those areas tailored for specific application to the water sector, as contrasted to the generic template shown in Exhibit 4.2.

Water Challenges Facing the Security-Assessment Team

The basic human need for water and the concern for maintaining a safe water supply are driving factors for water-infrastructure protection. Public perception regarding the safety of the nation's water supply is also significant, as is the safety of people who reside or work near water facilities.

In order to set security-assessment priorities, the security-assessment team should focus on the types of infrastructure attacks that could result in significant human casualties and property damage or widespread economic consequences. In general, there are four areas of primary concentration:

- Physical damage or destruction of critical assets including intentional release of toxic chemicals
- Actual or threatened contamination of the water supply
- Cyber attack on information-management systems or other electronic systems
- Interruption of services from dependent infrastructure sectors

To address these potential threats, the security-assessment team requires additional focus on threat information in order to direct investments toward enhancement of corresponding protective measures. This also requires increased monitoring and analytical capabilities to augment detection of biological, chemical, or radiological contaminants that could be intentionally introduced into the water supply. The S^3E **Security Assessment Methodology** defines these and other needs.

Approaches to emergency response and the handling of security incidents at water facilities vary according to state and local policies and procedures. With regard to the public reaction associated with actual or perceived contamination, it is essential that the security-assessment team review the support capabilities and response efforts of local, state, and federal departments and agencies.

The security-assessment team must also examine other external dependencies. **The heaviest dependence is on the energy sector.** For example, running pumps to move water and wastewater and the operation of drinking-water and wastewater-treatment plants require large amounts of electricity. To a lesser extent, the water sector also depends on the transportation system for supplies of water-treatment chemicals, on natural-gas pipelines for the energy used in some operational activities, and on the telecommunications sector. The security-assessment team evaluates how these dependencies integrate to enhance water security, including contingency-planning schemes.

APPLYING THE S^3E SECURITY ASSESSMENT METHODOLOGY TO THE WATER SECTOR

The security-assessment team should use its collective expertise to:

- Identify the important mission and functions of the water sector [**Task 1—Operational Environment**] and incorporate the development of all business operations and continuity expectations. Examination focuses on water services and the customer base to determine the business contributions to the community, its culture, and business objectives. This includes but is not necessarily limited to the steps indicated in Exhibits 11.3 through 11.6.

Adequate Pressure for Fire Protection & Other Safety Uses	Corporate Image & Core Values
	Mission & Services
Adequate Volumetric Water Supply	Performance Objectives
Areas of Responsibility	Public Safety & Public Confidence
Business Incentives & Investment	Water Quality [especially potable water]

Exhibit 11.3 Identify Water Enterprise Mission Goals and Objectives

General Public, Critical Customers	Medical, Firefighters
Government, Military	Regional & International
Industrial & Business	Size of Population Served
Law Enforcement	Transportation

Exhibit 11.4 Identify Water Enterprise Customer Base

29 CFR 1910.36 - Means of Egress	Fair Credit Reporting Act [15 U.S.C. 1681] [As Amended]
29 CFR 1910.151 - Medical Services and First Aid	Family Education Rights and Privacy Act 1974 [As Amended]
29 CFR 1910.155 - Fire Protection	Federal Polygraph Protection Act [As Amended]
18 U.S.C. 2511 - Technical Surveillance	
21 U.S.C. 801 - Drug Abuse and Control	Freedom of Information Act [U.S.C. 552] [As Amended]
26 U.S.C. 7201 - Tax Matters	
29 U.S.C. 2601 - Family and Medical Leave Act	National Labor Relations Act [29 U.S.C. 151] [As Amended]
31 U.S.C. 5322 - Financial Transaction Reports	Interagency Agreements
	Internal Revenue Service
41 U.S.C. 51 - U.S. Contractor Kickbacks	Labor Management Relations Act
41 U.S.C. 701 - Drug Free Workplace Act	Memorandum of Agreement
46 U.S.C. 1903 - Maritime Drug Enforcement	Memorandum of Understanding
	Oil Pollution Act 1990 [As Amended]
50 U.S.C. 1801 - Foreign Intelligence Surveillance Act	Omnibus Crime Control and Safe Streets Act, 1968 [As Amended]
Americans with Disabilities Act [42 U.S.C. 12101]	Policy Coordination Committee [PCC]
Architectural Barrier Act 1968 [As Amended]	Privacy Act [5 U.S.C. 522e] [As Amended]
	Rehabilitation Act, 1973 [As Amended]
Bioterrorism Preparedness and Response Act, 2002 [As Amended]	Safe Drinking Water Act 1974 [As Amended]
Carriers, Suppliers, Vendors, Insurance	State and Local Industry Administrators and Authorities
Civil Rights Act 1964 [U.S.C. 200e] [As Amended]	State and Local Emergency Planning Officials and Committees
Commerce Commission	
Comprehensive Drug Abuse Prevention Act 1970 [As Amended]	Uniform Trade Secrets Act
	U.S. Department of Commerce
Comprehensive Environmental Response, Compensation and Liability Act 1980 [As Amended]	U.S. Department of Homeland Security [DHS]
	U.S. Department of Justice [DOS]
Controlled Substance Act [As Amended]	U.S. Patriot Act, 2002 [As Amended]
Drug Free Workplace Act 1988 [As Amended]	U.S. Postal System
	U.S. Environmental Protection Agency
Emergency Planning & Right-to-know Act 1986 [As Amended]	World Health Organization [WHO]
Employee Polygraph Protection Act 1988 [29 U.S.C. 2001] [As Amended]	

Exhibit 11.5 Identify Water Enterprise Commitments

Dispersed Locations	Infrastructure Sharing
Facility Land Use	Neighboring Facilities
Facility Layout & Location	Terrain & Climate

Exhibit 11.6 Characterize Configuration of Water Enterprise Facilities and Boundaries

- Describe the water-utility configuration, operations, and other elements [**Task 2—Critical Assessment**] by examining conditions, circumstances, and situations relative to safeguarding public health and safety and to reducing the potential for disruption of a reliable water system. Focus is on an integrated evaluation process to determine which assets need protection in order to minimize the impact of threat events. The security-assessment team identifies and prioritizes water facilities, processes, and assets critical to water-service objectives that might be subject to malevolent acts or disasters that could result in a potential series of undesired consequences. The process takes into account the impacts that could substantially disrupt the ability of the water utility to provide safe water services and strives to reduce security risks associated with the consequences of significant events. This includes but is not necessarily limited to the tasks shown in Exhibits 11.7 and 11.8.

- Review the existing design-basis threat profile [**Task 3—Threat Assessment**] and update it as applicable to the client. If a design-basis threat profile has not been established, then the security-assessment team needs to develop one. Both the security-assessment team and the client must have a clear understanding of undesired events and consequences that may impact water services to the community and business-continuance planning goals. Emphasis is on prioritizing water threats through the process of identifying methods of attack, past or recent events, and types of disasters that might result in significant consequences; the analyses of trends; and the assessment of the likelihood of an attack occurring. Focus is on threat characteristics, capabilities, and target attractiveness. Social-order demographics used to develop trend analysis in order to support framework decisions are also considered. Under this task the security-assessment team should also validate previous water-assessment studies and update conclusions and recommendations into a broader framework of strategic planning. Potential undesirable events include but are not necessarily limited to the tasks shown in Exhibits 11.9 through 11.13.

Administrative & Aircraft/Fleet Facilities	Motors & Pumps
Aqueducts, Aquifers, Wells, Dams	Operational Centers & Control Rooms
Assembly Yards	Regulator Stations & Storage Tanks
Board/Conference Rooms	Physical Infrastructure
Cable/Communications Rooms & Centers	Pipelines & Branching Systems
Chemical Treatment & Storage	Population & Population Throughputs
Chlorination & Pump Stations	Primary & Backup Redundancy Systems
Communications	SCADA & Other Control Systems
Contamination Test Laboratories	Security Systems
Control/Isolation Valve Systems	Shops & Laboratories
Crisis Management Centers	Suppliers, Couriers, Vendors
Critical Dependencies	Time-Sensitive Processes
Distribution & Hydraulic Systems	Training Facilities
Early Warning Systems	Transmission Systems
Electrical & Mechanical Rooms	Visitors, Mail & Material Delivery
HAZMAT & High Value Storage Areas	Warehouse Receiving/Shipping
Hours of Operation	Warehouse Storing/Staging
Infrastructure Sharing	Water & Wastewater Treatment Facilities
IT Network Centers	Water Test Laboratories
Intake Stations & Reservoirs	Water Trouble Board Centers
Microwave & Telecommunication Sites	Water Treatment Processes
Mission, Service Capabilities	

Exhibit 11.7   Critical Assessment of Water Enterprise Facilities, Assets, Operations, Processes, and Logistics

Most critical facilities based on value system	Least critical facilities based on value system
Most critical assets based on value system	Least critical assets based on value system

Exhibit 11.8   Prioritize Critical Water Enterprise Assets in Relative Importance to Business Operations

Armed Attack	Inability to perform routine transactions
Arson	Intentional opening/closing of control/isolation valves and gates
Assassination, kidnapping, assault	
Barricade/hostage incident	Intentional release of stored chemicals into treated water
Bomb threats and bomb incidents	
Chemical, Biological, Radiological [CBR] contamination of assets, supplies, equipment, and areas	Institutional corruption, fraud, waste, abuse
	Intentional shutdown of electrical power
	Loss of key personnel
Computer-based extortion	Misuse/damage of control systems
Computer system failures and viruses	Misuse of the water supply chain
Computer theft	Misuse of wire and electronic communications
Cyber attack on SCADA or other systems	Theft or destruction of dangerous chemicals
Cyber stalking, software piracy	Phishing, hacking, computer crime
Damage/destruction of control/isolation valve systems	Power failure
	Robbery and burglary
Damage/destruction of interdependency systems resulting in production shutdown	Sabotage
	Sexual abuse, sexual exploitation
Damage/destruction of pipelines and cabling	Social engineering
Damage/destruction of pump stations and motors	Software piracy
	Storage of criminal information on the system
Damage/destruction of service capability	Unlawful access to stored wire and electronic communications and transactional records access
Delay in processing transactions	
Disable manufacturing, production, distribution and delivery processes	
	Unlawful wire and electronic communications interception and interception of oral communications
Disable water pretreatment, treatment, or distribution processes	
Eavesdropping	Telecommunications failure
Economic espionage censorship	Terrorism
Email attacks	Torture
Encryption breakdown	Treason, sedition, and subversive activities
Explosion	Unlawful access to stored communications
FAX security	Use of cellular/cordless telephones to disrupt network communications or in support of other crimes
Hacker attacks	
Hardware failure, data corruption	
Homicide	Voice mail system
Inability to access system or database	Wide-scale disruption and disaster
	Wide-scale evacuation

Exhibit 11.9 Determine Types of Malevolent Acts that could Reasonably Cause Water Enterprise Undesirable Events

Bribery, graft, conflicts of interest	Online pornography and pedophilia
Civil disorders, civil rights violations	Planned production shutdown or outage
Conspiracy	Maintenance activity
Embezzlement, extortion and threats	Racketeering
Historical rate increases	Racketeer influenced and corrupt
Demonstrations and strikes	organizations
Human error	Rape and sodomy
Industrial accidents and mishaps	Release and misuse of personal
Inefficiencies and waste	information
Ineffective hiring practices	Resource constraints and competing
Ineffective personnel and vehicle controls	priorities
Ineffective safety and security plans	Shrinkage, shelf-life, spoilage
Ineffective supervision and training	Sudden unexpected layoffs
Ineffective use of resources and time	Supplier, courier disruptions
Ineffective visitor controls	Tracking effects of outage overtime
Ignorance and indifference	Tracking effects of outage across related
Insensitive terminations	resources and dependent systems
Loss of competitiveness	Theft of trade secrets, intellectual
Loss of proprietary information	propriety, confidential information,
Mail fraud	patents, and copyright infringements
Natural disasters	Trespassing and vandalism
Obscenity	Unauthorized transfer of American
Obstruction of justice	technology to rogue foreign states
	Unidentified materials or wrong deliveries

Exhibit 11.10 Assess Other Disruptions Impacting Water Operations

Activist groups, subversive groups or cults	State-sponsored/independent terrorist groups
Disenfranchised individuals, hackers	Terrorist organizations posing as
Legitimate entities serving as conduits for terrorist financing	legitimate entities using the system for criminal purposes
Organized criminal groups	Vandals, lone wolves, insiders

Exhibit 11.11 Identify Category of Water Enterprise Perpetrators

Cost to repair	Loss of life [customers and employees]
Compromise of public confidence	Loss or disruption of communications
Dealing with unanticipated crisis	Loss or disruption of information
Disabling of or loss of key personnel	Loss or interruption of treated water
Duration of loss of fire protection	Loss of untreated water sources or ability to tap water sources
Duration of loss of water potability	
Economic loss	Number of critical customers impacted
Illness [customers and employees]	Number of users impacted
Impact on regional economic base	Production shutdown
Impact on utility ratepayers	Reduced ability to distribute water
Inability of customers to conduct business	Slowdown/interruption of delivery services
Interruption or inability to store water	
Local civil unrest	Temporary closure of financial institutions

Exhibit 11.12 Assess Initial Impact of Water Enterprise
Loss Consequence

Consequence analysis	Economic conditions
Social/political status	Social demographics

Exhibit 11.13 Assess Initial Likelihood of Water Enterprise Threat
Attractiveness and Likelihood of Malevolent Acts Occurring

- Examine and measure the effectiveness of the security system **[Task 4—Evaluate Program Effectiveness].** The security-assessment team evaluates business practices, processes, methods, and existing protective measures to assess their level of effectiveness against vulnerability and risk, and to identify the nature and criticality of business areas and assets of greatest concern to water operations. Focus is on vulnerability and vulnerable access points where penetration may be accomplished and how. Emphasis is placed on the physical characteristics of vulnerable points, accessibility to the location, and the effectiveness of protocol obstacles an adversary would likely have to overcome to reach a critical water asset. These include but are not necessarily limited to the tasks depicted in Exhibits 11.14 through 11.19.

Alert Notification System	Reporting Security Incidents
Emergency Planning	Roles & Responsibilities
Functional Interfaces	Security Awareness
Operations, Processes, Protocols	Security Plans, Policies & Procedures

Exhibit 11.14 Evaluate Existing Water Enterprise Security Operations and Protocols [P_{E1}]

Dependency Programs	Qualifications & Training
Management Team	Staffing, Skills, Experience
Organization & Composition	Supervision
Operational Capabilities	

Exhibit 11.15 Evaluate Existing Water Enterprise Security Organization [P_{E1}]

Agency for Toxic Substances and Disease Registry	Computer Security Institute
	Critical Infrastructure Protection Board [PCIPB]
American Chemistry Council	Defense Central Investigation Index [DCII]
American Electronics Association [AEA]	El Paso Intelligence Center [EPIC]
American Institute of Certified Public Accountants	EPA National Security Research Center
	EPA Office of Air and Radiation
American National Standards Institute	EPA Office of Prevention, Pesticides and Toxic Substances
American Psychological Association	
American Society for Industrial Security	EPA Office of Solid Waste and Emergency Response
American Society for Testing and Materials	
	EPA Water Security Division
American Society for Testing Materials, Vaults	EPA Water Security Team
American Water Works Association	Factory Mutual Research Corporation
American Water Works Association Research Foundation	Federal, state, and local agency responders
	Illuminating Engineers Society of North America
Association of Metropolitan Sewage Agencies	
Centers for Disease Control and Prevention [CDC]	Institute for a Drug Free Workplace
	Institute of Electrical and Electronic Engineers [IEEE]
Commerce Commission	

Exhibit 11.16 Evaluate Existing Water Enterprise Interface and Relationship with Partner Organizations [P_{E1}]

International Association of Chiefs of Police	National Water Awareness Technology Evaluation Research and Security Center
International Association of Professional Security Consultants	National Water Sample Test Laboratories
International Computer Security Association	National White Collar Crime Center [NWCCC]
International Criminal Police Organization [INTERPOL]	Occupational Safety and Health Review Commission [OSHRC]
International Electronics Supply Group [IESG]	Office of Civilian Radioactive Waste Management
International Parking Institute	
International Organization for Standardization [ISO]	Office of National Drug Control Policy
	Policy Coordination Committee [PCC]
Interstate Commerce Commission [ICC]	Private Security Advisory Council
Joint Terrorism Task Force [JTTF]	Regional Information Sharing System [RISS]
Law Enforcement Support Center [LESC]	Middle Atlantic-Great Lakes Organized Crime Law Enforcement Network [MAGLOCLEN]
Library of Congress, Congressional Research Service	Mid-States Organized Crime Information Center [MOCIC]
Medical Practitioners and Medical Support Personnel	New England State Police Information Network [NESPIN]
Multi-lateral Expert Groups	Regional Organized Crime Information Center [ROCIC]
Mutual Aid Associations	
National Advisory Commission on Civil Disorder	Rocky Mountain Information Network [RMIN]
National Crime Information Center [NCIC]	Western States Information Network [WSN]
National and State Public Health Associations	Safe Manufactures National Association
National Drinking Water Advisory Council	The Terrorism Research Center
National Electrical Manufacturing Association [NEMA]	U.S. Central Intelligence Agency [CIA]
	U.S. Coast Guard
National Fire Protection Association [NFPA]	U.S. Department of Commerce [DOC]
National Information Infrastructure [NII]	U.S. Department of Homeland Security [DHS]
National Infrastructure Protection Center	U.S. Department of Justice [DOJ]
National Institute of Law Enforcement and Criminal Justice	U.S. Environmental Protection Agency [EPA]
National Institute of Standards and Technology [NIST]	U.S. Federal Bureau of Investigation [FBI]
	U.S. Government Accounting Office [GAO]
National Insurance Crime Bureau [NICB] Online	U.S. Postal System
	Underwriters Laboratories [UL]
National Labor Relations Board	Water Information Sharing and Analysis Centers [ISAC]
National Parking Association	
National Safety Council	World Health Organization [WHO]

Exhibit 11.16 *(continued)*

Assessment Capability	Modem & Internet Access
Badge Controls	Performance Standards & Vulnerabilities
Barrier/Delay Systems	Personnel & Vehicle Access Control Points
Chemical & Other Vendor Deliveries	Physical Protection System Features
Communications Capability	Post Orders & Security Procedures
Cyber Intrusions, Firewalls, Other Protection Features	Response Capability
Display & Annunciation Capability	Security Awareness
Evacuation Response Plans	Security Training/Exercises/Drills
Intrusion Detection Capability	System Capabilities & Expansion Options
Lock & Key Controls, Fencing & Lighting	System Configuration & Operating Data

Exhibit 11.17 Evaluate Existing Water SCADA and Security System Performance Levels [P_{E1}]

Deception actions	Likely approaches and escape routes
Delay-penetration obstacles	Target selection and alternative targets
Interception analysis	Time-delay response analysis

Exhibit 11.18 Define the Water Enterprise Adversary Plan, Distractions, Sequence of Interruptions, and Path Analysis

Closeout Actions	Integrated Responders
Containment	Production Shutdown
Divert Distribution	Tactics & Techniques
First & Second Responders	

Exhibit 11.19 Assess Water Enterprise Effectiveness of Response and Recovery [P_{E1}]

Business Continuity Planning	Loss Consequences
Day-to-Day Planning	Performance Expectations
Emergency Preparedness Planning	Rank Order Facilities & Assets
Hiring Practices	Vulnerability Analysis

Exhibit 11.20 Analyze Effectiveness of Water Security Strategies and Operations [P_{E1}]

- Recognize the importance of bringing together the right balance of people, information, facilities, operations, processes, and systems [**Task 5—Program Analyses**] to deliver a cost-effective solution. The security-assessment team develops an integrated approach to measurably reduce security risk by reducing vulnerability and consequences through a combination of workable solutions that bring real tangible enterprise value to the table. Emphasis is placed on assessing the current status and level of protection provided against desired protective measures to reasonably counter the threat. Operational, technical, and financial constraints are examined, as well as future expansion plans and constraints. These include,but are not necessarily limited to the tasks shown in Exhibits 11.20 through 11.26.

Compromise of public confidence	Loss of life [customers and employees]
Cost to repair	Loss or disruption of communications
Dealing with unanticipated crisis	Loss or disruption of information
Disabling of or loss of key personnel	Loss or interruption of treated water
Duration of loss of fire protection	Loss of untreated water sources or ability to tap water sources
Duration of loss of water potability	
Economic loss	Number of critical customers impacted
Illness [customers and employees]	Number of users impacted
Inability of customers to conduct business	Reduced ability to distribute water
Impact on regional economic base	Slowdown/interruption of delivery services
Impact on utility ratepayers	
Interruption or inability to store water	Production shutdown
Local civil unrest	Temporary closure of financial institutions

Exhibit 11.21 Refine Previous Analysis of Water Enterprise Undesirable Consequences that can Affect Functions

Consequence analysis	Economic conditions
Social/political status	Social demographics

Exhibit 11.22 Refine Previous Analysis of Water Enterprise Likelihood of Malevolent Acts of Occurrence

Creation and/or revision or modification of sound business practices and security policies, plans, protocols, and procedures.	Development, revision, or modification of interdependency requirements.
Creation and/or revision or modification of emergency operation plans including dependency support requirements.	SCADA/security system upgrades to improve detection and assessment capabilities.

Exhibit 11.23 Analyze Selection of Specific Risk-Reduction Actions Against Current Risk, and Develop Prioritized Plan for Water Enterprise Mitigation Solutions [P$_{E2}$]

Reasonable and prudent mitigation options	Mirrors business culture image

Exhibit 11.24 Develop Short- and Long-Term Water Enterprise Mitigation Solutions

Effectiveness calculations	Residual vulnerability consequences

Exhibit 11.25 Evaluate Effectiveness of Water Enterprise Developed Mitigation Solutions and Residual Vulnerability [P$_{E2}$]

Practical, cost-effective recommendations	Reasonable return on investments

Exhibit 11.26 Develop Cost Estimate for Short- and Long-Term Water Enterprise Mitigation Solutions

PREPARING THE WATER SECURITY-ASSESSMENT REPORT

Chapter 10 provides a recommended approach to reporting the security-assessment results.

WATER-SECTOR INITIATIVES

Efforts to enhance water surety include a variety of initiatives. Appendix B summarizes some significant actions taken or currently underway by the U.S. Department of Homeland Security and industry.

REFERENCES

1. "The National Strategy for the Physical Protection of Critical Infrastructure and Key Assets" [February 2003]

APPENDIX A: A HISTORICAL OVERVIEW OF SELECTED TERRORIST ATTACKS, CRIMINAL INCIDENTS, AND INDUSTRY MISHAPS WITHIN THE WATER SECTOR

The author has compiled the information in this appendix in a chronological sequence to share the historical perspective of incidents with the reader interested in conducting further research. The listing, along with the main text of Chapter 2, presents a capsule review of terrorist activity and other criminal acts perpetrated around the world in the water sector and offers security practitioners specializing in particular regions of the world a quick reference to the range and level of threat.

Source: *The information is a consolidated listing of events, activities, and news stories as reported by the U.S. Department of Homeland Security, U.S. Department of State, and various other government agencies. Contributions from newspaper articles and news media reports are also included. They are provided for a better understanding of the scope of terrorism and other criminal activity and further research for the interested reader.*

JANUARY 2006 Los Angeles, CA, United States.
Five persons cut through fences and breached doors at one of the reservoirs owned and operated by the Department of Water and Power. Bags of salt were thrown into the reservoir, but attempts to tamper with system valves were not successful. The FBI, Los Angeles Police Department, Los Angeles Sheriff's Department, and Los Angeles Port Authority divers responded to the incident.

30 JULY 2005 California, United States.
A security fence is being erected along California's Nacimiento Dam in an attempt to stave off a terrorist attack to the hydroelectric plant there. Nacimiento Lake is in northern San Luis Obispo County, CA, but is owned and operated by Monterey County, CA. Officials said they are building the $110,000 fence after preparing a security analysis required by the Federal Energy Regulatory Commission following the September 11 terrorist attacks.

23 JANUARY 2005 Nairobi, Kenya.
Rival tribes from the Kikuyu and Masai clashed over access to water rights. A total of 14 people died in several confrontations.

28 DECEMBER 2004 Blackfoot, ID, United States.
At the Blackfoot Wastewater Treatment Plant someone purposefully altered a gas line at the plant, causing a leak in the building that could have destroyed the entire plant. So much gas was leaking inside the

building that all the oxygen had been pushed out. An employee discovered the leak and shut off the gas line before damage could occur, averting a potential disaster. An explosion could have left the city without water for months. The FBI and local police cleared present employees of suspicion.

20 NOVEMBER 2004 Albuquerque, NM, United States.
FBI joined search for break-in of Escondido Reservoir. Water delivery from the reservoir was stopped immediately after intrusion alarms at the gate and entrance were triggered. The 6-million-gallon reservoir, like all of the city's 47 reservoirs, is enclosed in a tank and serves about 6,000 city residents. The residents normally served by the Escondido Reservoir received water from another of the city's reservoirs until water samples were tested and cleared.

30 SEPTEMBER 2004 Baghdad, Iraq.
Three car bombs in the southern suburbs outside a newly opened sewage plant killed 35 children among a death toll of 41. At least 130 were wounded.

2003 Baghdad, Iraq.
Sabotage/bombing of main water pipeline in Baghdad.

2003 United States.
Al Qaeda threatens American water systems via call to Saudi Arabian magazine. Al Qaeda does not "rule out the poisoning of drinking water in American and Western cities."

23 SEPTEMBER 2003 Mecosta County, Michigan, United States.
Incendiaries were left at a pumping station supplying a water-bottling plant owned by Nestle Waters North America. Incendiaries failed to ignite and were removed from the station without incident. The Earth Liberation Front [ELF] claimed responsibility.

30 MAY 2003 Guamalito, Colombia.
Terrorists attacked a section of the Cano Limon-Covenas oil pipeline, spilling nearly 7,000 barrels of crude oil into the Cimitarra creek, a major source of drinking water for more than 5,000 people, causing extensive environmental damage, and leaving families without drinking water. Ecopetrol of Colombia and a consortium of West European and U.S. companies jointly own the pipeline. No group claimed responsibility, although both the Revolutionary Armed Forces of Colombia [FARC] and National Liberation Army [ELN] terrorist groups have attacked this

pipeline previously.

2002 Bogota, Colombia.
Colombian rebels damaged a gate valve in the dam that supplies most of Bogota's drinking water.

2002 Rome, Italy.
Italian police arrest four Moroccans who are allegedly planning to contaminate Rome's water-supply system with a cyanide-based chemical, targeting buildings that included the United States Embassy. Ties to al Qaeda are suggested.

2002 Winter Park, CO, United States.
The Earth Liberation Front [ELF] threatened the water supply of the town of Winter Park, CO. Previously, this group claimed responsibility for the destruction of a ski lodge in Vail, CO, that threatened lynx habitat.

2001 Israel.
Palestinians destroyed water-supply pipelines to West Bank settlements of Yitzhar and Kibbutz Kisufim. The Agbat Jabar refugee camp near Jericho was disconnected from its water supply after Palestinians looted and damaged local water pumps. Palestinians accused Israel of destroying a water cistern, blocking water-tanker deliveries, and attacking materials for a wastewater-treatment project.

2000 Queensland, Australia.
Queensland police arrested a man for using a computer and radio transmitter to take control of the Maroochy Shire Wastewater System to release sewage into parks, rivers, and property.

1999 Lusaka, Zambia.
Terrorist bomb blast destroyed the main water pipeline, cutting off water for the city of Lusaka, which has a population of 3 million.

1998 Roosevelt Dam, Arizona, United States.
A 12-year-old computer hacker broke into the SCADA computer system that runs Arizona's Roosevelt Dam, gaining complete control of the dam's massive flood gates. The cities of Mesa, Tempe, and Phoenix, AZ, are downstream of this dam. No damage was done.

1993 Iraq.
To quell opposition to his government, Saddam Hussein reportedly poisoned and drained the water supplies of southern Shiite Muslims, the Ma'dan.

1991 Irar, Kuwait.
The allied coalition targeted Baghdad's water-supply and sanitation systems.

1988 Angola, South Africa.
Cuban and Angolan forces launched an attack on Calueque Dam via land and air. Considerable damage was inflicted on the dam wall; the power supply to the dam was cut. The water pipeline to Owamboland was destroyed.

1984 The Dalles, Oregon, United States.
Members of the Rajneeshee religious cult contaminated a city water-supply tank in The Dalles, Oregon, using salmonella. A community outbreak of over 750 cases occurred in a country that normally reports fewer than five cases per year.

1975 Angola, South Africa.
South African troops moved into Angola to occupy and defend the Ruacana hydropower complex, including the Gove Dam on the Kunene River. The goal was to take possession of and defend the water resources of southwestern Africa and Namibia.

1964 Guantanamo Bay, Cuba.
The Cuban government ordered the water supply cut off to the United States Naval Base at Guantanamo Bay.

1951 North Korea.
North Korea released flood waves from the Hwachon Dam, damaging floating bridges operated by UN troops in the Pukhan Valley.

1948 Israel.
Arab forces cut off West Jerusalem's water supply in the first Arab-Israeli war.

1939–1942 China.
Japanese chemical- and biological-weapons activities reportedly included tests against military and civilian targets by lacing water wells and reservoirs with typhoid and other pathogens.

1907–1913 Los Angeles, CA, United States.
The Los Angeles Valley aqueduct/pipeline suffered repeated bombings in an effort to prevent diversions of water from the Owens Valley to Los Angeles.

1863 United States.
Union General Ulysses S. Grant cut levees in the Civil War campaign against Vicksburg.

1672 The Netherlands.
In defense of a French attack upon the Netherlands, the Dutch opened dikes and flooded the country, creating a watery barrier that was virtually impenetrable.

720–705 BCE Armenia.
After a successful campaign against the Halidians of Armenia, Sargon II of Assyria destroyed the Halidians' intricate irrigation network and flooded their land.

2500 BCE Umma.
In a dispute over the Gu'edena region, **Urlama, King of Lagash, diverted water from this region to boundary canals, drying up boundary ditches to deprive Umma of water.** His son cut off the water supply to Girsu, a city in Umma.

APPENDIX B: UNITED STATES GOVERNMENT WATER-SECTOR INITIATIVES

Government at all levels, the private sector, and concerned citizens across the country have established an important partnership and commitment to address the threat posed by those who wish to harm the United States. Critical infrastructure owners and operators are assessing their vulnerabilities and increasing their investment in security. State and municipal governments across the country continue to take important steps to identify and assure the protection of assets and services within their jurisdictions. Federal departments and agencies are working closely with industry to take stock of key assets and facilitate protective actions, while improving the timely exchange of important security-related information. The Department of Homeland Security [DHS] is working closely with key public- and private-sector entities to implement protection initiatives.

Water-infrastructure-protection initiatives are guided both by the challenges that the water sector faces and by the Bioterrorism Act. Additional protection initiatives include efforts to:

- **Identify high-priority vulnerabilities and improve site security.** The Environmental Protection Agency [EPA], in concert with the Department of Homeland Security [DHS], state and local governments, and other water-sector leaders, is working to identify processes and technologies to better secure key points of storage and distribution, such as dams, pumping stations, chemical-storage facilities, and treatment plants. The EPA and the DHS continue to provide tools, training, technical assistance, and limited financial assistance for research on vulnerability-assessment methodologies and risk-management strategies.
- **Improve sector monitoring and analytic capabilities.** The EPA continues working with sector representatives and other federal agencies to improve information on contaminants of concern and to develop appropriate monitoring and analytical technologies and capabilities.
- **Improve sector-wide information exchange and coordinate contingency planning.** The DHS and the EPA continue working with the water-sector coordinator and the Water Information Sharing Analysis Center [ISAC] to coordinate timely information on threats, incidents, and other topics of special interest to the water sector. The DHS and

the EPA are also working with the sector and the states to standardize and coordinate emergency-response efforts and communications protocols.

- **Work with other sectors to manage unique risks resulting from interdependencies.** The DHS and the EPA are working with cross-sector working groups to develop models for integrating priorities and emergency-response plans in the context of interdependencies between the water sector and other critical infrastructures.

Chapter 12

THE ENERGY SECTOR

This chapter outlines the:
- Importance of the energy sector to the economic security of the nation
- Energy-sector contributions to the economic security of America
- Energy-sector attractiveness to terrorist and criminal elements
- Energy-sector vulnerabilities
- Tailoring the S^3E Security Methodology to the energy sector
- Energy-sector challenges facing the security-assessment team
- Applying the Security Assessment Methodology to the energy sector
- Preparing the Energy Security Assessment Report
- Historical overview of selected energy incidents
- U.S. government energy initiatives

IMPORTANCE TO THE ECONOMIC SECURITY OF THE NATION

Energy drives many of the sophisticated processes at work in American society today. It is essential to our economy, national defense, and quality of life. The energy sector is commonly divided into two segments in the context of critical infrastructure protection: electricity and oil and natural gas. The electric industry serves almost 130 million households and institutions. The United States consumed nearly 3.6 trillion kilowatt-hours in 2001. Worldwide, there were 438 nuclear reactors in use at the end of the year 2000, generating electricity in 33 countries. About 16 percent of all electricity in the world is produced by nuclear power. The share of nuclear power in Europe is 35 percent. France is the largest single user of nuclear power, producing about 77 percent of all nuclear power in the European Union. In contrast, nuclear power in the United States only accounts for 22 percent of its total energy, generating 31 percent of the world total. [1]

Some of the larger and more symbolic dams are major components of other critical infrastructure systems that provide water and electricity to large populations, cities, and agricultural complexes. There are approximately 80,000 dam facilities identified in the National Inventory of Dams. Most are small, and their failure would not result in significant property damage or loss of life. The federal government is responsible for roughly 10 percent of the dams whose failure could cause significant property damage or have public health and safety consequences. The remaining critical dams belong to state or local governments, utilities, and corporate or private owners. [2]

Within the United States, oil and natural-gas facilities and assets are widely distributed, consisting of more than 300,000 producing sites, 4,000 offshore platforms, more than 600 natural-gas-processing plants, 153 refineries, more than 1,400 product terminals, and 7,500 bulk stations. [3]

Contributions to Economic Security

Electricity
Almost every form of productive activity, whether in businesses, manufacturing plants, schools, hospitals, or homes, requires electricity. Electricity is also necessary to produce other forms of energy, such as refined oil. Were a widespread or long-term disruption of the power grid to occur, many of the activities critical to the economy and national defense—including those associated with response and recovery—would

be impossible. The North American electric system is an interconnected, multinodal distribution system that accounts for virtually all the electricity supplied to the United States, Canada, and a portion of Baja California Norte, Mexico. The physical system consists of three major parts: generation, transmission and distribution, and control and communications.

Generation assets include hydroelectric dams, fossil-fuel plants, and nuclear-power plants:

- **Hydro plants** usually are located on a river and include a dam that forms a lake. The dam provides a source of energy to power turbine generators. When streams and lakes are navigable, locks may be provided to permit the passage of boats through the dam. These locks are normally operated by or through an agreement with the U.S. Army Corps of Engineers. The U.S. Coast Guard enforces the navigable waterways.

- In **fossil plants** boilers are fired by coal, oil, gas, lignite, or other fuel that produces steam, which turns the generators to produce electricity that is then stepped up through a transformer and passed through a switching yard to the transmission system.

- In **nuclear plants** material fission within the reactor produces the heat to convert water to steam.

- **Other sources** of energy include wind, solar, and fuel cells. These devices are generally used in small numbers, normally in association with larger energy systems. They require individualized security measures for their protection.

Transmission and distribution systems link areas of the national grid. Most power systems are interconnected through **transmission systems** linked to the national power grid, allowing utilities to facilitate distribution of power during periods of emergency. A serious power shortage or outage on one system could affect one or more other systems. To minimize these situations, devices used to isolate troubled areas are attractive targets to terrorists and require unique protective measures. The transmission system is vulnerable to interference from natural disasters such as rain and ice storms, high winds and falling trees, earthquakes, and lightning; man-made interferences such as automobile accidents, and plane crashes; and deliberate tampering, sabotage, or damage.

Distribution systems manage and control the distribution of electricity into homes and businesses. The distribution system provides energy to the customer base via feeders, transformers, vaults, and pipelines and ends at the customer meter. The system is vulnerable to natural and man-made interferences, including theft of service and diversion of energy.

Control and **communications systems** operate and monitor critical infrastructure components.

Supervisory-control and data-acquisition [SCADA] systems and other **management-information systems [MIS]** including **security systems** are the nerve center of the energy system. SCADA systems monitor and track operational system transactions and activity. They maintain a system in a stable condition. Management-information systems provide other direct support data. Security systems monitor and track the operational integrity of critical assets and areas.

In addition to these components, the electric infrastructure also comprises ancillary facilities and systems that guarantee fuel supplies necessary to support electricity generation, some of which involve the handling of hazardous materials. The electricity sector also depends heavily on other critical infrastructures such as telecommunications and transportation for power generation.

After New York's power blackout in 1965, the industry established the North American Electric Reliability Council [NERC] to develop guidelines and procedures for preventing similar incidents. NERC is a nonprofit corporation composed of 10 regional reliability councils, whose voluntary membership represents all segments of the electricity industry, including public and private utilities from the U.S. and Canada. Through NERC, the electricity sector coordinates programs to enhance security for the electricity industry. [4]

The electricity sector is highly regulated even as the industry is being restructured to increase competition. The Federal Energy Regulatory Commission [FERC] and state utility regulatory commissions regulate some of the activities and operations of certain electricity-industry participants.

Oil and Natural Gas
The oil and natural-gas industries are tightly integrated. The oil infrastructure consists of five general components:

- Oil production
- Crude-oil transport
- Refining
- Transport and distribution
- Control and other external support systems

Oil and natural-gas production include:

- Exploration
- Field development
- On- and offshore production
- Field collection systems
- Supporting infrastructures

Crude-oil transport includes:

- 160,000 miles of pipelines
- Storage terminals
- Ports and ships

The refinement infrastructure consists of about 150 refineries that range in size and production capabilities from 5,000 to over 500,000 barrels per day. [5]

Transport and distribution of oil includes:

- Pipelines
- Trains
- Ships
- Ports
- Terminals and storage
- Trucks
- Retail stations

The natural-gas industry consists of three major components:

- Exploration and production
- Transmission
- Local distribution

The U.S. produces roughly 20 percent of the world's natural-gas supply. There are 278,000 miles of natural-gas pipelines and 1,119,000 miles of natural-gas distribution lines in the U.S. [6]

Distribution includes:

- Storage facilities
- Gas processing

- Liquid-natural-gas facilities
- Pipelines
- City gates
- Liquefied-petroleum-gas storage facilities

City gates are distribution-pipeline nodes through which gas passes from interstate pipelines to a local distribution system. Natural-gas storage refers to underground aquifers, depleted oil and gas fields, and salt caverns.

The pipeline and distribution segments of the oil and natural-gas industries are highly regulated. Oversight includes financial, safety, and siting regulations. The exploration and production side of the industry is less regulated but is affected by safety regulations and restrictions concerning property access.

Pipelines

The United States has a vast pipeline industry consisting of many hundreds of thousands of miles of pipelines, many of which are buried underground. These lines move a variety of substances such as crude oil, refined petroleum products, and natural gas. Pipeline facilities already incorporate a variety of stringent safety precautions that account for the potential effects a disaster could have on surrounding areas. Moreover, most elements of pipeline infrastructures can be quickly repaired or bypassed to mitigate localized disruptions. Destruction of one or even several of their key components would not disrupt the entire system. As a whole, the response and recovery capabilities of the pipeline industry are well proven, and most large control-center operators have established extensive contingency plans and backup protocols. [7]

Pipelines that transport oil and gas supplies are components of the transportation-sector critical infrastructure and are regulated by the Department of Transportation [DOT] for safety purposes. They are included under the energy sector for assessment purposes only.

Nuclear Power

Nuclear-power plants are an important component of the energy sector's critical infrastructure. They represent about 20 percent of our nation's electrical generation capacity. The U.S. has 104 commercial nuclear reactors in 31 states. For 25 years, federal regulations have required that these facilities maintain rigorous security programs to withstand an attack of specified adversary strength and capability. Nuclear-power plants are also among the most physically hardened structures in the country, designed to withstand extreme events such as hurricanes, tornadoes, and earthquakes. Their reinforced engineering design provides inherent protection through such features as robust containment buildings, redundant safety systems, and sheltered spent-fuel storage facilities. The security at nuclear-power plants has been enhanced significantly in the aftermath of the 9/11 attacks. All plants remain at heightened states of readiness, and specific measures have been implemented to enhance physical security and to prevent and mitigate the effects of a deliberate release of radioactive materials. Steps have been taken to enhance surveillance, provide for more restricted site access, and improve coordination with law-enforcement and

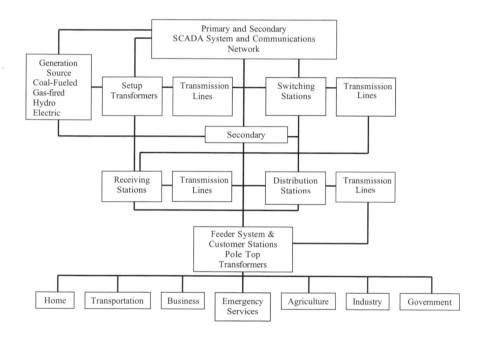

CONSEQUENCE SUMMARY

- A break anywhere in the system would cause loss of power requiring backup systems to kick in.
- A break or misuse of control units anywhere in the system may cause loss of system control.
- Contamination at certain points in the system may cause unacceptable consequences.
- Loss of communications may cause delays in detecting, assessing, responding to outages.
- A cyber attack may cause loss of system control.
- An attack may cause economic impact and widespread apprehension.

Exhibit 12.1 Typical Power-Utility Configuration

military authorities. In addition to these augmented security measures, all nuclear-power plants have robust security and emergency-response plans in place to further ensure public health and safety in the unlikely event of a malicious act and/or radioactive release.

The commercial nuclear-power industry is regulated by the Nuclear Regulatory Commission [NRC]. Security requirements for operating commercial nuclear-power plants are established in Title 10, Code of Federal Regulations [10 CFR]. Most security requirements are contained in 10 CFR parts 26, 50, 73, Appendices B and C, and published NRC guidelines and standards. The industry has an excellent security-assessment methodology, which is designated as **safeguards information**. It is not available to the public, and for security reasons the commercial nuclear-power industry is not a topic of this book.

A Prime Terrorist Target

Over the years terrorists have used energy systems as targets for bargaining demands, causing private corporations and governments to take extraordinary actions. A historical overview of selected terrorist attacks, criminal incidents, and industry mishaps within the energy sector is displayed in Appendix A.

Energy-Sector Vulnerabilities

Although the energy sector has not been attacked by terrorists in the United States, the potential risk is more than conjecture. Energy, particularly oil and gas pipelines, is a preferred target elsewhere in the world, especially in South America, Africa, and the Middle East. Al Qaeda has explicitly expressed its interest in attacking the American energy infrastructure, particularly the U.S. pipeline system. Some observations of attractiveness to the energy sector include but are not necessarily limited to the following:

- The energy sector faces significant vulnerabilities not found in other sectors—assets are spread throughout the nation with little definition to boundaries.
- Energy facilities have aging, expensive equipment, and some are difficult to replace, with long lead times for procurements.
- Aging facilities were not designed with threat or high-security needs in mind, and many have multiple built-in, single-point failures and mini-

mal redundancy systems to sustain continued operations. Modernization and upgrade programs have resulted in piecemeal solutions. Only the most modern facilities are advancing new concepts.

- Open and broadly dispersed transmission and distribution in remote, uncontrolled environments are not protected.

- Energy systems lack adequate minimal real-time monitoring capabilities and have underprotected SCADA, information-management, and security systems.

- At many utilities, the responsibility for security is an additional duty generally assigned to a safety manager, senior engineer, and/or administrative supervisor who has little or no security experience and less time to devote to security matters. In many locations the security organization lacks visibility, clear expectations for performance, and support from executive management. In other locations security staffing is insufficient, and training programs are in need of revitalization. Where contract guard services are employed, the contract vehicle often falls short of incorporating all the essential performance standards to provide adequate security organization, services, and qualifications.

Sufficient energy [and its distribution] is not only critical to sustaining day-to-day living and business needs; it also supports first responders during emergency and recovery operations in the event of a crisis. Moreover, it supports the nation's transportation systems [airports, seaports, railways, trucking, and mass-transport infrastructures], public health [medical centers and research laboratories], banking and financial institutions, agriculture [production, processes, and distribution], the defense industrial base [defense contractors supporting the military infrastructure], as well as the entire government infrastructure, including national and homeland security.

America's pipelines transport millions of gallons of crude oil and refined-petroleum products such as diesel fuel, gasoline, jet fuel, anhydrous ammonia, and carbon dioxide each day. About 1.2 million miles of liquid and gas pipelines snake through American communities, much of the system connecting oil from the Gulf of Mexico with refineries and distribution centers in the Northeast. Most of the pipeline system is underground, but some sections are not. About half of the 800-mile Alaskan pipeline is above ground. **The pipeline system is a key component of the many interconnected and interdependent critical infrastructure systems in America.** For example, oil provides power to generate electricity, which in turn supports the nation's transportation, manufacturing, and banking systems, and so on. Thus, a terrorist attack on the energy sector at key points could have a profound cascading effect on the nation's economic security. [9]

Latin America remains a major trouble spot for Americans and American business interests. Colombia stands out as a special problem because of the entrenched positions of the drug cartels, particularly since the Cali cartel and the well-organized guerrilla forces of the National Liberation Army are operating in the country. Among the regular guerrilla targets are the petroleum pipelines between fields in the southeast and the port of Covenas on the Caribbean. These are operated by a number of firms including British Petroleum and Oxy Colombia, a subsidiary of the United States firm Occidental Petroleum.

In an effort to strengthen defenses of the production fields and the pipelines, both British and United States companies have entered into agreements with the Colombian government to privately fund the training and assignment of petroleum protection duties to troops of the regular Colombian armed forces. Oxy Colombia was reported by *The New York Times* to have spent $7 million [up from a previous $3.9 million] for an expanded force, including two new platoons of 80 soldiers. The British firm has reportedly agreed to spend between $54 and $60 million to fund a battalion of 500 troops and 150 officers, including a mobile strike force to police a new 550-mile pipeline.

It has been estimated that guerrilla operations cost all the oil companies operating in the country $140 million annually, although the oil firms deny making direct extortion payments. In addition, a special war tax levied by the government cost the industry another $250 million a year. Over the past ten years, 1.4 million barrels of crude oil have been lost in attacks on the pipelines. Among the security threats is the targeting of civilian security guards. The murders of several have led to the resignations of many more.

TAILORING THE S^3E SECURITY-ASSESSMENT METHODOLOGY FOR THE ENERGY SECTOR

Applying the investigative principles outlined in Part I, the security-assessment team can use the tailored version of the S^3E **Security Assessment Methodology Model** shown on page after next as a roadmap to conducting security assessments for the energy sector.

The exhibit on the next page highlights in grey shaded area those areas tailored for specific application to the energy sector as contrasted to the general template shown in Exhibit 4.2.

Energy Challenges Facing the Security-Assessment Team

Electricity

The electricity sector is highly complex, and its numerous component assets and systems span the North American continent. Many of the sector's key assets, such as generation facilities, key substations, and switchyards, present unique protection challenges for the security-assessment team. The security-assessment team must keep abreast of industry structural changes currently taking place within the sector that may alter the parameters of the security assessment and its approach.

Another challenge for the security-assessment team is effective, sector-wide communications. The owners and operators of the electric system are a large and heterogeneous group. Industry associations serve as clearing houses for industry-related information, but not all industry owners and operators belong to such organizations. Data needed to perform thorough analyses on the infrastructure's interdependencies is not readily available. A focused analysis of time-phased effects of one infrastructure on another, including loss of operations metrics, would help identify dependencies and establish protection priorities and strategies.

For certain transmission and distribution facilities, providing redundancy and increasing generating capacity yield greater reliability of electricity service. However, this approach faces several challenges. Long lead times, possible denials of rights-of-way, state and local siting requirements, "not-in-my-backyard" community perspectives, and uncertain rates of return when compared to competing investment needs are hurdles that may prevent owners and operators of electricity facilities from investing sufficiently in security and service assurance measures.

Building a less vulnerable grid represents another option for protecting the national electricity infrastructure. Work is ongoing to develop a national R&D strategy for the electricity sector. Additionally, FERC has developed R&D guidelines, and the Department of Energy's [DOE] National Grid Study contains recommendations focused on enhancing physical and cyber security for the transmission system.

Dams

Under current policies and laws, dam owners are largely responsible for the safety and security of their own structures. Hence there are no institu-

Strategic Planning
Tasks 1, 2, & 3
Operational Environment
Critical Assessment
Threat Assessment

Program Effectiveness
Task 4
Infrastructure Interdependency
Evaluation [P_{E1}]

Operational Environment Task 1
Enterprise Goals
Corporate Image, Core Values, Areas of
Responsibility, Performance Objectives,
Business Incentives, Investment
Commitments
Federal Statutes, State Laws, Municipal
Codes, Confidentiality Laws, Civil
Common Law, Willful Torts, Contractual
Obligations, Agreements, Governing
Authorities, Voluntary Commitments
Characterization
Mission, Location, Layout/Land Use,
Neighboring Facilities, Terrain, Climate,
Resources, Operations/Processes
Operations – Logistics
Visitors, Customers, Vendors; Mail &
Material Delivery, Receiving, Shipping;
Staging, Storing, Distribution & Delivery;
Time-Sensitive Processes;
Communications

Enterprise
Roles & Responsibilities
Operations, Processes, Protocols
Functional Relationships
Security Awareness
Reporting Security Incident
Alert Notification System

Security Capabilities
Management Team
Organization & Composition
Staffing, Skills, Experience
Qualifications & Training
Operational Capabilities
Dependency Programs

Critical Assessment Task 2
Resources, Operations, Functions
Administrative, Aircraft/Fleet Facilities
Board/Conference Rooms
Cable/Communications Rooms
Electrical/Mechanical Rooms
Physical Infrastructure
Telecommunications Centers
IT Network Centers
Training Facilities
Warehouses, Loading Docks
SCADA & Security Systems
Infrastructure Sharing
Critical Interdependencies
Primary, Backup, Redundant Systems
All Energy Groups
HAZMAT, High Value Storage Areas
Pipe, Valve, Storage, Tunnel Systems
Shops & Laboratories
Microwave/Telecommunications Sites
Electricity & Pipelines
Assembly Yards, Open Storage
Receiving, Converter Stations
Coal-Fueled & Gas-fired Power Plants
Generators & Shafts
Hydro Electric & Nuclear Power Plants
Transmission Lines, Distribution Systems
Voltage Transformers & Couplings
Dams
Abutments, Drain Pipes, Reservoirs
Earth/Rock Fill, Pumps & Valves
Gated/Service, Emergency Spillways
Embankments & Hydraulic Components
Outlet Conduits, Pipes & Towers
Penstocks & Flood Gates
Oil & Natural Gas Pipelines
Citygates, Depleted Fields
Petroleum Gas Storage Facilities
Processing Units

Barriers & Delay Measures
Terrain, Approaches, Fences
Barrier, Walls, Gates
Detection Measures
Interior/Exterior Sensors
Special Purpose Sensors
Tamper Systems
Access Control Measures
People/Vehicle/ Delivery Controls
Screening, Verification
Assessment Measures
Perimeter Security, Lighting
Posts and Patrols, CCTV
Communications Media
Radio, Telephone, PA,
Hardwire, Wireless
Fiber Optic
Information Network
Security Control Center[s]
Annunciation & Display
Event Storage, Reports
Alarm Integration
Information Networks
Infrastructure
Primary, Secondary, Backup,
Redundancy

Response & Recovery
Tactics, Techniques
First & Second Responders
Integrated Responders
Production Shutdown
Grid Distribution
Closeout Actions

Threat Assessment Task 3
International, National, Regional, Local
Range & Levels of Threats
Design Basis Threat Profile, Probability
Threat Categories, Capabilities,
Methods & Techniques

Quality Assurance/Quality Control Process

Exhibit 12.2 The S^3E Security-Assessment Methodology
for the Energy Sector

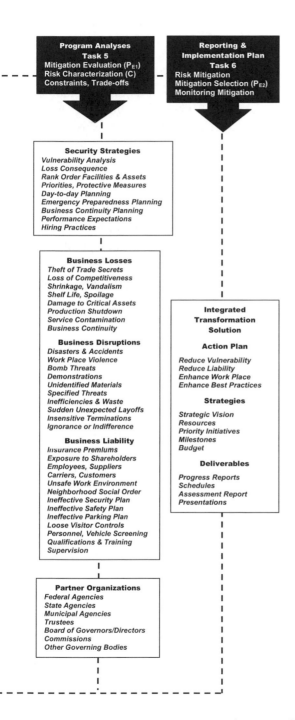

Program Analyses
Task 5
Mitigation Evaluation (P_{E1})
Risk Characterization (C)
Constraints, Trade-offs

**Reporting &
Implementation Plan**
Task 6
Risk Mitigation
Mitigation Selection (P_{E2})
Monitoring Mitigation

Security Strategies
Vulnerability Analysis
Loss Consequence
Rank Order Facilities & Assets
Priorities, Protective Measures
Day-to-day Planning
Emergency Preparedness Planning
Business Continuity Planning
Performance Expectations
Hiring Practices

Business Losses
Theft of Trade Secrets
Loss of Competitiveness
Shrinkage, Vandalism
Shelf Life, Spoilage
Damage to Critical Assets
Production Shutdown
Service Contamination
Business Continuity

Business Disruptions
Disasters & Accidents
Work Place Violence
Bomb Threats
Demonstrations
Unidentified Materials
Specified Threats
Inefficiencies & Waste
Sudden Unexpected Layoffs
Insensitive Terminations
Ignorance or Indifference

Business Liability
Insurance Premiums
Exposure to Shareholders
Employees, Suppliers
Carriers, Customers
Unsafe Work Environment
Neighborhood Social Order
Ineffective Security Plan
Ineffective Safety Plan
Ineffective Parking Plan
Loose Visitor Controls
Personnel, Vehicle Screening
Qualifications & Training
Supervision

**Integrated
Transformation
Solution**

Action Plan

Reduce Vulnerability
Reduce Liability
Enhance Work Place
Enhance Best Practices

Strategies

Strategic Vision
Resources
Priority Initiatives
Milestones
Budget

Deliverables

Progress Reports
Schedules
Assessment Report
Presentations

Partner Organizations
Federal Agencies
State Agencies
Municipal Agencies
Trustees
Board of Governors/Directors
Commissions
Other Governing Bodies

Exhibit 12.2 *(continued)*

tional security standards to apply, and resources available to protect dam property vary greatly from one category to the next. Additionally, the distributed nature of dam ownership also complicates assessment of the potential consequences of failure for certain categories of dams. The design and location of dams make them inherently vulnerable to attack. Given these realities, the need to develop more comprehensive mechanisms for assessing and managing risks to dams is clear.

Oil and Natural Gas

Protection of critical assets requires both heightened security awareness and investment in protective equipment and systems. One serious challenge for the security-assessment team is the lack of metrics to determine and justify corporate security expenditures. In the case of natural disasters or accidents, there are well-established methods for determining risks and cost-effective levels of investments in protective equipment, systems, and methods for managing risk [e.g., insurance]. It is not clear what levels of security and protection are appropriate and cost-effective to meet the risks of terrorist attack. Due to public-safety requirements related to their operations and facilities, the oil and natural-gas industries fortunately have substantial protection programs already in place.

Pipelines

Pipelines are not independent entities but rather integral parts of industrial and public-service networks. Loss of a pipeline could impact a wide array of facilities and industrial factories [both upstream and downstream] that depend on reliable fuel delivery to operate. Several hundred thousand miles of pipeline span the country, and it is not realistic to expect total security for all facilities. As such, the security-assessment team must focus its protection efforts on infrastructure components whose impairment would have significant effects on the energy markets and the economy as a whole. Determining what to protect and when to protect it are factors in cost-effective infrastructure protection. During periods of high demand such as the winter months, pipeline systems typically operate at peak capacity and are more important to the facilities and functions they serve. Many of the products that pipelines deliver are inherently volatile. Hence, their protection is a significant issue.

The security-assessment team must also keep abreast of the pipeline industry's increasing interdependencies, as part of the transportation sector, with the energy and telecommunications sectors, necessitating cooperation with other critical infrastructures during protection and response planning. A comprehensive security assessment should address the difficulties in evaluating the broader implications of an attack on their critical facilities. These interdependencies call for cross-sector coordination to be truly responsive to national concerns. The assessment process must also examine where pipelines cross state, local, and international jurisdictions. The number and variety of stakeholders create a confusing and sometimes conflicting array of regulations and security programs for the industry to manage, especially with respect to the ability of pipeline facilities to recover, reconstitute, and reestablish service quickly after a disruption.

Nuclear Power

Losing the capabilities of a single nuclear-power plant may have only a minor impact on overall electricity delivered within the context of our robust national power grid. Nevertheless, a terrorist attack on any nuclear facility would be considered a significant security event. In an unlikely worst-case scenario, a successful terrorist strike against a nuclear facility could result in the release of radioactive material. Even if radioactive material were not released, widely held misconceptions of the potential consequences of an attack on a nuclear facility could have significant negative impact. The NRC is currently performing a detailed design-basis threat and vulnerability analysis for nuclear-power plants to help identify additional security enhancements that may be warranted. Additional prudent measures should be examined to help strengthen the defensive posture of these facilities.

Applying the S³E Security Assessment Methodology to the Energy Sector

The diversity and complexity of the energy sector require a further breakout for each industry. The template provided below can easily be designed for that purpose. The security-assessment team should use its collective expertise to:

- Identify the important mission and functions of the energy utility **[Task 1—Operational Environment]** and incorporate the development of all business operations and continuity expectations. Examination focuses on services and the customer base to determine the business contributions to the community, its culture, and business objectives. This includes but is not necessarily limited to the identification of business goals and objectives, customer base, commitments, and facilities and boundaries, as shown in Exhibits 12.3 through 12.6.

Adequate Power Production & Distribution	Corporate Image & Core Values
Adequate Power Supply & Demand Loads	Performance Objectives
Areas of Responsibility	Public Safety & Public Confidence
Business Incentives & Investment	

Exhibit 12.3 Identify Energy Mission Business Goals and Objectives

General Public, Critical Customers	Medical, Firefighters
Government, Military	Regional & International
Industrial & Business	Size of Population Served
Law Enforcement	Transportation

Exhibit 12.4 Identify Energy Customer Base

- Describe the energy-utility configuration, operations, and other elements [**Task 2—Critical Assessment**] by examining conditions, circumstances, and situations relative to safeguarding public health and safety and to reducing the potential for disruption of services. Focus is on an integrated evaluation process to determine which assets need protection in order to minimize the impact of threat events. The security-assessment team identifies and prioritizes facilities, processes, and assets critical to energy service objectives that might be subject to malevolent acts or disasters that could result in a potential series of undesired consequences. The process takes into account the impacts that could substantially disrupt the ability of the utility to provide safe energy services and strives to reduce security risks associated with the consequences of significant events. This includes but is not necessarily limited to assessment of facilities, assets, operations, processes, and logistics; and prioritization of assets in relative importance to business operations, as shown in Exhibits 12.7 and 12.8.

- Review the existing design-basis threat profile [**Task 3—Threat Assessment**] and update it as applicable to the client. If a design-basis threat profile has not been established, then the security-assessment team needs to develop one. Both the security-assessment team and the client must have a clear understanding of undesired events and consequences that may impact energy services to the community and business continuance planning goals. Emphasis is on prioritizing energy threats through the process of identifying methods of attack,

10 CFR 45, 50, 73 - Nuclear Power Security	Federal Polygraph Protection Act [As Amended]
29 CFR 1910.36 - Means of Egress	
29 CFR 1910.151 - Medical Services and First Aid	Freedom of Information Act [U.S.C. 552] [As Amended]
29 CFR 1910.155 - Fire Protection	National Labor Relations Act [29 U.S.C. 151] [As Amended]
5 U.S.C. 522e - Federal Privacy Act of 1974	
18 U.S.C. 2511 - Technical Surveillance	Interagency Agreements
21 U.S.C. 801 - Drug Abuse and Control	International Atomic Emergency Agency
26 U.S.C. 7201 - Tax Matters	Internal Revenue Service
29 U.S.C. 2601 - Family and Medical Leave Act	Labor Management Relations Act
	Memorandum of Agreement
31 U.S.C. 5322 - Financial Transaction Reports	Memorandum of Understanding
	National, Regional, State, and Local Grid Planning Committees
41 U.S.C. 51 - U.S. Contractor Kickbacks	
41 U.S.C. 701 - Drug Free Workplace Act	Oil Pollution Act 1990 [As Amended]
46 U.S.C. 1903 - Maritime Drug Enforcement	Omnibus Crime Control and Safe Streets Act, 1968 [As Amended]
31 U.S.C. 5322 - Financial Transaction Reports	
	Policy Coordination Committee [PCC]
Architectural Barrier Act 1968 [As Amended]	Power Information Sharing and Analysis Centers [ISAC]
Bioterrorism Preparedness and Response Act, 2002 [As Amended]	
	Rehabilitation Act, 1973 [As Amended]
Carriers, Suppliers, Vendors, Insurance	State and Local Industry Administrators and Authorities
Civil Rights Act 1964 [U.S.C. 200e] [As Amended]	
	State and Local Emergency Planning Officials and Committees
Commerce Commission	
Comprehensive Drug Abuse Prevention Act 1970 [As Amended]	Uniform Trade Secrets Act
	U.S. Department of Commerce
Comprehensive Environmental Response, Compensation and Liability Act 1980 [As Amended]	U.S. Department of Homeland Security [DHS]
	U.S. Department of Defense [DOD]
	U.S. Department of Energy [DOE]
Controlled Substance Act [As Amended]	U.S. Department of Justice [DOJ]
Drug Free Workplace Act 1988 [As Amended]	
Emergency Planning & Right-to-Know Act 1986 [As Amended]	U.S. Environmental Protection Agency [EPA]
	U.S. Patriot Act, 2002 [As Amended]
Employee Polygraph Protection Act 1988 [29 U.S.C. 2001] [As Amended]	U.S. Postal System
	U.S. Nuclear Regulatory Commission [NRC]
Fair Credit Reporting Act [15 U.S.C. 1681] [As Amended]	

Exhibit 12.5 Identify Energy Commitments

Dispersed Locations	Infrastructure Sharing
Facility Layout & Location	Neighboring Facilities
Facility Land Use	Terrain & Climate

Exhibit 12.6 Characterize Configuration of Energy Enterprise Facilities and Boundaries

All Groups	
Administrative & Aircraft/Fleet Facilities	Microwave & Telecommunication Sites
Board/Conference Rooms	Microwave & Telecommunications Sites
Cable/Communications Rooms	Mission, Service Capabilities
Communications	Operational Centers & Control Rooms
Crisis Management Centers	Physical Infrastructure
Crisis Management Centers	Physical Infrastructure
Critical Dependencies	Pipe, Valve, Storage, Tunnel Systems
Early Warning Systems	Population & Population Throughputs
Electrical Trouble Board Centers	Primary & Backup, Redundancy Systems
Electrical/Mechanical Rooms	SCADA & Other Control Systems
HAZMAT, High Value Storage Areas	Security Systems
Hours of Operation	Shops & Laboratories
Infrastructure Sharing	Training Facilities
IT Network Centers	Warehouses, Loading Docks
Electricity and Pipelines	
Assembly Yards	Open Storage
Coal-Fueled & Gas-fired Power Plants	Receiving, Converter Stations
Generators & Shafts	Transmission Lines, Distribution Systems
Hydro Electric & Nuclear Power Plants	Voltage Transformers & Couplings
Dams	
Abutments, Drain Pipes, Reservoirs	Gated/Service, Emergency Spillways
Earth/Rock Fill, Pumps & Valves	Outlet Conduits, Pipes & Towers
Embankments & Hydraulic Components	Penstocks & Flood Gates
Oil & Natural Gas Pipelines	
Citygates, Depleted Fields	Processing Units
Petroleum Gas Storage Facilities	

Exhibit 12.7 Critical Assessment of Energy Enterprise Facilities, Assets, Operations, Processes, and Logistics

Most critical facilities based on value system	Least critical facilities based on value system
Most critical assets based on value system	Least critical assets based on value system

Exhibit 12.8 Prioritize Critical Energy Enterprise Assets in Relative Importance to Business Operations

past or recent events, and types of disasters that might result in significant consequences; the analysis of trends; and the assessment of the likelihood of an attack occurring. Focus is on threat characteristics, capabilities, and target attractiveness. Social-order demographics used to develop trend analysis in order to support framework decisions are also considered. Under this task the security-assessment team should also validate previous energy-assessment studies and update conclusions and recommendations into a broader framework of strategic planning. Potential undesirable events include but are not necessarily limited to malevolent acts, other disruptions, categories of perpetrators, initial impact of loss consequences, and likelihood of threat occurrence, as shown in Exhibits 12.9 through 12.13.

- Examine and measure the effectiveness of the security system **[Task 4—Evaluate Program Effectiveness]**. The security-assessment team evaluates business practices, processes, methods, and existing protective measures to assess their level of effectiveness against vulnerability and risk and to identify the nature and criticality of business areas and assets of greatest concern to energy operations. Focus is on vulnerability and vulnerable access points where penetration may be accomplished and how. Emphasis is placed on the physical characteristics of vulnerable points, accessibility to the location, and the effectiveness of protocol obstacles an adversary would likely have to overcome to reach a critical power asset. These include but are not necessarily limited to security operations and protocols, existing security organization, relationships with partner organizations, SCADA and security-system performance, adversary path analysis, and response and recovery effectiveness, as shown in Exhibits 12.14 through 12.19.

- Recognize the importance of bringing together the right balance of people, information, facilities, operations, processes, and systems **[Task 5—Program Analyses]** to deliver a cost-effective solution. The security-assessment team develops an integrated approach to measurably reduce security risk by reducing vulnerability and consequences through a combination of workable solutions that bring real,

Armed attack	Loss of key personnel
Arson	Misuse of energy supply chain
Assassination, kidnapping, assault	Theft or destruction of chemicals
Barricade/hostage incident	Misuse/damage of control systems
Bomb threats and bomb incidents	Misuse of wire and electronic communications
Chemical, Biological, Radiological [CBR] contamination of assets, supplies, equipment, and areas	Phishing, hacking, computer crime
	Power failure
	Rape and sodomy
Computer-based extortion	Sabotage
Computer system failures and viruses	Sexual abuse, sexual exploitation
Computer theft	Social engineering
Cyber attack on SCADA or other systems	Software piracy
Cyber stalking, software piracy	Storage of criminal information on the system
Damage/destruction of interdependency systems resulting in production shutdown	Unlawful access to stored wire and electronic communications and transactional records access
Damage/destruction of control/isolation valve systems	
Damage/destruction of pipelines and cabling	Telecommunications failure
Damage/destruction of service capability	Terrorism
Damage of feeders, towers, pipelines	Torture
Damage of generation capability	Tracking effects of outage overtime
Delay in processing transactions	Tracking effects of outage across related resources and dependent systems
Disable manufacturing, production, distribution and delivery processes	Theft of trade secrets, intellectual propriety, confidential information, patents, and copyright infringements
Email attacks	
Encryption breakdown	Treason, sedition, and subversive activities
Explosion	Unauthorized transfer of American technology to rogue foreign states
Embezzlement, extortion and threats	
FAX security	Unidentified materials or wrong deliveries
Hacker attacks	Unlawful access to stored communications
Hardware failure, data corruption	Unlawful wire and electronic communications interception and interception of oral communications
Homicide	
Human error	
Inability to access system or database	Use of cellular/cordless telephones to disrupt network communications or in support of other crimes
Inability to perform routine transactions	
Intentional opening/closing of control/isolation values and gates	
	Voice mail system
Intentional release of dangerous chemicals	Wide-scale disruption and disaster
Intentional shutdown of electrical power	Wide-scale evacuation

Exhibit 12.9 Determine Types of Energy Enterprise Malevolent Acts that could Reasonably Cause Undesirable Events

Bribery, graft, conflicts of interest	Loss of proprietary information
Civil disorders, civil rights violations	Mail fraud
Conspiracy	Maintenance activity
Eavesdropping	Natural disasters
Economic espionage censorship	Obscenity
HAZMAT spill	Obstruction of justice
Historical rate increases	Online pornography and pedophilia
Industrial accidents and mishaps	Planned production shutdown or outage
Inefficiencies and waste	Racketeering
Ineffective hiring practices	Racketeer influenced and corrupt
Ineffective personnel and vehicle controls	organizations
Ineffective safety and security plans	Resource constraints and competing
Ineffective supervision and training	priorities
Ineffective use of resources and time	Release and misuse of personal
Ineffective visitor controls	information
Ignorance and indifference	Robbery and burglary
Insensitive terminations	Shrinkage, shelf-life, spoilage
Institutional corruption, fraud, waste, abuse	Sudden unexpected layoffs
	Supplier, courier disruptions
Loss of competitiveness	Trespassing and vandalism

Exhibit 12.10 Assess Other Disruptions Impacting Energy Enterprise Operations

Activist groups, subversive groups or cults	State-sponsored or independent terrorist groups
Disenfranchised individuals, hackers	
Legitimate entities serving as conduits for terrorist financing	Terrorist organizations posing as legitimate entities using the system for criminal purposes
Organized criminal groups	
	Vandals, lone wolves, insiders

Exhibit 12.11 Identify Category of Energy Perpetrators

Compromise of public confidence	Inability to transfer power
Cost to repair	Local civil unrest
Dealing with unanticipated crisis	Loss of energy sources or ability to generate energy
Disabling of or loss of key personnel	
Duration of loss of emergency services	Loss of life [customers and employees]
Duration of loss of energy source	Loss or disruption of communications
Economic loss	Loss or disruption of information
Illness [customers and employees]	Number of critical customers impacted
Impact on regional economic base	Number of users impacted
Impact on utility ratepayers	Production shutdown
Inability of customers to conduct business	Slowdown/interruption of delivery services
Inability to distribute power, oil or gas	
Inability to store or transfer oil and gas	Temporary closure of financial institutions

Exhibit 12.12 Assess Initial Impact of Loss Consequence for the Energy Enterprise

Consequence analysis	Social demographics
Economic conditions	Social/political status

Exhibit 12.13 Assess Initial Likelihood of Energy Enterprise Threat Attractiveness and Likelihood of Malevolent Acts of Occurrence

Alert Notification System	Reporting Security Incidents
Emergency Preparedness Planning	Roles & Responsibilities
Functional Interfaces	Security Awareness
Operations, Processes, Protocols	Security Plans, Policies & Procedures

Exhibit 12.14 Evaluate Existing Entergy Enterprise Security, Operations, and Protocols $[P_{E1}]$

Dependency Programs	Qualifications & Training
Management Team	Staffing, Skills, Experience
Operational Capabilities	Supervision
Organization & Compositions	

Exhibit 12.15 Evaluate Existing Security Organization $[P_{E1}]$

Agency for Toxic Substances and Disease Registry	National Information Infrastructure [NII]
American Electronics Association [AEA]	National Infrastructure Protection Center
American Institute of Certified Public Accountants	National Institute of Law Enforcement and Criminal Justice
American National Standards Institute	
American Psychological Association	National Institute of Standards and Technology [NIST]
American Society for Industrial Security	
American Society for Testing and Materials	National Insurance Crime Bureau [NICB] Online
American Society for Testing Materials, Vaults	National Labor Relations Board
Centers for Disease Control and Prevention [CDC]	National Parking Association
Commerce Commission	National Safety Council
Computer Security Institute	National White Collar Crime Center [NWCCC]
Critical Infrastructure Protection Board [PCIPB]	Occupational Safety and Health Review Commission [OSHRC]
Defense Central Investigation Index [DCII]	
El Paso Intelligence Center [EPIC]	Office of Civilian Radioactive Waste Management
EPA National Security Research Center	Office of National Drug Control Policy
EPA Office of Air and Radiation	Policy Coordination Committee [PCC]
EPA Office of Prevention, Pesticides and Toxic Substances	Private Security Advisory Council
EPA Office of Solid Waste and Emergency Response	Regional Information Sharing System [RISS]:
Factory Mutual Research Corporation	Middle Atlantic-Great Lakes Organized Crime Law Enforcement Network [MAGLOCLEN]
Federal Power Commission [FPC]	
Federal, state, and local agency responders	Mid-States Organized Crime Information Center [MOCIC]
Illuminating Engineers Society of North America	
Institute for a Drug Free Workplace	New England State Police Information Network [NESPIN]
Institute of Electrical and Electronic Engineers [IEEE]	
International Association of Chiefs of Police	Regional Organized Crime Information Center [ROCIC]
International Association of Professional Security Consultants	
	Rocky Mountain Information Network [RMIN]
International Computer Security Association	Western States Information Network [WSN]
International Criminal Police Organization [INTERPOL]	Safe Manufactures National Association
	The Terrorism Research Center
International Electronics Supply Group [IESG]	U.S. Central Intelligence Agency [CIA]
International Parking Institute	U.S. Coast Guard
International Organization for Standardization [ISO]	U.S. Customs
Interstate Commerce Commission [ICC]	U.S. Department of Commerce [DOC]
Joint Terrorism Task Force [JTTF]	U.S. Department of Defense [DOD]
Law Enforcement Support Center [LESC]	U.S. Department of Energy [DOE]
Library of Congress, Congressional Research Service	U.S. Department of Homeland Security [DHS]
	U.S. Department of Justice [DOJ]
Medical Practitioners and Medical Support Personnel	U.S. Environmental Protection Agency [EPA]
	U.S. Federal Bureau of Investigation [FBI]
Multi-lateral Expert Groups	U.S. Government Accounting Office [GAO]
Mutual Aid Associations	U.S. Postal System
National Advisory Commission on Civil Disorder	U.S. Nuclear Regulatory Commission [NRC]
National American Electric Reliability Council [NAER]	Underwriters Laboratories [UL]
National Crime Information Center [NCIC]	
National Electrical Manufacturing Association [NEMA]	Water Information Sharing and Analysis Centers [ISAC]
National Fire Protection Association [NFPA]	World Health Organization [WHO]

Exhibit 12.16 Evaluate Energy Enterprise Interface and Relationship with Partner Organizations [P_{E1}]

Assessment Capability	Modem & Internet Access
Badge Controls	Performance Standards & Vulnerabilities
Barrier/Delay Systems	Personnel & Vehicle Access Control Points
Chemical & Other Vendor Deliveries	Physical Protection System Features
Communications Capability	Post Orders & Security Procedures
Cyber Intrusions, Firewalls, Other Protection Features	Response Capability
Display & Annunciation Capability	Security Awareness
Evacuation Response Plans	Security Training/Exercises/Drills
Intrusion Detection Capability	System Capabilities & Expansion Options
Lock & Key Controls, Fencing & Lighting	System Configuration & Operating Data

Exhibit 12.17 Evaluate Existing Energy Enterprise SCADA and Security-System Performance Levels [P_{E1}]

Deception actions	Likely approaches and escape routes
Delay-penetration obstacles	Target selection and alternative targets
Interception analysis	Time-delay response analysis

Exhibit 12.18 Define the Energy Enterprise Adversary Plan, Distractions, Sequence of Interruptions, and Path Analysis

Closeout Actions	Integrated Responders
Containment	Production Shutdown
Divert Distribution	Tactics and Techniques
First and Second Responders	

Exhibit 12.19 Assess Effectiveness of Energy Enterprise Response and Recovery [P_{E1}]

Business Continuity Planning	Loss Consequences
Day-to-Day Planning	Performance Expectations
Emergency Preparedness Planning	Rank Order Facilities & Assets
Hiring Practices	Vulnerability Analysis

Exhibit 12.20 Analyze Effectiveness of Energy Enterprise Security Strategies and Operations [P_{E1}]

Compromise of public confidence	Inability to transfer power
Cost to repair	Local civil unrest
Dealing with unanticipated crisis	Loss of energy sources or ability to generate energy
Disabling of or loss of key personnel	
Duration of loss of emergency services	Loss of life [customers and employees]
Duration of loss of energy source	Loss or disruption of communications
Economic loss	Loss or disruption of information
Illness [customers and employees]	Number of critical customers impacted
Impact on regional economic base	Number of users impacted
Impact on utility ratepayers	Production shutdown
Inability of customers to conduct business	Slowdown/interruption of delivery services
Inability to distribute power, oil or gas	
Inability to store or transfer oil and gas	Temporary closure of financial institutions

Exhibit 12.21 Refine Previous Analysis of Undesirable Consequences that can Affect Energy Enterprise Functions

Consequence analysis	Social demographics
Economic conditions	Social/political status

Exhibit 12.22 Refine Previous Analysis of Likelihood of Energy Enterprise Threat Attractiveness and Malevolent Acts of Occurrence

Creation and/or revision or modification of sound business practices and security policies, plans, protocols, and procedures.	Development, revision, or modification of interdependency requirements.
Creation and/or revision or modification of emergency operation plans including dependency support requirements.	SCADA/security system upgrades to improve detection and assessment capabilities.

Exhibit 12.23 Analyze Selection of Specific Risk Reduction Actions Against Current Risks, and Develop Prioritized Plan for Energy Enterprise Risk Reduction [P_{E1}]

Mirrors business culture image	Reasonable and prudent mitigation options

Exhibit 12.24 Develop Short- and Long-Term Energy Enterprise Mitigation Solutions

Effectiveness calculations	Residual vulnerability consequences

Exhibit 12.25 Evaluate Effectiveness of Developed Energy Enterprise Mitigation Solutions and Residual Vulnerability [P_{E2}]

Practical, cost-effective recommendations	Reasonable return on investments

Exhibit 12.26 Develop Cost Estimate for Short- and Long-Term Energy Enterprise Mitigation Solutions

tangible enterprise value to the table. Emphasis is placed on assessing the current status and level of protection provided against desired protective measures to reasonably counter the threat. Operational, technical, and financial constraints are examined, as well as future expansion plans and constraints. These include but are not necessarily limited to analysis of security strategies and operations, refinement of previous consequence analysis, refinement of event attractiveness and likelihood of occurrence, analysis of and planning for risk reduction, development of short- and long-term solutions, evaluation of solutions and remaining vulnerability, and development of cost estimates, as shown in Exhibits 12.10 through 12.26.

PREPARING THE ENERGY SECURITY-ASSESSMENT REPORT

Chapter 10 provides a recommended approach to reporting the security-assessment results.

U.S. GOVERNMENT ENERGY INITIATIVES

Appendix B summarizes some initiatives taken by government and industry to enhance the security integrity of the power sector.

REFERENCES

1. "The National Strategy for the Physical Protection of Critical Infrastructure and Key Assets" [February 2003]
2. Ibid.
3. Ibid.
4. U.S. Government Accounting Office [2002]
5. "The National Strategy for the Physical Protection of Critical Infrastructure and Key Assets" [February 2003]
6. Ibid.
7. Ibid.
8. Ibid.
9. Ibid.

APPENDIX A: A HISTORICAL OVERVIEW OF SELECTED TERRORIST ATTACKS, CRIMINAL INCIDENTS, AND INDUSTRY MISHAPS WITHIN THE ENERGY SECTOR

The author has compiled the information in this appendix in a chronological sequence to share the historical perspective of incidents with the reader interested in conducting further research. The listing, along with the main text of Chapter 2, presents a capsule review of terrorist activity and other criminal acts perpetrated around the world and offers security practitioners specializing in particular regions of the world a quick reference to range and level of threat.

Source: *The information is a consolidated listing of events, activities, and news stories as reported by the U.S. Department of Homeland Security, U.S. Department of State, and various other government agencies. Contributions from newspaper articles and news media reports are also included. They are provided for a better understanding of the scope of terrorism and other criminal activity and further research for the interested reader.*

23 OCTOBER 2006 Venezuela.
Hezbollah threatens to sabotage the petroleum base of the United States in Latin America.

3 OCTOBER 2006 Abuja, Nigeria.
Suspected militants raided a residential compound for ExxonMobil contractors in Eket in the eastern delta, killing two Nigerian security guards and abducting five foreign workers. The attack came two days after 25 Shell contractors were abducted in neighboring Rivers state.

20 MAY 2006 Buenaventura, Colombia.
Colombia's main port city was left without electricity after left-wing FARC rebels attacked its power installation. The incident came after at least two days of grenade attacks, which left at least 24 people injured. Police said they had arrested children who were used in the grenade attacks.

20 MAY 2006 Veazie, ME, United States.
Copper-wire theft causes power disruptions. Persons broke into all four substations of Bangor Hydro-Electric C, breaking locks and tampering with wiring, making the electrical system unsafe in addition to the cable theft. Power loss was experienced throughout the area for about six hours.

12 MAY 2006 Nigeria.
An American executive for a Houston-based oil firm was targeted for assassination as gas pipelines were destroyed by terrorists.

15 MAY 2006 Lagos, Nigeria.
Up to 200 people burned to death after a fuel pipeline on the outskirts of Lagos exploded when thieves tried to tap into it to steal oil. Nigerian President Olusegun Obasanjo ordered a police investigation and increased protection for other pipelines.

12 SEPTEMBER 2005 Los Angeles, CA, United States.
Major sections of the city lost power when lines were accidentally cut by crews.

13 AUGUST 2005 London, United Kingdom.
The U.S. Department of Homeland Security warned that road oil tankers are being earmarked by terrorists for strikes against selected targets in London and three American cities.

28 JULY 2005 Queensland, Australia.
A bomb threat shut a large petroleum fuel refinery in Australia. Seven hundred workers were evacuated by police when an anonymous caller threatened to blow up the Caltex Australia site. A thorough search by a police bomb squad failed to find any traces of explosives. The refinery, the fifth largest in Australia, was commissioned in 1965 and has a crude oil throughput of 4.4 million gallons per day.

22 JULY 2005 Sana, Yemen.
Clashes over the doubling of fuel prices between Yemen security forces and demonstrators raged for 48 hours and left at least 39 people dead.

17 JUNE 2005 Basra, Iraq.
Three "pirates" boarded a supertanker at this Iraqi port, and another group boarded a supertanker as it waited to take on crude oil. Damage and losses were minimal in both cases as attackers fled on approach of investigating vessels.

30 MAY 2005 Awka, Nigeria.
Six people died when a southern Nigerian pipeline caught fire as a gang illegally siphoned fuel.

28 APRIL 2005 Somalia.
Pirates free hostages of LPG carrier off Somalia. Seventeen crew members of an unnamed Panamanian-flagged liquefied-petroleum-gas

[LPG] carrier taken hostage by armed pirates off Somalia were freed unharmed after a two-week ordeal. The LPG carrier was hijacked 135 miles off the east coast of Somalia by a gang in speedboats and later taken into Somalian territorial waters, where it remained anchored off the coast. The vessel was lured into a trap using a distress flare.

29 MARCH 2005, Omaha, NE, United States.

Someone broke into the Omaha Public Power District substation and cut two 13,800-volt power lines. One of the power lines was a back-up circuit for part of downtown Omaha's power. Omaha police believe the person who cut the lines is seriously burned. When power employees arrived on the scene, they found a saw and smelled burned flesh. Authorities believe the person who cut the line may have been one of homeless persons who frequently visit the area of the substation and may have been trying to steal the lines for the copper that's inside.

16 MARCH 2005 Portland, OR, United States.

Drug addicts are reported to be stealing power lines to feed their methamphetamine habits, and Portland General Electric [PGE] says it could drive up Northwest power costs. Police arrested four people with 600 to 800 pounds of stolen power lines. Detectives say the suspects took them from a PGE substation, planned to sell it for scrap, and use the money for methamphetamine. PGE says the value of the copper used to make power lines has recently skyrocketed, so thieves are desperate to steal it. Some even risk their lives by trying to swipe live lines. It is illegal for scrap yards to knowingly accept stolen metal. However, there is no law requiring yards to ask where the seller acquired the metal.

11 MARCH 2005 Washington, DC, United States.

Hackers caused no serious damage to systems that feed the nation's power grid, but their untiring efforts have heightened concerns that electric companies have failed to adequately fortify defenses against a potential catastrophic strike. The fear is that in a worst-case scenario, terrorists or others could engineer an attack that sets off a widespread blackout and damages power plants, prolonging an outage. The Federal Energy Regulatory Commission and others in the industry said the companies' computer security is uneven. The biggest threat to the grid, analysts said, may come from power companies using older equipment that is more susceptible to attack.

2 MARCH 2005 Hopatcong Township, NJ, United States.

Someone cut through a fence and tried to sabotage a portion of the Central Power & Light Company's electrical grid. The company suspected union involvement, since the utility's union went on strike three months before. Oil valves were opened on two transformers at the Marble Hill substation. The act of vandalism caused an estimated several hundred gallons of oil to drain from the units. Crews were sent before power was lost at the substation, which serves customers that include a hospital and a water-treatment plant.

26 FEBRUARY 2005 Baghdad, Iraq.
Saboteurs blow up an oil pipeline in northern Iraq.

15 FEBRUARY 2005 Allegheny, PA, United States.
Six Greenpeace activists were sentenced to jail terms ranging from 5 to 30 days for climbing a smokestack at a coal-fire power plant in protest of President Bush's energy policy. The protesters cut a hole in the fence that surrounds Allegheny Energy's Hatfield Ferry Power Station, then climbed the 700-foot smokestack and unfurled a 2,500-square-foot banner. The six pleaded guilty to misdemeanor charges of reckless endangerment, failure to disperse, disorderly conduct, and defiant trespass.

30 JANUARY 2005 Orlando, FL, United States.
Utilities combat power thieves. Rising energy prices, tighter competition, and the desire to keep bystanders from being electrocuted have led utilities nationwide to fight back with growing fervor. They're hiring full-time investigative teams and running nighttime surveillance. Energy thieves cost U.S. utilities between $4 billion and $6 billion yearly. To stem the losses, the nation's investor-owned utility companies hire an average of 10 full-time investigators each to root out theft, according to the Edison Electric Institute, a trade association for U.S. shareholder-owned electric companies. Utilities prosecute about 10 percent of cases.

23 JANUARY 2005 Washington, DC, United States.
Report examines liquid-natural-gas tanker risks. When an explosion flattened a liquefied-natural-gas [LNG] plant in Algeria, killing 30 workers, communities in coastal towns in New England, Alabama, and California became concerned. The Algerian inferno undermined industry arguments that the modern era of LNG transport is inherently safe. It also became a rallying point for groups fighting proposed new LNG terminals in U.S. towns. To many energy experts, fear of a devastating LNG fire from an accident or terrorist attack is the toughest obstacle facing the industry. A report by Sandia National Laboratories—the most comprehensive examination of LNG tanker risks to date—concluded that terrorists have the capability to tear a hole in a tanker, causing damage to the

environment as well as people. LNG imports are widely acknowledged to be crucial in meeting future natural-gas needs. Yet public concern about safety led more than a half-dozen communities to reject an LNG import terminal or rally against a proposed facility.

19 JANUARY 2005 Chicago, IL, United States.

A barge carrying hundreds of thousands of gallons of a glutinous petroleum byproduct exploded, caught fire, and sank in a ship channel on Chicago's southwest side. A boiler on the barge apparently exploded, igniting the clarified slurry oil. Initial estimates indicated that the barge was carrying about 13,000 barrels, or more than 500,000 gallons.

17 JANUARY 2005 Nigeria.

Shell's Nigerian subsidiary restarted oil production at five platforms shut down the previous month by protests in the troubled Niger Delta, but a pipeline leak cut output elsewhere. Meanwhile, a weeks-old leak in the Trans-Niger pipeline in the Egbema area of the Niger Delta cut oil production by 44,000 barrels a day. Unidentified people had punctured the pipeline at several points. Thieves who routinely break open pipelines to steal oil—siphoning off up to 10 percent of Nigeria's 2.5 million-barrels-per-day output—pose another threat to oil multinationals in the region. Nigeria is Africa's largest oil exporter and the fifth largest supplier of crude to the U.S.

13 JANUARY 2005 Nigeria.

Protesting villagers forced the Nigerian subsidiary of Royal Dutch/Shell Group of Companies to cut oil output in the volatile Niger Delta, hampering efforts by the Anglo-Dutch oil company to recover from earlier protests. They followed a string of oil-platform invasions by hundreds of villagers in December in the Niger Delta, where poverty-stricken residents complained they were not benefiting from the riches pumped from their soil. Shell had been negotiating with the villagers to reach an agreement allowing production to restart. Including the shutdowns, unrest forced Nigerian oil producers to cut off 230,000 barrels per day of oil output, or nearly 10 percent of its capacity. Nigeria is Africa's top oil exporter and the fifth largest supplier of oil to the United States.

6 JANUARY 2005 Kearney, NE, United States.

Vandals turn off a gas-pipeline valve to about 150 homes and businesses in frigid temperatures.

28 DECEMBER 2004 Nevada, United States.

Eight high-voltage transmission lines servicing the northern region and Reno area were sabotaged. Officials reported the collapse of any one tower could possibly bring down a string of other towers.

19 DECEMBER 2004 Saudi Arabia.

The Saudi branch of al Qaeda called for "all mujahideen in the Arabian Peninsula" to target "oil resources that do not serve the nation of Islam." The statement urged al Qaeda members and sympathizers around the Arab world to unite "to strike all the foreign targets in the Arabian Peninsula and attack all the infidels' havens everywhere." The statement came four days after al Qaeda leader Osama bin Laden urged Islamic militants to stop Westerners from obtaining Middle Eastern oil, saying such a blow would be fatal to the West.

16 DECEMBER 2004 Gary, IN, United States.

Utility workers investigating a power outage found cut fencing and an unexploded homemade bomb in a Northern Indiana Public Service Company substation.

7 DECEMBER 2004 Canada

An antiglobalization group calling itself the Initiative for Internationalist Resistance claimed responsibility for sabotaging Hydro-Quebec. Explosives apparently were used on a tower carrying high-voltage electricity to the U.S. The incident coincided with a visit from President George W. Bush to Canada. The 450-kilovolt lines transmit electricity from Quebec's northern James Bay region to the New England power pool in the Boston area. The line is the only one that exports electricity from Quebec.

7 DECEMBER 2004 Russia.

Terrorists blew up a main gas pipeline in the southern Russian region of Dagestan. The blast was the fourth in a year to hit the Dagestani section of the Mozdok-Gasimagomed pipeline, which with its arteries stretches 750 miles. No group immediately took responsibility, but Chechen rebel leaders previously had threatened and been blamed for attacks against pipelines, electricity towers, and other infrastructure in Russia.

7 DECEMBER 2004 NIGERIA.

Hundreds of protesters ended their three-day siege of oil platforms in the oil-rich Niger Delta, paddling away in canoes and boats after Royal Dutch-Shell Companies and Chevron Texaco Corporation agreed to talks. The three-day protest shut down production of 90,000 barrels a day.

6 DECEMBER 2004 Nigeria.

Hundreds of protesters besieged two oil platforms run by Royal Dutch/Shell Group companies and Chevron Texaco Corporation in Nigeria's southern oil region, shutting down production of 90,000 barrels of oil a day. The protesting villagers from Kula community invaded the oil-pumping facilities owned by Shell in the Ekulama oil fields and another belonging to Chevron Texaco at Rober-Kiri Island in the swamps of the oil-rich delta. Shell pumps 70,000 barrels daily from the two facilities, while Chevron Texaco pumps 20,000 barrels daily from its own station. Nigeria, at 2.5 million barrels a day, is Africa's leading oil exporter, the world's seventh-largest exporter, and the fifth biggest source of U.S. oil imports

4 DECEMBER 2004 Quebec, Canada.

Experts investigated suspicious material found near a damaged hydroelectric tower that carries power to the United States. The site in Saint-Herménégilde is nine miles from the U.S. border. The investigation was referred to the police force's terrorism unit.

3 DECEMBER 2004 Madrid, Spain.

Bomb explosions hit five gas stations in the Madrid, Spain area after the ETA, a Basque separatist group, phoned in a warning of the attack. The blasts caused no injuries, as the bombs were small and the areas had been evacuated in time. The blasts came as hundreds of thousands of Madrid residents were pouring out of town for a long bank holiday. Four major roads out of Madrid were blocked. The explosions marked a significant return to violence for ETA, which had been relatively inactive since the March 11 train bombings that killed 191 people in Madrid.

26 NOVEMBER 2004 Philadelphia, PA, United States.

Some oil deliveries to refineries on the U.S. East Coast were halted when the Coast Guard closed the Delaware River to clean up an oil spill near Philadelphia. The Cypriot-flagged *Athos I* leaked 30,000 gallons of crude oil en route to Citgo Petroleum Corp.'s Paulsboro, NJ, plant. Seven refineries that account for 75 percent of East Coast oil-processing capacity rely on the 150-mile-long river to receive oil and ship cargoes of fuel, asphalt, and chemicals.

14 OCTOBER 2004 Pewaukee, WI, United States.

Four men were reported taking photos and shooting videos outside the headquarters of a transmission company. At the same time, two of its transmission towers collapsed after the bolts were removed.

13 OCTOBER 2004 Los Angeles, CA, United States.

DWP gets anti-terror training. Two years after Los Angeles, CA, Department of Water and Power [DWP] security workers asked for better training, the Los Angeles Police Department [LAPD] launched a program to increase antiterrorism instruction. In addition, the DWP is instituting a number of programs to improve security, including the daily testing of water supplies, installation of security cameras, increased helicopter surveillance, increased size of the DWP security forces, and stricter background checks.

11 OCTOBER 2004 Philadelphia, PA, United States.
A fake bomb at electrical tower forced the closure of a major highway connecting Philadelphia with its western suburbs for several hours. A box made to look like an explosive but later determined to be a hoax was discovered at the foot of an electric transmission tower. An FBI spokesperson did not know if the incident was related to the sabotage of an electrical-transmission tower in Wisconsin on October 10.

6 OCTOBER 2004 Moundsville, WV, United States.
Security was stepped up at American Electric Power's [AEP] Kammer-Mitchell power plant in Moundsville, WV, after suspicious vehicles were spotted on a roadway taking photographs of the facility. The plant manager believed those involved were from a group of environmental activists, not terrorists.

28 SEPTEMBER 2004 New Caney, TX.
Police connected a pipeline explosion to an August vandalism case near Porter, TX. Evidence pointed to one or more suspects behind the explosion and the ensuing six-hour fire. No one was injured, and nearby houses and other structures were not affected. Damage to construction equipment, electricity-transmission lines, and the six-inch pipeline itself were estimated to be over $1 million dollars. Evidence showed that a track hoe and a bucket truck at the explosion site were tampered with.

28 SEPTEMBER 2004 Nigeria.
Nigerian rebels fighting for the sovereignty of the oil-producing Niger Delta told oil companies in the world's seventh largest exporter to shut production before they begin an "all out war." The Niger Delta People's Volunteer Force advised all foreigners to leave the area, which pumps all of Nigeria's 2.3 million-barrels-per-day output. Companies feared a repeat of the previous year's rebellion, which forced them briefly to shut 40 percent of production. Shell evacuated more than 200 staff from the fighting. The communiqué further advised "all foreign embassies [to] withdraw their citizens from the Niger Delta until the resolution of the fundamental issues." Five multinational oil companies produce Nigerian oil. The country is the fifth largest supplier to the U.S.

27 SEPTEMBER 2004 Asia, the Middle East, and Africa.

Terrorists and insurgents stepped up attacks on oil and gas operations in an effort to disrupt jittery energy markets, destabilize governments, and scare off foreign workers. The attacks were most intense in Iraq but also occurred in previous months in Indonesia, Pakistan, India, Russia, and Nigeria. In many cases, the attacks were by terrorists or rebels seeking to cause economic disruption or to steal oil to finance their operations. They came as world oil production stretched close to its limit. Analysts viewed them as a factor adding pressure on oil prices. High prices were cited by the Federal Reserve as a contributor to slowed U.S. economic growth. Analysts believe that al Qaeda favors economic targets because they create the immediate impact of the attack and economic repercussions and can potentially destabilize the country where the attack occurs.

26 SEPTEMBER 2004 Lagos, Nigeria.

Eight officers from a Russian oil tanker were arrested and accused of trying to smuggle 1,100 tons of crude oil after their ship's disappearance following its earlier apprehension by the Nigerian Navy.

25 SEPTEMBER 2004 Spain.

Spanish police blew up a bomb placed on an electricity pylon and hunted for another device after ETA Basque guerrillas said they had planted them in what appeared to be a new tactic to hit the power infrastructure. A caller claiming to represent ETA telephoned a Basque newspaper saying the separatist group had put bombs on two high-tension electricity-cable pylons run by grid operator Red Electrica. Police quickly found one on a pylon near the French border in the Basque province of Guipuzcoa. A source at Red Electrica did not expect the incident to have much impact on the electricity flow from France, a key power supplier for Spain. ETA is considered a terrorist group by Spain.

15 SEPTEMBER 2004 Irun, Spain.

Four explosive devices were detonated on an electricity tower. The blast damaged the foundations of the tower but did not manage to bring the structure down or affect the electricity supply. Officials disclosed that ammonal had been used in the devices, each of which contained between one and two pounds. They also stressed that the type of explosive material matched that used by the armed organization ETA in the attacks carried out during the summer in Cantabria, Asturias, and Galicia.

10 AUGUST 2004 Istanbul, Turkey.

Two bombs exploded within a half hour of each other at a liquefied-petroleum gas plant in the outskirts of the city, causing damage but no casualties.

9 AUGUST 2004 Nigeria.

A year of bloodletting and unrest killed more than 1,000 in the oil-rich Niger Delta. Tensions over oil revenues aggravated ethnic strife. Kidnappings and sabotage escalated, forcing costly shutdowns by companies pumping crude in the oil-rich swamps of the volatile Niger Delta. Major oil companies hope to double production in West Africa's Gulf of Guinea, estimated to hold up to 10 percent of the world's oil reserves. The United States, Europe, and Asia are increasingly looking to the region's oil as an alternative to crude from the Middle East. The Nigerian subsidiary of San Ramon, CA-based ChevronTexaco Corp. is among the companies hit hardest by Nigeria's worsening oil-related violence, suffering an estimated $750 million in costs from sabotage to its wells, pipelines, and other facilities since March 2003.

4 AUGUST 2004 Athens, Greece.

A homemade bomb exploded near an electrical substation outside Athens, causing damage but no injuries. Authorities were not immediately clear if there was a link between the blast and the Olympics, which opens the next week amid unprecedented security. The device, which police said was made with a cooking-gas canister and a triggering fuse, exploded in the bathroom of a building near a substation in the town of Metamorphosi, 6 miles north of Athens. No electrical facilities were damaged, a government official said. According to the official, the culprits broke the bathroom window and dropped the device inside.

30 JULY 2004 Sao Paulo, Brazil.

Brazilians protesting the country's land-ownership policies seized land at a power plant in southern Sao Paulo State owned by Duke Energy Corp., the largest U.S. utility owner. About 1,500 followers of the Brazilian Landless Workers Movement arrived in seven buses, three trucks, and 38 cars to set up tents and shacks at the Taquarucu power plant in preparation to invade a nearby farm. Rural workers prepared the invasion for the past two weeks and gathered people from 12 settlements in the Pontal de Paranapanema region. There were 230 land invasions in Brazil in the first half of the year, surpassing the 222 invasions in all of 2003.

28 JULY 2004 Washington, DC, United States.

The government has warned repeatedly that anti-American groups may be planning attacks on U.S. refineries, but some experts question whether refineries really would be high-priority targets. Experts say refineries pose difficulties for would-be attackers because the typical U.S. facility is a sprawling, well-protected complex that would require careful targeting and some measure of luck to score a big hit. An attack on a single refinery would affect the markets and cause alarm, but putting a serious dent in U.S. refining supply would be a daunting task, experts said, claiming other parts of the U.S. energy-supply chain are more vulnerable to attack, such as portions of oil and natural-gas pipelines above ground in unpopulated areas.

22 JULY 2004 Limerick Township, PA, United States.

Exelon Corp. presented plans to Limerick Township, PA, to build a larger training center and install seven guard towers at its Limerick Generating Station, part of mandated security upgrades ordered by the Nuclear Regulatory Commission [NRC]. The NRC issued the regulations in 2003 to address concerns about security at the nation's nuclear-power plants in light of the terrorism threat. The 84,000-square-foot training center would be located in a parking lot adjacent to the learning center, just northeast of the second cooling tower. The guard towers will be strategically placed throughout the facility to improve security. Several township supervisors believed Exelon asked for an excessive amount of waivers from township ordinances.

19 JULY 2004 Texas City, TX, United States.

Law enforcement sought a man seen taking pictures of two refineries in Texas City, TX, located on the Texas Gulf coast about 30 miles south of Houston. The town has three refineries including the largest U.S. plant operated by BP Plc. [the third-largest U.S. refinery], processing 470,000 barrels of crude oil per day. U.S. refinery security officials said their security guards regularly report people observing or taking pictures of refineries.

7 JULY 2004 Geismar, LA, United States.

The U.S. Coast Guard ordered the closure of two facilities in Louisiana that handle natural-gas liquids [NGL] and petroleum products for failure to comply with new maritime-security codes.

27 JUNE 2004 Guam.

Numerous power outages reported, hundreds head to shelters, flooded vehicles and stranded residents being helped in Guam. A state of emergency was declared for Tropical Storm Tingting. The Ugum Water Treatment Plant was taken offline, and the Guam Waterworks Authority [GWA] reported a clogged intake at the Ugum River as the cause of the shutdown. Operation crews worked to address numerous feeder and circuit outages in small pocket areas of Guam.

17 JUNE 2004 Saudi Arabia.

Escalating sabotage against pipelines in Iraq heightened fears that terrorists are planning a wholesale assault on energy targets throughout the region and are taking aim at the world's largest oil supplier— Saudi Arabia. The head of Saudi Arabia's government oil monopoly remains confident that the industry is well protected. However, independent experts warn that an attack on any of Saudi Arabia's major facilities could cripple world oil supplies. Experts warn that the country's pipelines, oil wells, refineries, and export terminals are enticing targets for al Qaeda, whose operatives in Saudi Arabia are threatening to launch a devastating attack. Extensive terrorist attacks on oil targets in Saudi Arabia and Iraq would be tantamount to an energy Pearl Harbor, forcing severe shortages and boosting prices in the United States and other countries heavily dependent on imported oil. The United States gets more than 50 percent of its oil from foreign suppliers.

17 June 2004 Wisconsin, United States.

Joseph Konopka, who calls himself "Dr. Chaos," was sentenced to 23 years in prison for conspiring to knock out power lines, burn buildings, and damage computers in Wisconsin. Authorities said more than 50 acts in various Wisconsin counties affected more than 30,000 power customers and caused more than $800,000 in damages. Konopka was the self-appointed leader of a loose affiliation called "The Realm of Chaos," which recruited youths to engage in property damage. Konopka was serving a prison term for taking two bottles of cyanide from an abandoned chemical warehouse and hiding them in the Chicago subway in 2001.

15 JUNE 2004 Iraq.

Terrorists sabotaged a pipeline in southern Iraq, prompting a stoppage of all oil exports through Iraq's two offshore terminals in the Persian Gulf. The previous week, a number of pipeline attacks occurred over a three-day period, with the major attack on the line between Kirkuk and Turkey. After those attacks, Iraq's interim prime minister reported that Iraq lost more than $200 million in revenues.

2 JUNE 2004 Amarillo, TX, United States.

A cause had not been determined for an hour-long complete power outage at a nuclear-weapons plant near Amarillo, TX. The blackout affected the entire Pantex plant, America's only nuclear-weapons assembly and disassembly facility, although backup power kicked in very quickly. It was the most severe power outage ever at the plant—a depository of large amounts of radioactive materials—but officials said security was never compromised.

30 MAY 2004 Khobar, Saudi Arabia.

A group of Islamic extremists seized foreign oil workers in the Persian Gulf city of Khobar, 250 miles northeast of Riyadh. About 50 hostages were rescued, but one American and 21 other people were killed, including citizens from at least 10 countries. The ringleader of the gunmen was captured. Three terrorists escaped, but two were killed a few days later. A group allied with al Qaeda asserted responsibility for the attack, driving a wedge between Saudi Arabia and the U.S. It was the fourth time in 13 months that Islamic extremists launched deadly attacks on foreign targets in Saudi Arabia.

29 MAY 2004 Washington, DC, United States.

Attacks on oil facilities in Iraq and Saudi Arabia resulted in price rises several times over two months.

24 MAY 2004 Washington, United States.

The 400-mile Olympic pipeline through Western Washington was shut down Monday, May 24, after a small section of a test line exploded and caught fire. A shutdown of more than a few days could lead to an increase in gas prices in the region. The pipeline, the main supplier of fuel for Seattle-Tacoma International Airport, will remain shut down until state and federal investigators determine the cause of the leaks that occurred Sunday and decide that the line is safe, Olympic spokesperson Lee Keller said.

24 DECEMBER 2003 Chongqing, China.

A huge explosion at a natural-gas field in southwest China kills more than 190 people.

28 NOVEMBER 2003 Warri, Nigeria.

Foreign oil workers were kidnapped in Nigeria and a ransom sought by militants—the fourth kidnapping in one month.

30 OCTOBER 2003 Sacramento, CA, United States.

Local authorities reported that an individual was observed removing bolts from high-voltage transmission towers.

22 OCTOBER 2003 Wasco, OR, United States.

State police found bolts missing from two high-power electrical towers.

August 2003 Malacca Straits.
Gunmen took control of a Malaysian-flagged tanker carrying 1,000 tons of fuel oil from Singapore to Penang. The attackers, whom Malaysian police believed were affiliated with GAM, demanded a ransom of $100,000 against the ship's crew of nine but later released them after the ship owner paid $50,000.

14 August 2003 United States and Canada.
Power outage caused total blackouts at 21 power plants throughout the eastern United States and part of Canada. A communiqué released by al Qaeda claimed responsibility, but lightning was blamed as the official cause.

16 July 2003 La Pesquera, Colombia.
Military officials reported that either ELN or FARC terrorists bombed a section of the Cano Limon-Covenas oil pipeline, owned by Colombian and U.S. oil companies, causing an unknown amount of damage.

9 June 2003 Lima, Peru.
Approximately 60 terrorists kidnapped 71 workers employed by Techint Group, an Argentine company building a natural-gas pipeline in southeastern Peru. The kidnapped group consisted of 64 Peruvians, four Colombians, two Argentines, and a Chilean. A rescue operation freed all the hostages on 11 June, but the terrorist escaped. Shining Path was responsible.

30 May 2003 Guamalito, Colombia.
Terrorists attacked a section of the Cano Limon-Covenas oil pipeline, spilling nearly 7,000 barrels of crude oil into the Cimitarra creek, a major source of drinking water for more than 5,000 people and causing extensive environmental damage, leaving families without drinking water. Ecopetrol of Colombia and a consortium of West European and U.S. companies jointly own the pipeline. No group claimed responsibility, although both FARC and ELN terrorists had attacked this pipeline previously.

15 May 2003 Karachi, Pakistan.
Nineteen small bombs exploded at Shell, an Anglo-Dutch-owned company, stations and at two Caltex, a subsidiary of U.S. giant Caltex, **petrol stations**, injuring seven persons. The small bomb-firecrackers fitted with timing devices were packed into boxes, placed in garbage bins, and appeared aimed to scare. The group Muslim United Army claimed responsibility in a faxed letter to the newspaper *Dawn*.

2 May 2003 Maracaibo, Venezuela.

A car bomb exploded, damaging surrounding buildings, including a local office of the U.S. oil company Chevron Texaco. The explosion occurred outside the home of controversial cattle livestock producer, Antonio Melian. Mr. Melian is a leading activist in Zulia State, and he had been the center of opposition government debate in the wake of the two-month nationwide labor stoppage that failed to bring down the Chavez Frias government.

26 March 2003 Narwal, Kashmir.

A bomb placed inside the engine of an empty oil tanker parked outside a fuel storage area exploded and caught fire, killing one person and injuring six others.

15 Feb 2003 Saravena, Colombia.

Military officials reported that either ELN or FARC terrorists bombed a section of the Cano Limon-Covenas oil pipeline, causing an unknown amount of damage. The pipeline is owned by Colombian and U.S. oil companies.

6 February 2003 Arauquita, Colombia.

Military officials reported either ELN or FARC terrorists bombed a section of the Cano Limon-Covenas oil pipeline, causing an unknown amount of oil to spill. The pipeline is owned by Colombian and U.S. oil companies.

22 January 2003 Arauquita, Colombia.

Military officials reported either the ELN or FARC terrorists bombed a section of the Cano Limon-Covenas oil pipeline, causing an unknown amount of damage. The pipeline is owned by Colombian and U.S. oil companies.

6 October 2002 Mukalla, Yemen.

A barge loaded with explosives rammed into French supertanker *MN Limburg,* killing 1 crewmember and spilling 90,000 barrels of oil into the Gulf of Aden. An al Qaeda affiliate, the Aden-Abyan Islamic Group in Yemen, claimed responsibility. The organizer, Abd al Rahman al Nashiri, was also responsible for the *U.S.S. Cole* attack.

8 October 2001 Pennsylvania, United States.

Authorities announced a credible threat to Three Mile Island, the site of America's worst nuclear accident, which occurred in 1979. The threat closed down two nearby airports for four hours, and military aircraft were sent to patrol the area. By the next morning, the threat was dismissed and the alert cancelled.

4 OCTOBER 2001 Alaska, United States.

A drunken local resident pierced a 25-year old pipeline with a bullet fired from a .338 caliber rifle. The single shot resulted in the release of more than 285,000 gallons of crude oil across two acres of tundra forest. It shut down the Trans-Alaskan Pipeline, which carries one million gallons of oil per day, for more than three days.

MARCH 2000 Manila, Philippines.

Terrorists attacked the Department of Energy in the central Philippines to protest rising oil prices. The ABB was responsible.

APRIL 1999 Phnom Penh, Cambodia.

Authorities arrested five CFF members for plotting to blow up a fuel depot outside Phnom Penh with antitank weapons.

18 OCTOBER 1998 Antioquia, Colombia.

A bomb exploded on the Ocensa pipeline in Antioquia Department, killing approximately 71 persons and injuring at least 100 others. The explosion caused major damage when the spilled oil caught fire and burned nearby houses in the town of Machuca. The Colombia State Oil Company Ecopetrol and a consortium including U.S., French, British, and Canadian companies jointly own the pipeline. The ELN claimed responsibility.

18 JULY 1998 Pastaza Province, Ecuador

The Indigenous Defense Front for Pastaza Province [FDIP] kidnapped three employees of an Ecuadorian pipeline maker subcontracted by a U.S. oil company in Pastaza Province. The group accused the company of causing environmental damage in its oilfield developments. It released the hostages over the next two days.

25 MARCH 1998 Cupiagua, Colombia.

At the British Petroleum oil field in Cupiagua, a bomb blast injured one U.S. citizen and two British workers. At least one bomb was placed near the oil workers' sleeping trailers and detonated around midnight. Police blame the attack on the National Liberation Army.

28 APRIL 1986 Soviet Union.

A power surge at the Chernobyl Nuclear Power Station led to an explosion that created the world's greatest nuclear accident. The explosion burned the core of the reactor, releasing radioactive materials into the atmosphere for 10 days. Timely and reliable response by Soviet officials was nonexistent. The accident was not announced until three days after it occurred. Inhabitants of a town just 3 km away from the disaster site were not evacuated until 60 hours after they were contaminated. A village 30 km from the plant was evacuated a week later. Rainwater contaminated with radioactive material fell as far away as Britain and Scandinavia. The Dnierper River, which provides drinking water to 35 million people, including those living in Kiev, was contaminated. Runoffs from watersheds in the Ukraine were contaminated up to 4 years after the accident. The land surrounding the plant remained uninhabited for a long time.

28 MARCH 1979 Pennsylvania, United States.

A chain of events at Three Mile Island Nuclear Power Station caused the loss of cooling water, resulting in a near meltdown and the worst nuclear accident in the United States. No radiation was leaked into the atmosphere, but the reactor was a total loss. The site took 10 years to clean at a cost of more than $1 billion dollars.

FEB 1972 Germany and Holland

Terrorists blew up a German electrical installation and a Dutch gas plant. Black September was responsible.

APPENDIX B: ENERGY-SECTOR INITIATIVES

The electricity industry has a history of taking proactive measures to assure the reliability and availability of the electricity system. Individual enterprises also work actively within their communities to address public-safety issues related to their systems and facilities. Since September 11, 2001, the sector has reviewed its security guidelines and initiated a series of intra-industry working groups to address specific aspects of security. It has created a utility-sector security committee at the chief-executive-officer level to enhance planning, awareness, and resource allocation within the industry.

Electricity

Additional electricity-sector protection initiatives include efforts to:

- **Establish a partnership with the NERC.** The sector as a whole, with NERC as the sector coordinator, has been working in collaboration with DOE since 1998 to assess its risk posture in light of the new threat environment, particularly with respect to the electric system's dependence on information-technology networks. In the process the sector has created an awareness program that includes a "Business Case for Action" for industry senior executives, a strategic reference document entitled "An Approach to Action for the Electric Power Sector," and security guidelines related to physical and cyber security.
- **Share and manage security information.** With respect to managing security information, the sector has established an indication, analysis, and warning program that trains utilities on incident reporting and alert-notification procedures. The sector has also developed threat-alert levels for both physical and cyber events, which include action-response guidelines for each alert level. The industry has also established an Information Sharing Analysis Center [ISAC] to gather incident information, relay alert notices, and coordinate daily briefs between the federal government and electric-grid operators around the country.
- **Identify equipment-stockpile requirements.** Department of Homeland Security [DHS] and the DOE will work with the electricity sector to inventory components and equipment critical to electric-system operations and to identify and assess other approaches to enhance restoration and recovery including standardizing equipment and increasing component interchangeability.

- **Reevaluate and adjust nationwide protection planning, system restoration, and recovery in response to attacks.** The electric-power industry has an excellent process and record of reconstitution and recovery from disruptive events. Jointly, industry and government need to evaluate this system and its processes to support the evolution from a local and regional system to an integrated national-response system. The DHS and the DOE will work with the electricity sector to ensure that existing coordination and mutual-aid processes can effectively and efficiently support protection, response, and recovery activities as the structure of the electricity sector continues to evolve.

- **Develop strategies to reduce vulnerabilities.** The DHS and the DOE will work with state and local governments and the electric-power industry to identify the appropriate levels of redundancy of critical parts of the electric system as well as requirements for designing and implementing redundancy in view of the industry's realignment and restructuring activities.

- **Develop standardized guidelines for physical security program.** The DHS and the DOE will work with the energy sector to define consistent criteria for criticality, standard approaches for vulnerability and risk assessments for critical facilities, and physical security training for electricity-sector personnel.

Dams

Dam initiatives to overcome protection challenges include the following efforts:

- **Develop risk-assessment methodologies for dams.** The DHS, in cooperation with appropriate federal, state, and local government representatives and private-sector dam owners, has agreed to design risk-assessment methodologies for dams and develop criteria to prioritize the dams in the national inventory to identify structures requiring enhanced security evaluation and protection focus.

- **Develop protective action plans.** The DHS, together with other appropriate departments and agencies, has agreed to establish an intergovernmental working group to explore appropriate protective actions for the nation's critical dams.

- **Establish a sector ISAC.** The DHS has agreed to work with other appropriate public and private sector entities to establish an information and warning structure for dams similar to the ISAC model in use within other critical infrastructure sectors.

- **Institute a national dam-security program.** The DHS and other appropriate departments and agencies, such as the Association of State Dam Safety Officials and the United States Society of Dams, have agreed to collaborate to establish a nationwide security program for dams.
- **Develop emergency action plans.** The DHS, together with other appropriate departments and agencies, will identify the areas downstream from critical dams that could be affected by dam failure and develop appropriate population and infrastructure protection and emergency action plans.
- **Develop technology to provide protective solutions.** The DHS, together with other appropriate departments and agencies, has agreed to explore new protective technology solutions for dams. Technology solutions hold significant promise for the identification and mitigation of water-borne threats. For example, technical options might include deploying sensors, barriers, and communications systems to reduce the possibility of an unauthorized craft or device entering a critical zone located near a navigational dam.

Oil and Natural Gas

Oil- and natural-gas-sector initiatives include the following efforts to:

- **Plan and invest in research and development for the oil and gas industry to enhance robustness and reliability.** Utilizing the federal government's national scientific and research capabilities, the DHS and the DOE will work with oil- and natural-gas-sector stakeholders to develop an appropriate strategy for research and development to support protection, response, and recovery requirements.
- **Develop strategies to reduce vulnerabilities.** The DHS and the DOE will work with state and local governments and industry to identify the appropriate levels of redundancy of critical components and systems as well as requirements for designing and enhancing reliability.
- **Develop standardized guidelines for physical security programs.** The DHS and the DOE will work with oil and natural-gas industry representatives to define consistent criteria for criticality, standard approaches for vulnerability and risk assessments for various facilities, and physical security training for industry personnel.

- **Develop guidelines for measures to reconstitute capabilities of individual facilities and systems.** The DHS and the DOE will convene an advisory task force of industry representatives from the sector, construction firms, equipment suppliers, oil-engineering firms, state and local governments, and federal agencies to identify appropriate planning requirements and approaches.
- **Develop a national system for locating and distributing critical components in support of response and recovery activities.** The DHS and the DOE will work with industry to develop regional and national programs for identifying spare-parts requirements, notifying parties of their availability, and distributing them in an emergency.

Pipelines

Historically, individual enterprises within this sector have invested in the security of their facilities to protect their ability to deliver oil and gas products. Representatives from major entities within this sector have examined the new terrorist risk environment. As a result, they have developed a plan for action, including industry-wide information sharing. In addition to industry efforts, the DOT has developed a methodology for determining pipeline-facility criticality and a system of recommended protective measures that are synchronized with the threat levels of the Homeland Security advisory system. Additional pipeline protection initiatives include efforts to:

- **Develop standard reconstitution protocols.** The DHS, in collaboration with the DOE, the DOT, and industry, is working to identify, clarify, and establish authorities and procedures as needed to reconstitute facilities as quickly as possible after a disruption.
- **Develop standard security-assessment and threat-deterrent guidelines.** The DHS, in collaboration with the DOE and the DOT, is working with state and local governments and the pipeline industry to develop consensus security guidance on assessing vulnerabilities, improving security plans, implementing specific deterrent and protective actions, and upgrading response and recovery plans for pipelines.
- **Work with other sectors to manage risks resulting from interdependencies.** The DHS, in collaboration with the DOE and the DOT, is working with cross-sector working groups to develop models for integrating protection priorities and emergency-response plans.

Nuclear-Power Plants

To overcome protection challenges, the U.S. government has set a strategy to:

- **Coordinate efforts to perform standardized vulnerability and risk assessments.** The NRC and the DHS are working with owners and operators of nuclear-power plants to develop a standard methodology for conducting vulnerability and risk assessments.

- **Establish common processes and identify resources needed to augment security at nuclear-power plants.** The NRC and the DHS are working in concert with plant owners, operators, and appropriate local, state, and federal authorities to develop a standard process for requesting external security augmentation at nuclear-power plants during heightened periods of alert and in the event of an imminent threat.

- **Criminalize the carrying of unauthorized weapons or explosives into nuclear facilities.** The NRC, in coordination with the DHS, is pursuing legislation to make the act of carrying an unauthorized weapon or explosive into a nuclear-power plant a federal crime.

- **Enhance the capabilities of nuclear-power-plant security forces.** The NRC, in coordination with the DHS, is pursuing legislation authorizing security guards at licensed facilities to carry and use more powerful weapons. It will also assist the industry in developing standards and implementing additional training in counterterrorist techniques for private security forces.

- **Seek legislation to apply sabotage laws to nuclear facilities.** The NRC, in coordination with the DHS, is pursuing legislation to make federal prohibitions on sabotage applicable to nuclear facilities and their operations.

- **Enhance public outreach and awareness.** The NRC and the DHS are working with plant owners, operators, and appropriate local and state authorities to enhance programs for public outreach, awareness, and emergency preparedness.

Chapter 13

THE TRANSPORTATION SECTOR

This chapter outlines the:

- Criticality of the transportation sector to the economy of the nation
- Transportation-sector contributions to economic security
- Transportation-sector attractiveness to terrorist and criminal elements
- Transportation-sector vulnerabilities
- Tailoring the S^3E Security Methodology to the transportation sector
- Transportation-sector challenges facing the security-assessment team
- Applying the security-assessment methodology to the transportation sector
- Preparing the Transportation Security Assessment Report
- Historical overview of selected transportation incidents
- U.S. government transportation initiatives

THE ECONOMY AND NATIONAL SECURITY

Critical transportation systems crisscross the nation and extend beyond our borders to move millions of passengers and tons of freight each day, making them both attractive targets to terrorists and difficult to secure. Protecting these systems is further complicated by the need to balance security with the expeditious flow of people and goods through a vast, interconnected network of diverse key modes including aviation, highway, motor carrier [trucking], motor coach [intercity buses], maritime, pipeline, rail [passenger and freight], and transit [buses, subways, ferry-boats, and light rail]. The nation's transportation systems are inherently open environments designed to move people and commerce quickly to their destinations—they move over 30 million tons of freight and 1.1 billion passenger trips each day. The diversity and size of the transportation sector make it vital to our economy and national security. [1]

The United States shares a 5,525-mile border with Canada and a 1,989-mile border with Mexico. The maritime border includes 95,000 miles of shoreline and navigable waterways, as well as a 3.4-million-square-mile exclusive economic zone. There are 361 seaports and over 500 international airports dispersed throughout the nation. All people and goods legally entering the U.S. are processed through these air, land, or sea ports of entry. Each year, more than 500 million people legally enter the country. Some 330 million are non-citizens; more than 85 percent enter via land borders, often as daily commuters. [2]

Virtually every community in America is connected to the global transportation network by seaports, airports, highways, pipelines, railroads, and waterways. Protecting and promoting the efficient and reliable flow of people, goods, and services while preventing terrorists from using transportation conveyances or systems to deliver implements of destruction are daunting challenges. The diversity and size of the transportation sector make it vital to the economy and national security, including military mobilization and deployment. Together the various transportation industries provide free mobility of our population and fast movement of goods. [3]

Contribution to Economic Security

World trade, completely dependent on the transportation sector, accounts for $12.5 trillion in merchandise value. The U.S. is the world's leading importer and exporter of goods, with about $2 trillion in merchandise value [4].

Aviation

U.S. air carriers transport millions of passengers every day and at least twice as many bags and other cargo [5]. The aviation infrastructure is vast, consisting of thousands of entry points. It also has symbolic value, representing the freedom of movement that Americans value so highly as well as the technological and industrial prowess that has made the United States a world power. The nation's aviation system consists of two main parts:

- Airports and the associated assets needed to support their operations, including the aircraft that they utilize
- Aviation command, control, communications, and information systems needed to support and maintain safe use of our national airspace

Maritime Traffic

Seaports play a critical role in the U.S. economy. The U.S. Bureau of Transportation reports that in 2001 the nation's ports imported goods valued in excess of $523 billion and exported goods valued at about $196 billion. The number of containers shipped to the U.S. is expected to grow from more than eight million annually today to more than 20 million by 2017. [6]

Goods valued at over $1.4 billion dollars move through U.S. ocean ports every day, translating into millions of containers annually. Every year more than 7,500 ships make 51,000 calls on U.S. ports. They carry the bulk of 890 tons of goods, including 8 million containers, 175 billion gallons of oil and other fuels, and 7 million passengers, contributing an excess of $1 trillion to the nation's annual gross domestic product. [7]

The U.S. maritime border consists of 95,000 miles of shoreline and 3.4 million square miles of exclusive economic zone. There are more than 1,000 harbor channels and 25,000 miles of inland, intracoastal, and coastal waterways serving 300 ports within more than 3,700 terminals that handle passenger and cargo movements. There are 361 public ports in the U.S., and the largest 50 account for over 90 percent of the cargo tonnage. Twenty-five U.S. ports account for over 98 percent of all container shipments. Ninety-five percent of our nation's international trade moves by water, with more than 16,000 containers arriving at U.S. ports every day and half of all goods entering the United States arriving by oceangoing cargo containers. Cruise ships visiting foreign destinations embark from 16 U.S. ports. Ninety percent of the U.S. cargo arrives by ocean. [8]

Seaports are made up of a complex web of assets ranging from trucks, containers, terminal equipment, ships, and cranes to third-party suppliers and personnel to actual facilities with dynamic operational environments and specific business processes and objectives. The maritime-shipping infrastructure includes seaports and their associated assets; ships and passenger transportation systems; coastal and inland waterways; locks, dams, and canals; and the network of railroads and pipelines that connects these waterborne systems to other transportation networks.

Most ports have diverse waterside facilities that are owned, operated, and accessed by various entities. State and local governments control some port-authority facilities, while others are owned and operated by private corporations. Most ships are privately owned and operated. Cargo is stored in terminals at ports and loaded onto ships or other vehicles that pass through on their way to domestic and international destinations. The Department of Defense (DOD) has also designated certain commercial seaports as strategic seaports, which provide facilities and services needed for military deployment.

Rail and Public Transportation

The U.S. rail system includes all passengers and rail owners/operators. It includes light-rail systems, intercity passenger systems, commuter-rail operations, as well as subway systems nationwide. During every hour of every day, trains traverse the United States, linking producers of raw materials to manufacturers and retailers and employees to businesses. They carry mining, manufacturing, and agricultural products; liquid chemicals and fuels; and consumer goods. They transport 40 percent of intercity freight—a much larger portion than is moved by any other single mode of transportation. About 20 percent of that freight is coal—a critical resource for the generation of electricity. [9]

More than 20 million intercity travelers use the rail system annually. Forty-five million passengers ride trains and subways operated by local transit authorities. Each year passengers take approximately 9.5 billion trips on public transit. Mass-transit systems—designed to be publicly accessible—carry more passengers in a single day than air or rail transportation. Most are owned and operated by state and local agencies. A city relies on its mass-transit system to serve a significant portion of its workforce, in addition to being a means of evacuation in case of emergency. Protection of mass-transit systems is, therefore, an important requirement. [10]

In 2001, over 83 million tons of hazardous materials were shipped by rail in the United States across a 170,000-mile rail network that extends through every major city as well as thousands of small communities. Hazardous materials are defined by 49 U.S.C.5103 as any substance that is capable of posing an unreasonable risk to health, safety, and property when transported. It includes substances such as ammonia, hazardous wastes from chemical-manufacturing processes and medical facilities, and elevated-temperature materials such as molten aluminum and flammable, radioactive, and other materials. The DOT hazmat regulations classify hazardous materials into nine classes. The classification system helps communicate the dangers of these materials to emergency responders and transportation workers. The nine classes of hazardous materials are [11]:

- Class 1—Explosives
- Class 2—Gases
- Class 3—Flammable liquids
- Class 4—Flammable solids
- Class 5—Oxidizing substances and organic peroxides
- Class 6—Poisonous and infectious substances
- Class 7—Radioactive materials
- Class 8—Corrosives
- Class 9—Miscellaneous materials

The Department of Transportation (DOT) estimates that there are over 800,000 shipments of hazardous materials daily by all modes of transportation in quantities varying from several ounces to many thousands of gallons. [12] Together with other agencies, it is responsible for the following:

- In addition to the DOT, several federal agencies have authority over certain aspects of rail shipments of hazardous materials. These include the Department of Homeland Security [DHS], Environmental Protection Agency [EPA], the Department of Labor's Occupational Safety and Health Administration [OSHA], the Nuclear Regulatory Commission [NRC], the Department of Energy [DOE], and Department of Defense [DOD].
- Two administrations within the DOT—the Research and Special Program Administration [RSPA] and the Federal Railroad Administration [FRA]—have responsibilities for developing regulations pertaining to the transportation of hazardous materials and rail safety. Under the Homeland Security Act of 2002, the Department of Transportation shares responsibility with the Transportation Security Administration [TSA], within the DHS, for rail security.

▸ The FRA oversees the safety of track, signal, and train controls; motor power and equipment; operating practices; highway-rail-grade-crossing safety; and hazardous materials.

▸ Within the DHS, the TSA (created in the immediate aftermath of the terrorist attacks of September 11, 2001) has focused primarily on aviation issues but is now responsible for the security of all modes of transportation including rail.

▸ The U.S. Coast Guard has the responsibility for preventing spills from vessels and waterfront facilities. It also serves as the federal on-scene coordinator under the National Contingency Plan for oil or hazardous substances released in the coastal zone.

- The EPA has authority for implementing and enforcing legislation governing the protection of public health and the environment against chemical and other polluting discharges and for abating and controlling pollution when spills occur. The EPA also identifies substances and quantities that qualify as extremely hazardous and participates with other agencies in responding to hazardous-material transportation incidents involving radioactive materials.

- OSHA promulgates standards to protect the safety and health of employees. It also establishes hazardous-material training and safety requirements for emergency responders through its general industry standards, including its hazardous-waste operations and emergency-response standards.

- Although the DOT regulates the transportation of nuclear materials, including spent fuel, as hazardous materials, the NRC also regulates the transportation of nuclear material by its licensees. The primary role of the NRC is the establishment of packaging standards for fissile and other radioactive materials exceeding certain limits. The NRC certifies spent-fuel casks and other radioactive-material packaging designs that meet these standards and requires its licensees to use certified casks for transport. The NRC also plays a significant role through safety and security requirements and through inspection and enforcement.

- The responsibilities of the DOE regarding spent nuclear fuel are related to its role as an operator of nuclear facilities, including its role in developing the proposed Yucca Mountain Repository. The DOE's Office of Civilian Radioactive Waste Management is responsible for shipping spent nuclear fuel and oversees nuclear-waste-fund activities related to the Yucca Mountain Repository, which include the transportation of spent nuclear fuel. Both the DOE and the NRC have authority to approve packages suitable for transport. The DOE's authority is for defense or DOE-owned materials, while the NRC's authority is for shipments by its licensees.

• The DOD's Military Traffic Management Command oversees the shipments of DOD hazardous materials by rail companies and ensures that they are shipped according to the DOD's safety and security standards. Through its Military Traffic Management Command [MTMC], the DOD contracts with U.S. rail companies for the shipment of arms, ammunition, explosives, and other hazardous materials. The Department of the Navy and the DOE each ship radioactive materials, including high-level spent nuclear fuel. From 1997 to 2001, MTMC shipped 728,000 tons of hazardous materials by rail, which represents a very small percentage of the 459 million tons of all hazardous materials shipped by rail during the same time period. The Naval Nuclear Propulsion Program, a joint organization within the Departments of the Navy and Energy, ships naval spent nuclear fuel from shipyards to the DOE's Idaho National Engineering and Environmental Laboratory for examination and temporary storage. According to program data, spent nuclear fuel from nuclear-powered warships accounts for approximately 0.05 percent of all spent nuclear fuel in the United States. The DOE ships its own radioactive-waste material including low-level radioactive material, transuranic waste, and spent nuclear fuel.

Although rail moves only a small percentage of all hazardous materials, it is the predominant method of transportation for some types of materials, such as flammable solids.

Trucking and Busing

The trucking and busing industry is a fundamental component of the national transportation infrastructure. Without the sector's resources, the movement of people, goods, and services around the country would be greatly impeded. Components of this infrastructure include highways, roads, intermodal terminals, bridges, tunnels, trucks, buses, maintenance facilities, and roadway border crossings. More than 35 million trucks move in and out of U.S. seaports, warehouses, and distribution centers annually to deliver goods to the marketplace. [13]

Terrorist Target

Interdependencies exist between transportation and nearly every other sector of the economy. Consequently, a threat to the transportation sector may impact other industries that rely on it. Since the advent of major transportation systems in the 1800s, they have been vulnerable to attack from many quarters. Terrorists know that a disruption to one segment of a system can create a cascading effect across the rest of the network. Public transportation is used by millions of people daily. It is an attractive target for killing indiscriminately and in large quantities. No longer can corporations or private citizens ignore the reality of traveling in an increasingly dangerous world. A historical overview of selected terrorist attacks, criminal incidents, and industry mishaps within the transportation sector is displayed in Appendix A.

Vulnerabilities

The transportation sector is vast and complex, with multiple vulnerable points of entry. Differences in design, structure, and mission of transportation industries complicate the sector's overall protection framework. There is no way that a substantial portion of the sector can be protected all of the time, particularly if the attackers are prepared to die in the process. The size and breadth of the sector make it difficult to react to threats effectively or efficiently in all scenarios. This fact complicates protection efforts, but it also offers certain mitigating potential in the event of a terrorist attack. The time at which goods and materials are most vulnerable to loss is when they are in the transport and distribution mode.

The single most important threat to the transportation sector is a systematic planned threat by professional criminals and organized crime. The cost of transportation lost due to theft is $9 to $12 billion a year. Experts predict that losses will reach $20 billion dollars by 2010.

Threats to Aviation

Air piracy is not uncommon. The downing of civilian airliners, however, has not been limited to the placement of prepositioned explosives on board aircraft. The State Department reports that between 1978 and 1993 there were 25 incidents of MANPADS [man-portable precision-guided munitions] attacks upon civilian aircraft, which resulted in 536 fatalities. The attacks occurred in Angola, Soviet Georgia, Afghanistan, Sudan, Mauritania, Zimbabwe, Costa Rica, and Chad. [14]

A safe and secure civil-aviation industry is a critical component of the nation's overall security, physical infrastructure, and economic foundation. The threat of terrorism to aviation was significant throughout the 1990s; for instance, a plot to destroy 12 U.S. airliners was discovered and thwarted in 1995. The enormous size of U.S. airspace alone defies easy protection. Furthermore, given this country's hundreds of airports, thousands of planes, tens of thousands of daily flights, and the seemingly limitless ways terrorists or criminals can devise to attack the system, aviation is vulnerable on every front.

As the events of September 11 illustrated, aviation's vital importance to the U.S. economy and the freedom it provides our citizens make its protection an important national priority. Aviation faces several unique protection challenges. Its distribution and open access through thousands of entry points at home and abroad make it difficult to secure. Furthermore, components of the aviation infrastructure are not only attractive terrorist targets but also serve as potential weapons to be exploited.

Over 2 million passengers and their baggage are checked each day for articles that could pose threats to the safety of an aircraft and those aboard it. While there is more to airport security than screening, this screening of travelers and their luggage is the part of the security picture that impacts air travelers the most and is the most visible of security measures. Other security measures are less visible or known to the air traveler and generally are not discussed by aviation authorities. While the screening of passengers and their luggage has significantly improved during the last two decades, serious problems remain regarding the reliability and dependability of the Civil Aviation Security Program, and more work in this area needs to be done. [15]

Threats to Maritime Traffic

Protecting our seaports and waterways is no small task. The United States has jurisdiction over 3.5 million square miles of ocean and 95,000 miles of coastline. Some 7,500 ships with foreign flags make 51,000 calls on American ports each year. They carry the bulk of the approximately 890 million tons of goods that come into the country, including 7.8 million containers, 175 billion gallons of oil and other fuels, and hundreds of thousands of cruise ship passengers and crew members [16] . Some $1.35 trillion in imports and $1 trillion in exports are processed through our ports annually. The maritime industry pumps more than $750 billion into the U.S. economy and employs more than 13 million people. Given the importance of foreign trade to the U.S. economy, an attack that shuts down a major American port for even a few days could devastate the regional economy that it serves. [17]

Historically it has always been easier to appreciate the vulnerability of airports and airlines. There is a 30-year history of hijackings and terrorist killings. But we live in a new age where terrorists are more sophisticated and more determined than ever, better financed, and backed by a fanatical support group including some nations.

Shoring up harbors, ports, and waterfronts and safeguarding vessels pose many security concerns given their size, accessibility, and attractiveness as terrorist targets. Cargo and cruise ships present potentially desirable terrorist targets, given the potential for loss of life, ecological destruction, or disruption of commerce. In addition, ports often are not only gateways for the movement of goods but also industrial hubs and close to population centers, presenting other opportunities for terrorists bent on urban destruction. Al Qaeda's chief of maritime operations, Abdal Rahman al Nashiri, since captured, had developed a four-pronged strategy to attack vessels and seaports. These included ramming ships, blowing up medium-sized ports, attacking large vessels such as supertankers from the air using explosive-laden small aircraft, and attacking vessels with underwater demolition teams using mines or suicide bombs.

Al Qaeda is not the only terrorist group pursuing such targets and tactics. Members of the Indonesian terrorist group Hemaah Islamiah [IH] have been trained in sea-borne guerrilla tactics such as suicide diving capabilities and ramming, developed by the Sri Lankan group Liberation Tigers of Tamil Ealam.

Following the events of September 11, the United States addressed the General Assembly of the International Maritime Organization [IMO], an agency of the United Nations representing 162 nations dedicated to improving maritime safety, including combating acts of violence or crime at sea. In November 2001, the Commandant of the Coast Guard addressed the IMO, urging the body to consider an international scheme for port and shipping security. At the direction of IMO's Maritime Safety and Security Committee, a series of international-maritime-security work-group meetings were held to adopt comprehensive security resolutions to reduce risk to passengers, crew, and port personnel on board vessels, in port areas, and to the vessels and their cargo. In November 2002, one year later, the recommendations were adopted and incorporated as amendments to the 1974 International Convention for Safety of Life at Sea [SOLAS], and a new International Ship and Port Facility [ISPS] code, which was effective July 2004.

As a participant of the IMO and major contributor to the development of the security protocols of the ISPS code, the United States was also developing its own internal measures—the Maritime Transportation Security Act [MTSA] of 2002—to implement the security protocols of the new ISPS code. The intent of the MTSA, in part, is to help protect the nation's ports and waterways from terrorist attacks through a wide range of improvements. The MTSA, or Public Law 107-295, is largely administered by the United States Coast Guard, an agency within the Department of Homeland Security. The MTSA mandated major changes in the nation's approach to maritime security. A few notable differences exist between the MTSA and the ISPS are described below.

The MTSA extends its jurisdiction to facilities on the water that may be receiving a wide variety of domestic vessels, such as refineries located at nearby ports that produce highly volatile petrochemicals and convert crude oil into gasoline and heating oil.

The act calls for a comprehensive security framework for the nation's 361 major seaports, owners and operators of about 3,150 port facilities such as shipping terminals and offshore oil rigs, and 5,000 processing factories on waterways, coastlines, and harbors, such as power plants, refineries, or chemical facilities with hazardous materials. About 9,200 vessels such as cargo ships, ferries, tugs, and barges are also covered under the act. The act calls for owners and operators to conduct assessments and to develop security plans. The basic aim of such plans is to address potential vulnerabilities that could be exploited to kill people, cause environmental damage, or disrupt port operations and the economy by developing measures to mitigate these vulnerabilities. Company security plans were to be submitted to the Coast Guard for review and approval not later than December 2003 or self-certified that plans would be developed and implemented. Industry compliance with security plans for all owners and operations was expected by July 1, 2004—a schedule the Coast Guard decided to adopt that would align the United States with ongoing international improvements in maritime security being pursued by the IMO. Over 80 percent of major seaports, owners and operators and shipping companies met the initial target date for submitting their respective security plans to the Coast Guard. The remaining organizations negotiated extensions with the Cost Guard. The MTSA also mandates uniform biometric identification and background checks for maritime-transportation workers, which ISPS doesn't.

Cargo screening and monitoring are also important facets of port security. Ninety percent of the world's goods moves by cargo. Nearly 50 percent of the value of all U.S. imports arrives in 16 million containers through our nation's seaports. Security experts remain concerned about the potential for using the maritime transportation system as a conduit for smuggling terrorists, weapons of mass destruction, or other dangerous materials into the country. Before September 11, less than 3 percent of the containers were inspected. [18]

Security measures and technology innovations are leading the way toward increasing the integrity of shipment containers before they arrive in the United States as well as after they arrive in our ports. One interim initiative launched by the Customs Service—the Customs-Trade Partnership Against Terrorism—between the government and the private sector requires importers to take steps that will ensure tighter security of cargo, and, in turn, the government agrees to give the more secure, low-risk cargo a "fast lane" or "easy pass" through our ports of entry.

Ali M. Koknar [19] reports that sea piracy and terrorism are joining forces and creating troubled waters for the maritime industry with world-wide economic and security implications. Oil, natural gas, and other hazardous-cargo-laden vessels and passenger liners are potential targets for terrorists. Due to their hazardous cargo, these vessels could also serve as weapons in attacks directed at port facilities and other shore-side installations or against stationary navy or passenger ships.

According to Koknar, the Piracy Reporting Center at the International Maritime Bureau [IMB] reported that pirate attacks rose by 20 percent in 2003 to 445 incidents. These attacks are not only costly in terms of business losses; they are also deadly. Bounties range from $8 to $200 million per hijacking. Pirates killed 21 crew members on the high seas in 2003, more than twice the number of deaths reported in 2002. As piracy requires comprehensive planning, experience, and coordination with corrupt officials, Koknar contends that attacks on the high seas are carried out by no less than sophisticated syndicates, often with political or terrorist connections. When piracy is politically motivated, the global maritime industry may confront the same fate the airline industry did after 9/11.

Indonesia, Singapore, and Malaysia remain the most pirate-infested zones in the world, accounting for more than a quarter of the attacks, followed by the Indian subcontinent. African countries have seen an increase in piracy particularly in Nigeria, the Ivory Coast, Cameroon, and Tanzania. Local villagers along the coasts of Indonesia, Malaysia, and Africa welcome the pirate business and provide the perpetrators with food and shelter. In many countries in Southeast Asia, Latin America, and Africa, coast-guard operatives, corrupt drug agents, and other law-enforcement officials moonlight as pirates. Renegade members of British-trained Indonesian antipiracy squads still roam the Malacca Straits. The Strait of Malacca is a vital waterway for ships carrying petroleum products to Asia. As many as 50,000 ships use this waterway each year, and any blockage would force nearly half the world's fleet to sail further, generating a substantial increase in the need for vessel capacity, raising freight rates worldwide, and jolting the economies of China, Japan, and South Korea, which rely on imported energy. As a result of these threats, Japan, which imports 80 percent of its crude oil via the Strait of Malacca, has offered to help Indonesia beef up its coast guard. NATO operates a security mission code-named "Active Endeavor" in the Mediterranean. Up to eight navy vessels keep tabs on cargo flow in strategic locations. The program has helped reduce illegal immigration and smuggling, but the NATO program does not cover regions of the world where piracy and maritime terrorism incidents are most frequent.

Ships near Cape Horn in Africa are attacked frequently by the so-called Puntland Coast Guard, a pirate militia from a breakaway region of Somalia, which has built a lucrative practice of hijacking vessels for ransom. Indonesian authorities blame most of the piracy attacks in their region on the Free Aceh Movement, also known as Gerakan Aceh Merdeka [GAM], which is seeking to break off from Jakarta and set up an independent state. GAM has been fighting since 1975 for independence for the gas- and oil-rich region on the northern tip of Sumatra, about 1,100 miles northwest of Jakarta. Asian intelligence agencies believe the GAM is working with al Qaeda, which after the overthrow of the Taliban considered shifting its base from Afghanistan to Aceh. Ayman Al Zawahiri, the purported mastermind behind many of al Qaeda's attacks, was reported to have visited Aceh in June 2000. Since then, the number of crew kidnapping and ransom operations in Aceh by GAM militants has increased. [20]

In many cases, maritime terrorism goes beyond piracy for economic gain, with the focus more on disrupting operations and causing harm. Al Qaeda's chief of maritime operations, Abd al Rahman al Nashiri, is well known for his involvement in the *USS Cole* bombing and other deeds. During interrogation, Nashiri revealed that if warships became too difficult to approach, tourist ships would be targeted. Nashiri's capture led to the discovery that al Qaeda had also undertaken preparations to attack U.S. commercial vessels transiting the Straits of Gibraltar. Nashiri's comments were backed by a 190-page dossier, captured with him, which listed cruise liners sailing from Western ports as targets of opportunity. The dossier indicated that large cruise ships exceeding 140,000 gross tons and carrying more than 5,000 passengers continue to be desirable targets for terrorists. Cruise ships, which in the United States alone carry nearly seven million passengers each year, are considered prestigious targets because there is a perception that they are filled with wealthy Americans. For this reason, most American cruise lines stopped serving certain eastern Mediterranean ports after September 11.

Piracy and maritime insurance fraud perpetrated by pirate crews are estimated to cost around $16 billion annually, not taking into account the large number of uninsured or unreported vessel losses. The risk of piracy and politically motivated terrorist attacks on the high seas is deemed so serious that yacht insurers will not cover vessels that sail in the Red Sea. After the terrorist attack at Bandaranaike International Airport in Sri Lanka, Lloyd's of London Insurance imposed massive war-risk surcharges on all shipping to Sri Lanka. Other insurance brokers have more than tripled their premiums. Some are demanding that governments and ship owners pay bonds.

Since the terrorist attacks of September 11, 2001, the nation's 361 ports have increasingly been viewed as potential targets for future attacks for many reasons. There is concern about the potential for using the maritime system as a conduit for smuggling weapons of mass destruction or other dangerous materials into the country. The maritime-transport system is vulnerable to being used and/or targeted because it is open and porous enough that terrorists can enter and/or manipulate it for their own purposes. This is especially true of the container system, where the velocity of trade, the use of uniform containers, and the relative ease with which their contents can be willfully misrepresented offers many opportunities for terrorists, just as they do for drug and contraband smugglers. Bulk shipments also pose a danger because of the dangerous nature of some of their cargo. [21]

Only a portion of the more than 14 million containers that arrive in the U.S. every year are inspected. Only 17.5 percent of the high-risk cargo identified by customs and border protection was inspected overseas under the Container Security Initiative [CSI] program. Many security experts fear that the containers could easily be used to smuggle weapons, such as so-called dirty bombs, into the country. The placement of U.S. inspectors in foreign ports to check containers at the point of origin is a start, but the program is not meeting expectations. [22]

Experts warn that U.S. seaports could be tempting targets for terrorists bent on killing large numbers of people and disrupting the U.S. economy. Port, ferry, and cruise-ship terminals are often located in highly congested areas where large numbers of people live and work. A terrorist attack on our nation's seaports would be devastating, bringing the worldwide movement and processing of oceangoing cargo containers to a halt. Other sources of vulnerability come from the dependent network of linkages, such as 152,000 miles of rail, 460,000 miles of pipelines, and 45,000 miles of interstate highways that are used to bring products to the ports for shipment to and from the ports for delivery.

The system has already been targeted by pirates and criminals in various parts of the world with some success. The maritime sector also provides an attractive opportunity for groups who manipulate document-disclosure requirements to create a legitimate business front for a criminal enterprise.

Threats to Rail and Public-Transportation Systems

Rail security has become a matter of national security and is critical to protecting U.S. commerce and the safety of travelers. If the effect on air transportation resulting from the September 11 attacks is any indicator, then a terrorist attack on a major mass-transit system could have a significant regional and national economic impact.

Experts theorize that the most likely attacks on rail systems would involve conventional explosives and do not rule out suicide bombings such as those experienced in other countries. A less likely but more devastating scenario involves the use of a chemical agent such as sarin gas or biological agents such as anthrax or smallpox, which could be released into terminals, stations, subways, tunnels, and ventilation systems. Air currents above ground and those generated by moving trains below ground can spread gases and germs for miles within minutes, infecting a large number of people and contaminating vast areas along the travel route. Less deadly toxic industrial chemicals such as chlorine, phosgene, and cyanide are easier to obtain than sarin, smallpox, or anthrax and can kill or injure a smaller number of people. Terrorists can also derail any train or blow up a bridge or tunnel, killing many people and crippling a community's infrastructure for months or even years.

The greater risk associated with rail transport is the legitimate shipment of hazardous materials. Freight railways yearly transport millions of tons of hazardous materials across the continental U.S. that are essential to other sectors and public services. Much of this volume is carried on rail networks that travel through populated areas, increasing the concern that accidents or attacks on these shipments could have severe consequences. While the vast majority arrive safely at their destination, serious incidents involving these materials have the potential to cause widespread disruption, injury, or panic. Additionally, the proposed shipments of spent nuclear fuel at sites from 39 states across the country to the Yucca Mountain Repository have highlighted the need to safeguard hazardous materials against both accident and attack.

In response to the 9/11 terrorist attacks, industry and government have taken steps to improve the safety and security of the transportation of hazardous materials by rail. While much headway has been made in reducing the exposure of rail facilities to the risk of attack, the greatest challenges include the need for measures to better safeguard hazardous materials temporarily stored in rail cars while awaiting delivery to their ultimate destination—a practice commonly called "storage-in-transit"; the advisability of requiring companies to notify local communities of the type and quantities of materials stored in transit; and the appropriate amount of information rail companies should be required to provide local officials regarding hazardous-material shipments passing through their communities.

Other rail conditions that concern security experts are the rampant trespassing and vandalism in rail yards, unattended tank cars carrying hazardous materials, and train and equipment delayed or left unattended for extended periods of time.

Threats to Trucking and Busing

Each type of truck and bus transport has a different potential for attack:

- Transit buses are seen most likely to be used in attacking urban areas.
- Motor coaches are seen most likely to achieve a widespread attack across the nation.
- Schoolbuses could be used to paralyze a community by endangering the children aboard.
- Trucks carrying perishable goods could also disrupt the food and agriculture chain, including both retailers and customers.
- Trucks carrying high-value goods are seen most likely for attack by professional criminals and organized crime.
- Trucks carrying hazardous materials are seen most likely for attack or hijacking by terrorists to be used as a weapon against another target.

Truck cargo that spends a lot of time in transit and is parked in unattended areas with no supervision if the driver has to stop and spend the night is always an attractive target and becomes vulnerable not only to theft but to the adding of contraband to the shipment.

TAILORING THE S^3E SECURITY-ASSESSMENT METHODOLOGY TO THE TRANSPORTATION SECTOR

Applying the investigative principles outlined in Part I, the security-assessment team can use the tailored version of the S^3E **Security Assessment Methodology** model shown below as a road map to conducting security assessments within the transportation sector.

As the transportation sector is so diverse and complex, the S^3E **Security Assessment Methodology** model may require a further breakout for each industry. Exhibit 13.1 can be designed to provide industry-specific details.

The exhibit on the next page highlights in the grey shaded area those areas tailored for specific application to the transportation sector as contrasted to the generic template shown in Exhibit 4.2.

Transportation-Sector Challenges Facing the Security-Assessment Team

A daunting challenge for the security-assessment team is to determine the appropriate allocation of finite resources to manage risks while addressing specific threats and enhancing security across all transportation modes. Other specific challenges are discussed below.

Aviation Challenges

A major challenge that faces the security-assessment team is how best to close the back door of airports.

While attention and money are being focused on passenger and luggage screening, perimeters, cargo handling and processing, aircraft parking areas, maintenance-support facilities, and fuel-storage areas pose a greater vulnerability to the protection of airside operations. One theory for this neglect is that the airside of the airport does not directly impact on the traveler's flying experience. Other unique protection challenges that face the security-assessment team include:

- **Limited capabilities and available space.** Current detection equipment and methods are limited in number, capability, and ease of use. Suggestions to perform more manual physical searches of carry-on luggage and stored baggage do not sit well with the industry. New advanced screening technology is being developed, but more work in this area remains to be done.

Strategic Planning
Tasks 1, 2, & 3
Operational Environment
Critical Assessment
Threat Assessment

Program Effectiveness
Task 4
Infrastructure Interdependency
Evaluation [P$_{E_1}$]

Operational Environment Task 1
Enterprise Goals
Corporate Image, Core Values, Areas of Responsibility, Performance Objectives Business Incentives, Investment
Commitments
Federal Statutes, State Laws, Municipal Codes, Confidentiality Laws, Civil Common Law, Willful Torts, Contractual Obligations, Agreements, Governing Authorities, Voluntary Commitments
Characterization
Mission, Location, Layout/Land Use, Neighboring Facilities, Terrain, Climate, Resources, Operations, Processes
Operations – Logistics
Visitors, Customers, Vendors; Mail and Material Delivery, Receiving, Shipping; Staging, Storing, Distribution & Delivery; Time-Sensitive Processes; Communications

Enterprise
Roles & Responsibilities
Operations, Processes, Protocols
Functional Relationships
Security Awareness
Reporting Security Incident
Alert Notification System
Jurisdiction
Transfer Accountability

Security Capabilities
Management Team, Organization & Composition, Staffing, Skills, Experience, Qualifications & Training, Operational Capabilities Dependency Programs, Travel Routes, Schedules, Diversionary Tactics, Pre-Trip Inspections Call-in/Pick-up Protocols

Critical Assessment Task 2
Resources, Operations, Functions
Administrative, Aircraft/Fleet Facilities
Board/Conference Rooms
Cable/Communications Rooms
Electrical/Mechanical Rooms
Physical Infrastructure
Telecommunications Centers
IT Network Centers
Training Facilities
Warehouses, Loading Docks
SCADA & Security Systems
Infrastructure Sharing
Critical Interdependencies
Primary, Backup, Redundant Systems
All Transportation Groups
Generators, Pump Stations
HAZMAT, High Value Storage Areas
High Value/High Consequence Cargo
Overpass, Bridge & Tunnel Systems
Terminals, Shops & Laboratories
Microwave/Telecommunications Sites
Bonded Warehouses, Loading Docks
Aviation
Land/Air-Side Controls
Baggage Screening & Loading Areas
Gates, Taxi Ramps, Runways
Maritime
Port/Seaside Controls
Port Facility & Ship Security
Anchor Chain Collar, Pier/Anchorage
Beach/Shore Delivery Points
Meter Transfer Units, Temporary Storage
Boom & Netting
Discharge Points, Distribution Stations
Vessel Storage Bunkers, Reserve Supplies
Port/Seaside Operations
Trucking & Busing
Tankers, Tractor, Trailer & Compartments
Staging, Fuel Locations

Barriers & Delay Measures
Terrain, Approaches, Fences
Barrier, Walls, Gates
Detection Measures
Interior/Exterior Sensors
Special Purpose Sensors, Tamper Systems, Sensors, RFID Tags, GPS Tracking, Radiation Portal Monitors, Electronic Seals, Tamper Locks
Access Control Measures
People/Vehicle/Delivery Controls
Screening, Verification, Immigration & Passport Control, U.S. Customs Inspections, Manifests, Shipment Coding, Checkpoints, Controlled Deliveries, People & Conveyance/Cargo Screening, Declarations, Automatic ID System, 96/24 Hour Notification Rule
Assessment Measures
Perimeter Security, Lighting Posts and Patrols, CCTV
Communications Media
Radio, Telephone, PA, Hardwire, Wireless, Fiber Optic Information Network
Security Control Center[s]
Annunciation & Display, Event Storage, Reports, Alarm Integration Information Networks
Security Infrastructure
Primary, Secondary, Backup, Redundancy

Response & Recovery
Tactics, Techniques
First & Second Responders
Integrated Responders
Ground Transportation Modes
Redirect Traffic
Closeout Actions

Threat Assessment Task 3
International, National, Regional, Local Range & Levels of Threats Design Basis Threat Profile, Probability Threat Categories, Capabilities, Methods & Techniques

Quality Assurance/Quality Control Process

Exhibit 13.1 The S^3E Security-Assessment Methodology for the Transportation Sector

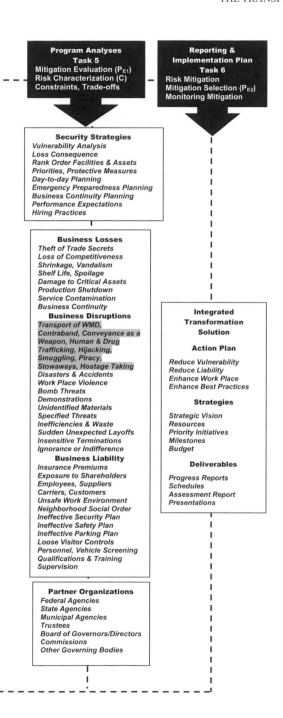

Program Analyses
Task 5
Mitigation Evaluation (P_{E1})
Risk Characterization (C)
Constraints, Trade-offs

Reporting &
Implementation Plan
Task 6
Risk Mitigation
Mitigation Selection (P_{E2})
Monitoring Mitigation

Security Strategies
Vulnerability Analysis
Loss Consequence
Rank Order Facilities & Assets
Priorities, Protective Measures
Day-to-day Planning
Emergency Preparedness Planning
Business Continuity Planning
Performance Expectations
Hiring Practices

Business Losses
Theft of Trade Secrets
Loss of Competitiveness
Shrinkage, Vandalism
Shelf Life, Spoilage
Damage to Critical Assets
Production Shutdown
Service Contamination
Business Continuity
Business Disruptions
Transport of WMD,
Contraband, Conveyance as a
Weapon, Human & Drug
Trafficking, Hijacking,
Smuggling, Piracy,
Stowaways, Hostage Taking
Disasters & Accidents
Work Place Violence
Bomb Threats
Demonstrations
Unidentified Materials
Specified Threats
Inefficiencies & Waste
Sudden Unexpected Layoffs
Insensitive Terminations
Ignorance or Indifference
Business Liability
Insurance Premiums
Exposure to Shareholders
Employees, Suppliers
Carriers, Customers
Unsafe Work Environment
Neighborhood Social Order
Ineffective Security Plan
Ineffective Safety Plan
Ineffective Parking Plan
Loose Visitor Controls
Personnel, Vehicle Screening
Qualifications & Training
Supervision

Integrated
Transformation
Solution

Action Plan

Reduce Vulnerability
Reduce Liability
Enhance Work Place
Enhance Best Practices

Strategies

Strategic Vision
Resources
Priority Initiatives
Milestones
Budget

Deliverables

Progress Reports
Schedules
Assessment Report
Presentations

Partner Organizations
Federal Agencies
State Agencies
Municipal Agencies
Trustees
Board of Governors/Directors
Commissions
Other Governing Bodies

Exhibit 13.1 *(continued)*

- **Time-sensitive cargo.** Just-in-time delivery of valuable cargo is essential for many businesses—any significant time delay in processing and transporting such cargo would negatively affect the U.S. economy.
- **Security versus convenience.** Maintaining security while limiting congestion and delays complicates the task of security and has important financial implications.
- **Accessibility.** Most airports are open to the public; their facilities are closed to public roadways for convenience and to streamline access for vehicles delivering passengers to terminals.

Another concern for the security-assessment team is the additional cost of increased security during sustained periods of heightened alert. Since September 11, 2001, airports across the country have in effect been working at surge capacity to meet the security requirements of the current threat environment.

Maritime-Traffic Challenges

The size, diversity, and complexity of this infrastructure make the inspection of all vessels and cargo that pass through our ports an extremely difficult undertaking.

Current inspection methods—both physical and technological—are limited and costly. As with other modes of transportation that cross international borders, the security-assessment team must balance the tension between efficient processing of cargo and passengers and adequate security. The biggest challenge is ensuring that legitimate cargo is not needlessly delayed.

Major portions of the maritime industry's operations are international in nature and governed by international agreements and multinational authorities such as the International Maritime Organization. Negotiation of maritime rules and practices with foreign governments lies within the purview of the Department of State. Often these international efforts involve extended negotiation timelines.

Much of the port system represents a significant protection challenge for the security-assessment team, particularly in the case of high-consequence cargo. The DOT and the DHS have been working together to develop physical and operational security guidelines for appropriate protective actions. The security-assessment team's solutions to increase the security of the maritime industry must also consider infrastructures subject to multi-agency jurisdictions and the international framework in which the industry operates.

Rail and Public-Transportation Challenges

Unlike airlines, where security checkpoints and screening equipment channel the movement of passengers and baggage, the rail and public-transportation system is designed to be easily accessible and is therefore harder to protect.

These systems often include confined spaces packed with people, luggage, and vehicles that are attractive targets for terrorists. The security-assessment team's security solutions to the container shipping challenge must recognize that in many cases commerce including essential national-security materials must continue to flow. **Stifling commerce to meet security needs simply swaps one consequence of a security threat for another.** In the event that a credible threat was to necessitate a shutdown, well-developed continuity of operations procedures can mitigate further unintentional negative consequences. For example, contingency planning can help determine how quickly commerce can be resumed; whether rerouting provides a measure of protection; or what specific shipments should be exempt from a shutdown, such as national-defense critical materials or perishable goods.

An additional area of concern for the security-assessment team is the marking of container cars to indicate the specific type of hazardous materials being transported. During an emergency response, placards on rail-cars help to alert first responders to hazardous materials they may encounter, but planners must take care to devise a system of markings that terrorists cannot easily decipher. This must be a national solution, not a local one.

Each city and region has a unique transit system, varying in size and design. No one security program or information-sharing mechanism will fit all systems. Despite these differences the security-assessment team, as a general rule, must apply basic planning factors that are relatively consistent from system to system.

Mass-transit systems were designed for openness and ease of public access, which makes the security-assessment team's task difficult in terms of deterring attacks over broad geographic areas. Solutions can result in significant financial commitments for owners and operators.

Trucking and Busing Challenges

Due to its heterogeneity in size and operations and the multitude of owners and operators nationwide, the trucking and busing infrastructure is highly resilient, flexible, and responsive to market demand.

For the same reasons, the sector is fractionated and regulated by multiple jurisdictions at state, federal, and sometimes local levels. The size and pervasive nature of the trucking and busing infrastructure pose significant protection challenges.

Transportation choke points [e.g., bridges and tunnels, intermodal terminals, border crossings, and highway interchanges] present unique protection challenges for the security-assessment team. Overall understanding of infrastructure choke points is limited and therefore common criteria for identifying them are difficult to establish. The security-assessment team must undertake a comprehensive, systematic effort to identify key assets, particularly those whose destruction or disruption would entail significant public health and safety consequences or economic impact.

Although many states have conducted risk assessments of their respective highway infrastructures, no true basis for comparison among them exists to determine relative criticality. Likewise, there is no coordinated mechanism for assessing choke-point vulnerabilities or conducting and evaluating risk-mitigation planning. A major reason for this lack of synchronization within the sector is a paucity of funds to promote communication among industry members and facilitate cooperation for joint protection-planning efforts. As a result, the sector as a whole has neither a coherent picture of industry-wide risks nor a set of appropriate security criteria on which to baseline its protection-planning efforts, such as which conditions constitute threats for the sector or standards for infrastructure protection or threat reduction. The sector's diverse and widely distributed constituency complicates this situation. The security-assessment team is left to explore new ground in establishing an initial baseline.

APPLYING THE S^3E SECURITY-ASSESSMENT METHODOLOGY TO THE TRANSPORTATION SECTOR

The security-assessment team should use its collective expertise to:

- Identify the important mission and functions of the transportation sector [**Task 1—Operational Environment**] and incorporate the development of all business operations and continuity expectations.

Areas of Responsibility	On Time Services
Business Incentives & Investment	Passenger & Cargo Safety
Corporate Image & Core Values	Performance Objectives
Mission & Services	Public Safety & Public Confidence

Exhibit 13.2 Identify Transportation Enterprise Mission Goals
& Objectives

General Public, Critical Customers	Medical, Firefighters
Government, Military	Regional & International
Industrial & Business	Size of Population Served
Law Enforcement	Transportation

Exhibit 13.3 Identify Transportation Sector Customer Base

Examination focuses on services and the customer base to determine the business contributions to the community, its culture, and business objectives. This includes but is not necessarily limited to the identification of goals and objectives, identification of customer base, identification of transportation commitments, and characterization of facilities and boundaries, as shown in Exhibits 13.2 through 13.5.

- Describe the transportation configuration, operations, and other elements [**Task 2—Critical Assessment**] by examining conditions, circumstances, and situations relative to safeguarding public health and safety and reducing the potential for disruption of services. Focus is on an integrated evaluation process to determine which assets need protection in order to minimize the impact of threat events. The security-assessment team identifies and prioritizes facilities, processes, and assets critical to transportation service objectives that might be subject to malevolent acts or disasters that could result in a potential series of undesired consequences. The process takes into account the impacts that could substantially disrupt the ability of the enterprise to provide safe transportation services and strives to reduce security risks associated with the consequences of significant events. This includes but is not necessarily limited to the assessment of facilities, assets, operations, processes, and logistics and prioritization of assets by importance to business operations, as shown in Exhibits 13.6 and 13.8.

- Review the existing design-basis threat profile [**Task 3—Threat Assessment**] and update it as applicable to the client. If a design-basis threat profile has not been established, then the security-assessment team needs to develop one. Both the security-assessment team and the client must have a clear understanding of undesired events and consequences that may impact transportation services to the community and business-continuance-planning goals. Emphasis is on prioritizing threats through: the process of identifying methods of attack, past or recent events, and types of disasters that might result

14 CFR 107 – Airport Security	Family Education Rights and Privacy Act 1974 [As Amended]
14 CFR 108 – Security Program	
14 CFR 109 – Indirect Air Carrier Security	Family and Medical Leave Act 1993 [As Amended]
14 CFR 129 – Operations: Foreign Air Carriers and Foreign Operators of U.S. Registered Aircraft Engaged in Common Carriage	
	Federal Polygraph Protection Act [As Amended]
19 CFR 4.30 – Security of Cargo in Unloading Areas	Federal Maritime Commission
19 CFR 19.40 – Container Stations and Control of Merchandise	Federal Privacy Act 1974 [As Amended]
	Freedom of Information Act [U.S.C. 552] [As Amended]
19 CFR 112 – Carriers, Cartmen and Lightermen	
29 CFR 1910.36 – Means of Egress	Foreign Corrupt Practices Act
29 CFR 1910.151 – Medical Services and First Aid	Immigration Control Act, 1986 [As Amended]
29 CFR 1910.155 – Fire Protection	Immigration and Nationality Act [As Amended]
49 CFR Part 85 – Cargo Security	International Atomic Emergency Agency
18 U.S.C. 659 – Theft of Goods [in] Interstate Commerce	National Labor Relations Act [29 U.S.C. 151] [As Amended]
18 U.S.C. 2511 – Technical Surveillance	Interagency Agreements
21 U.S.C. 801 – Drug Abuse and Control	Internal Revenue Service
26 U.S.C 7201 – Tax Matters	Labor Management Relations Act
29 U.S.C. 2601 – Family and Medical Leave Act	Memorandum of Agreement
31 U.S.C. 5322 – Financial Transaction Reports	Memorandum of Understanding
41 U.S.C. 51 – U.S. Contractor Kickbacks	Oil Pollution Act 1990 [As Amended]
41 U.S.C. 701 – Drug Free Workplace Act	Omnibus Crime Control and Safe Streets Act, 1968 [As Amended]
46 U.S.C. - 1903 – Maritime Drug Enforcement	
50 U.S.C. 1801 – Foreign Intelligence Surveillance Act	Policy Coordination Committee [PCC]
	Privacy Act [5 U.S.C. 522e] [As Amended]
Americans with Disabilities Act [42 U.S.C. 12101]	Rehabilitation Act, 1973 [As Amended]
Architectural Barrier Act 1968 [As Amended]	State and Local Industry Administrators and Authorities
Bioterrorism Preparedness and Response Act, 2002 [As Amended]	
	State and Local Emergency Planning Officials and Committees
Carriers, Suppliers, Vendors, Insurance	
Civil Aeronautics Board	Uniform Trade Secrets Act
Civil Rights Act 1964 [U.S.C. 200e] [As Amended]	U.S. Department of Commerce
Commerce Commission	U.S. Department of Defense [DOD]
Comprehensive Drug Abuse Prevention Act 1970 [As Amended]	U.S. DOD Military Traffic Management Command [MTMC]
Comprehensive Environmental Response, Compensation and Liability Act 1980 [As Amended]	U.S. Department of Energy [DOE]
	U.S. Department of Homeland Security [DHS]
Container Security Initiative [CSI] Program	U.S. Department of Justice [DOS]
Controlled Substance Act [As Amended]	U.S. Environmental Protection Agency [EPA]
Drug Free Workplace Act 1988 [As Amended]	U.S. Nuclear Regulatory Commission [NRC]
Emergency Planning & Right-to-Know Act 1986 [As Amended]	U.S. Occupational Safety and Health Administration [OSHA]
Employee Polygraph Protection Act 1988 [29 U.S.C. 2001] [As Amended]	U.S. Patriot Act, 2002 [As Amended]
	U.S. Postal System
Fair Credit Reporting Act [15 U.S.C. 1681] [As Amended]	U.S. Department of Transportation [DOT]
	World Health Organization [WHO]

Exhibit 13.4 Identify Transportation Commitments

Commercially Leased Facilities	Infrastructure Sharing
Dedicated site owned and operated by the enterprise	Neighboring Facilities
	Operated and maintained through reciprocal agreements with internal and external entities
Dispersed Locations	
Facility Land Use	Terrain & Climate
Facility Layout & Location	Types of assets at specific locations

Exhibit 13.5 Characterise Configuration of Transportation Enterprise Facilities and Boundaries

in significant consequences; the analyses of trends; and the assessment of the likelihood of an attack occurring. Focus is on threat characteristics, capabilities, and target attractiveness. Social-order demographics used to develop trend analysis in order to support framework decisions are also considered. Under this task the security-assessment team should also validate previous transportation-assessment studies and update conclusions and recommendations into a broader framework of strategic planning. Potential undesirable events include but are not necessarily limited to malevolent acts that could cause undesirable results, other disruptions to operations, types of perpetrators, impact of losses, and threat attractiveness, as shown in Exhibits 13.8 through 13.12.

- Examine and measure the effectiveness of the security system [**Task 4—Evaluate Program Effectiveness**]. The security-assessment team evaluates business practices, processes, methods, and existing protective measures to assess their level of effectiveness against vulnerability and risk and to identify the nature and criticality of business areas and assets of greatest concern to transportation operations. Focus is on vulnerability and vulnerable access points where penetration may be accomplished and how. Emphasis is placed on the physical characteristics of vulnerable points, accessibility to the location, and effectiveness of protocol obstacles an adversary would likely have to overcome to reach a critical transportation asset. These include but are not necessarily limited to an evaluation of existing operations, evaluation of existing security organization, evaluation of organization's relationship with partners, evaluation of SCADA, definition of adversary plan, and assessment of response and recovery plan, as described in Exhibits 13.13 through 13.18, evaluation of solutions and residual vulnerability,

Air operations area	Gates, taxi ramps, runways
Aircraft hangers	Ground support equipment parking areas
Aircraft repair and overhaul areas	Ground transportation parking
Aircrew waiting areas	Hi-jack, explosive ramp area
Airline support facilities	Hotel and Conference Center
Airport entrance	Import warehouses
Airport perimeter	Land/air side controls and operations
Airport personnel parking	Light power distribution systems
Arrival passenger waiting areas	Maintenance facilities
Banking and currency exchange	Manufacturing and shop areas
Baggage pickup areas	Passenger arrival areas
Baggage screening and loading areas	Passenger automobile parking
Cafeterias and restaurants	Passenger entrances
Cold storage	Passenger reception areas
Control tower	Passenger registry booths
Conveyance and compartments	Passenger terminal platforms
Departing passenger waiting areas	Private air fleet
Disembarking waiting areas	Retail stores and shops
Exhibition hall	Security facilities
Export cargo areas	Security operations area
Export warehousing and ramp areas	Special persons security
Flight simulator and training center	Telescoping gangways
Free trade zone area	Terminal building
Fuel tank farm	Vehicle maintenance areas
	VIP areas and ramp security
Rail Mode	
Security Check Points & Screening	Conveyance & Compartments
Rail Yards	Maintenance Yards
Maritime Mode	
Port/Sea Side Controls & Operations	Port Facility & Ship Security
Anchor Chain Collar, Pier/Anchorage	Beach/Shore Delivery Points
Meter Transfer Units, Temporary Storage	Boom & Netting Operations
Discharge Points, Distribution Stations	Vessel Stores, Bunkers & Reserve Supply
Truck and Busing Modes	
Tankers, Tractors	Trailer & Compartments
Staging	Fuel Locations
Safe Havens	Conveyance & Compartments

Exhibit 13.6 Critical Assessment of Transportation Facilities, Assets, Operations, Processes, and Logistics

Most critical facilities based on value system	Least critical facilities based on value system
Most critical assets based on value system	Least critical assets based on value system

Exhibit 13.7 Prioritize Critical Transportation Assets in Relative Importance to Business Operations

Armed attack	Misuse of wire and electronic communications
Arson	Peonage, slavery, and trafficking in persons including children
Assassination, kidnapping, assault	
Barricade/hostage incident	Phishing, hacking, computer crime
Bomb threats and bomb incidents	Piracy and Privateering
Chemical, Biological, Radiological [CBR] contamination of assets, supplies, equipment, and areas	Power failure
	Sabotage
	Rape and sodomy
Civil disorders, civil rights violations	Seamen and stowaways
Common carrier operation under the influence of alcohol or drugs	Sexual abuse, sexual exploitation
	Social engineering
Computer-based extortion	Software piracy
Computer system failures and viruses	Storage of criminal information on the system
Computer theft	Unlawful access to stored wire and electronic communications and transactional records access
Cyber attack on SCADA or other systems	
Cyber stalking, software piracy	Supplier, courier disruptions
Damage/destruction of interdependency systems resulting in production shutdown	Telecommunications failure
	Terrorism
Damage/destroy service capability	Torture
Damage/destruction of pipelines and cabling	Trafficking in contraband cigarettes and liquors, money laundering, cash smuggling, banned trading, and dangerous substances
Delay in processing transactions	
Email attacks	
Encryption breakdown	Trafficking in weapons of mass destruction and materials
Explosion	
Embezzlement, extortion and threats	Transportation for illegal sexual activity & related crimes
False impersonation, false statements	
FAX security	Theft or destruction of chemicals
Hacker attacks	Treason, sedition, and subversive activities
Hardware failure, data corruption	Trespassing and vandalism
HAZMAT spill	Unidentified materials or wrong deliveries
Hijacking and piracy	Unlawful access to stored communications
Importation, manufacture, distribution and storage of explosive materials	Unlawful wire and electronic communications interception and interception of oral communications
Inability to access system or database	
Inability to perform routine transactions	Use of cellular/cordless telephones to disrupt network communications or in support of other crimes
Intentional opening/closing of rail switching controls	
Intentional shutdown of electrical power	
Loss of key personnel	Voice mail system
Misuse/damage of control systems	Wide-scale disruption and disaster

Exhibit 13.8 Determine Types of Malevolent Acts That Could Reasonably Cause Transportation Undesirable Events

Bribery, graft, conflicts of interest	Maintenance activity
Conspiracy	Natural disasters
Customs violations	Obscenity
Eavesdropping	Obstruction of justice
Economic espionage censorship	Online pornography and pedophilia
Historical rate increases	Planned production shutdown or outage
Homicide	Racketeering
Human error	Racketeer influenced and corrupt
Industrial accidents and mishaps	organizations
Inefficiencies and waste	Release and misuse of personal
Ineffective hiring practices	information
Ineffective personnel and vehicle controls	Resource constraints and competing
Ineffective safety and security plans	priorities
Ineffective supervision and training	Robbery and burglary
Ineffective use of resources and time	Shrinkage, shelf-life, spoilage
Ineffective visitor controls	Sudden unexpected layoffs
Ignorance and indifference	Tracking effects of outage overtime
Insensitive terminations	Tracking effects of outage across related
Institutional corruption, fraud, waste,	resources and dependent systems
abuse	Theft of trade secrets, intellectual
Loss of competitiveness	propriety, confidential information,
Loss of proprietary information	patents, and copyright infringements
Mail fraud	Unauthorized transfer of American
	technology to rogue foreign states

Exhibit 13.9 Assess Other Disruptions Impacting
Transportation Operations

Activist groups or cults	Professional criminals
Disenfranchised individuals	State-sponsored, independent terrorist
Insiders	groups
Lone wolves	Vandals
Organized crime	

Exhibit 13.10 Identify Category of Transportation Perpetrators

Compromise of public confidence	Inability to transfer to secondary power
Dealing with unanticipated crisis	Local civil unrest
Disabling of or loss of key personnel	Loss of energy sources or ability to sustain productivity
Duration of loss of emergency services	
Duration of loss of recovery & rebuilding	Loss of life [customers and employees]
Economic revenue loss	Loss or disruption of communications
Illness [customers and employees]	Loss or disruption of information
	Number of critical customers impacted
Impact on regional economic base	Number of users impacted
Impact on utility ratepayers	Production shutdown
Inability of customers to conduct business	Slowdown and interruption of delivery services
Inability to deliver and distribute fuel	
Inability to store or transfer oil and gas	Temporary closure of financial institutions

Exhibit 13.11 Assess Initial Impact of Loss Consequences for the Transportation Enterprise

Consequence analysis	Social demographics
Economic conditions	Social/political status

Exhibit 13.12 Assess Initial Likelihood of Threat Attractiveness and Malevolent Acts of Occurrence for the Transportation Enterprise

Alert Notification System	Reporting Security Incidents
Emergency Preparedness Planning	Roles & Responsibilities
Functional Interfaces	Security Awareness
Operations, Processes, Protocols	Security Plans, Policies & Procedures

Exhibit 13.13 Evaluate Existing Transportation Security Operations and Protocols $[P_{E1}]$

Dependency Programs	Qualifications & Training
Management Team	Staffing, Skills, Experience
Operational Capabilities	Supervision
Organization & Composition	

Exhibit 13.14 Evaluate Existing Transportation Security Organization $[P_{E1}]$

Agency for Toxic Substances and Disease Registry	National Transportation Safety Board [NTSB]
American Electronics Association [AEA]	Office of Civilian Radioactive Waste Management
American Institute of Certified Public Accountants	Office of National Drug Control Policy
American National Standards Institute	Policy Coordination Committee [PCC]
American Psychological Association	Private Security Advisory Council
American Society for Industrial Security	Regional Information Sharing System [RISS]:
American Society for Testing and Materials	Middle Atlantic-Great Lakes Organized Crime Law
American Society for Testing Materials, Vaults	Enforcement Network [MAGLOCLEN]
Centers for Disease Control and Prevention [CDC]	Mid-States Organized Crime Information Center
Commerce Commission	[MOCIC]
Computer Security Institute	New England State Police Information Network
Critical Infrastructure Protection Board [PCIPB]	[NESPIN]
Civil Aeronautics Board	Regional Organized Crime Information Center
Civil Aviation Organization [ICAO]	[ROCIC]
Defense Central Investigation Index [DCII]	Rocky Mountain Information Network [RMIN]
El Paso Intelligence Center [EPIC]	Western States Information Network [WSN]
EPA National Security Research Center	Safe Manufactures National Association
EPA Office of Air and Radiation	The Terrorism Research Center
EPA Office of Prevention, Pesticides and Toxic	Transportation Information Sharing and Analysis
Substances	Centers [ISAC]
EPA Office of Solid Waste and Emergency Response	U.S. Central Intelligence Agency [CIA]
Factory Mutual Research Corporation	U.S. Coast Guard
Federal, state, and local agency responders	U.S. Customs
Illuminating Engineers Society of North America	U.S. Department of Agriculture
Institute for a Drug Free Workplace	U.S. Department of Commerce [DOC]
Institute of Electrical and Electronic Engineers [IEEE]	U.S. Department of Defense [DOD]
International Association of Chiefs of Police	U.S. DOD Military Traffic Management Command
International Association of Professional Security	[MTMC]
Consultants	U.S. Department of Energy [DOE]
International Computer Security Association	U.S. Department of Homeland Security [DHS]
International Criminal Police Organization [INTERPOL]	U.S. Department of Homeland Security Immigration
International Electronics Supply Group [IESG]	and Customs Enforcement [ICE]
International Maritime Organization [IMO]	U.S. Department of Justice [DOJ]
International Parking Institute	U.S. Department of State [DOS]
International Organization for Standardization [ISO]	U.S. Department of Transportation [DOT]
Interstate Commerce Commission [ICC]	U.S. Department of Transportation Federal Highway
Joint Maritime Information Element [JIMIE]	Administration
Joint Terrorism Task Force [JTTF]	U.S. DOT Security Administration
Law Enforcement Support Center [LESC]	U.S. Drug Enforcement Agency [DEA]
Library of Congress, Congressional Research Service	U.S. Environmental Protection Agency [EPA]
Medical Practitioners and Medical Support Personnel	U.S. Federal Bureau of Investigation [FBI]
Multi-lateral Expert Groups	U.S. Federal Railroad Administration [FRA]
National Institute of Standards and Technology [NIST]	U.S. Food and Drug Administration [FDA]
National Insurance Crime Bureau [NICB] Online	U.S. Government Accounting Office [GAO]
National Labor Relations Board	U.S. Postal System
National Parking Association	U.S. Nuclear Regulatory Commission [NRC]
National Safety Council	U.S. Treasury Department
National White Collar Crime Center [NWCCC]	Underwriters Laboratories [UL]
Occupational Safety and Health Review Commission	World Health Organization [WHO]
[OSHRC]	

Exhibit 13.15 Evaluate Transportation Interface and Relationship with Partner Organizations [P_{E1}]

96/24 Hour Notification Rules	Modem & Internet Access
Assessment Capability	Performance Standards & Vulnerabilities
Automated ID Systems	Personnel & Vehicle Access Control Points
Badge Controls	Physical Protection System Features
Barrier/Delay Systems	Post Orders & Security Procedures
Chemical & Other Vendor Deliveries	Radiation Portal Monitoring
Communications Capability	Response Capability
Cyber Intrusions, Firewalls, Other Protection Features	Security Awareness
Display & Annunciation Capability	Security Training/Exercises/Drills
Evacuation Response Plans	"Sensors, RFID tags, GPS Tracking"
Intrusion Detection Capability	System Capabilities & Expansion Options
Lock & Key Controls, Fencing & Lighting	System Configuration & Operating Data

Exhibit 13.16 Evaluate Existing Transportation SCADA and Security Performance Levels [P_{E1}]

Deception actions	Likely approaches and escape routes
Delay-penetration obstacles	Target selection and alternative targets
Interception analysis	Time-delay response analysis

Exhibit 13.17 Define the Transportation Adversary Plan, Distractions, Sequence of Interruptions, and Path Analysis

Containment	Integrated Responders
Closeout Actions	Production Shutdown
Divert Traffic	Tactics & Techniques
First & Second Responders	

Exhibit 13.18 Assess Effectiveness of Transportation Response and Recovery [P_{E1}]

Business Continuity Planning	Loss Consequences
Day-to-Day Planning	Performance Expectations
Emergency Preparedness Planning	Rank Order Facilities & Assets
Hiring Practices	Vulnerability Analysis

Exhibit 13.19 Analyze Effectiveness of the Transportation Enterprise Security Strategies and Operations [P_{E1}]

Compromise of public confidence	Inability to transfer to secondary power
Dealing with unanticipated crisis	Local civil unrest
Disabling of or loss of key personnel	Loss of energy sources or ability to sustain productivity
Duration of loss of emergency services	
Duration of loss of recovery & rebuilding	Loss of life [customers and employees]
Economic revenue loss	Loss or disruption of communications
Illness [customers and employees]	Loss or disruption of information
	Number of critical customers impacted
Impact on regional economic base	Number of users impacted
Impact on utility ratepayers	Production shutdown
Inability of customers to conduct business	Slowdown and interruption of delivery services
Inability to deliver and distribute fuel	
Inability to store or transfer oil and gas	Temporary closure of financial institutions

Exhibit 13.20 Refine Previous Analysis of Transportation Enterprise Undesirable Consequences That Can Affect Functions

Consequence analysis	Social demographics
Economic conditions	Social/political status

Exhibit 13.21 Refine Previous Analysis of Transportation Enterprise Threat Attractiveness and Likelihood of Malevolent Acts of Occurrence

Creation and/or revision or modification of sound business practices and security policies, plans, protocols, and procedures.	Development, revision, or modification of interdependency requirements.
Creation and/or revision or modification of emergency operation plans including dependency support requirements.	SCADA/security system upgrades to improve detection and assessment capabilities.

Exhibit 13.22 Analyze Selection of Specific Risk-Reduction Actions Against Current Risks, and Develop Prioritized Plan for Transportation Enterprise Mitigation Solutions [P_{E1}]

Reasonable and prudent mitigation options	Mirrors business culture image

Exhibit 13.23 Develop Short- and Long-Term Transportation Enterprise Mitigation Solutions

Effectiveness calculations	Residual vulnerability consequences

Exhibit 13.24 Evaluate Effectiveness of Developed Transportation Enterprise Mitigation Solutions and Residual Vulnerability [P_{E2}]

Practical, cost-effective recommendations	Reasonable return on investments

Exhibit 13.25 Develop Cost Estimate for Short- and Long-Term Transportation Enterprise Mitigation Solutions

- Recognize the importance of bringing together the right balance of people, information, facilities, operations, processes, and systems [**Task 5—Program Analyses**] to deliver a cost-effective solution. The security-assessment team develops an integrated approach to measurably reduce security risk by reducing vulnerabilities and consequences through a combination of workable solutions that bring real, tangible enterprise value to the table. Emphasis is placed on assessing the current status and level of protection provided against desired protective measures to reasonably counter the threat. Operational, technical, and financial constraints are examined as well as future expansion plans and constraints. These include but are not necessarily limited to analysis of security-operation effectiveness, refinement of consequences analysis, refinement of threat-attractiveness analysis, analysis of risk-reduction and mitigation solutions, development of short- and long-term mitigation solutions, and development of cost estimates, as described in Exhibits 13.19 through 13.25.

PREPARING THE TRANSPORTATION SECURITY-ASSESSMENT REPORT

Chapter 10 provides a recommended approach to reporting the security-assessment results.

U.S. GOVERNMENT TRANSPORTATION-SECTOR INITIATIVES

Appendix B summarizes some significant initiatives taken or currently underway by the U.S. Department of Homeland Security and industry to close the vulnerability gaps within the transportation sector.

REFERENCES

1. The National Strategy for the Physical Protection of Critical Infrastructure and Key Assets (February 2003).
2. Ibid.
3. Ibid.
4. U.S. Government Accounting Office (December 2005).
6. Ibid.
5. Ibid.
6. Ibid.
7. Ibid.
8. Ibid.
9. Ibid.
10. Ibid.
11. Ibid.
12. Ibid.
13. Ibid.
14. Ibid.
15. U.S. Government Accounting Office (December 2005)
16. Council of Foreign Relations (June 2004).
17. U.S. Government Accounting Office (December 2005).
18. Ibid.
19. Koknar, Ali M., *Terrorism on the High Seas* (Security Management, June 2004).
20. Ibid.
21. U.S. Government Accounting Office (December 2005).
22. Ibid.

APPENDIX A: A HISTORICAL OVERVIEW OF SELECTED TERRORIST ATTACKS, CRIMINAL INCIDENTS, AND INDUSTRY MISHAPS WITHIN THE TRANSPORTATION SECTOR

The author has compiled the information in this appendix in a chronlogical sequence to share the historical perspective with the reader interested in conducting further research. The listing, along with the main text of Chapter 2, presents a capsule review of terrorist activity and other criminal acts perpetrated around the world and offers security practitioners specializing in particular regions of the world a quick reference to range and level of threat.

Source: *The information is a consolidated listing of events, activities, and news stories as reported by the U.S. Department of Homeland Security, U.S. Department of State, and various other government agencies. Contributions from newspaper articles and news media reports are also included. They are provided for a better understanding of the scope of terrorism and other criminal activity and further research for the interested reader.*

20 NOVEMBER 2006 Frankfurt, Germany.
Police arrested 6 people attempting to sneak a bomb aboard a Frankfurt/Tel Aviv flight.

19 NOVEMBER 2006 Detroit, MI, United States.
Sisayethiticha Dinssa of Dallas Texas was arrested at Detroit Airport when $79,000 in cash, cyanide poison, and suspicious information on nuclear materials were found on his laptop computer. He is an unemployed U.S. citizen of African origin.

4 OCTOBER 2006 India.
Thirteen Indian airports were placed on high alert after the Central Industrial Security Force [CISF] received specific intelligence of a terrorist plan to hijack an airplane. Bomb-disposal staff and Special Forces were stationed at several airports.

3 OCTOBER 2006 Brindisi, Italy.
Two Turkish men hijacked a plane flying from Albania to Istanbul. The men forced the plane, which was carrying 133 passengers, to land in Brindisi, Italy, and surrendered to police, requesting political asylum.

25 SEPTEMBER 2006 Sri Lanka.
The Sri Lankan navy killed up to 70 Tamil Tiger rebels when 25 boats were attacked off the eastern coast. At least eight boats were sunk after the five-hour sea battle.

23 SEPTEMBER 2006 Nepal.

A WWF helicopter crashed in eastern Nepal, killing all 24 people onboard. The helicopter was carrying a group of conservationists, a Finnish diplomat, two Russian crew members, a U.S. aid worker, as well as Nepalese officials and reporters.

22 SEPTEMBER 2006 Lathen, Germany.

Twenty-three people were killed and 10 were injured when a high-speed magnetic train collided with a maintenance vehicle. The train was traveling at nearly 200 kilometers per hour.

11 JULY 2006 Bombay, India.

Seven bombs were simultaneously detonated on a crowded train. The death toll was 174; 493 were injured.

17 MARCH 2006 Ingushetia, Russia.

Two explosions damaged a railway line and destroyed a mobile-phone base station in the southern Russian province, which neighbors Chechnya. There were no victims in either incident.

17 AUGUST 2005 Baghdad, Iraq.

Three bombs ripped through a city bus station and nearby hospital emergency room. Over 40 people were killed.

11 AUGUST 2005 Oklahoma, United States.

A man, unaffiliated with any terrorist group, was detained at the city's airport after trying to carry a homemade bomb onto an aircraft.

AUGUST 2005 United States.

Federal cargo-inspection system found wanting. A system used by the Department of Homeland Security to help inspectors identify high-risk cargo coming into U.S. seaports needs improvement in order to better screen for weapons of mass destruction. In a summary report, the Department of Homeland Security's inspector general found deficiencies in an inspection system used by the Customs and Border Protection Bureau [CBP]. Called the Automated Targeting System, it is used by CBP inspectors at domestic and foreign ports to help identify high-risk cargo containers for inspection. About nine million containers arrive annually at U.S. seaports, making it impossible to physically inspect each of them without hampering the flow of commerce.

4 AUGUST 2005 Michigan, United States.

Security reacts after freighter skips check-in. Two countries, seven federal and local agencies, the U.S. Department of Homeland Security, and at least one dog reacted when a freighter motored past a checkpoint in Lake Erie. Sailing an Antiguan flag, with a Russian and Lithuanian crew and an American pilot at the wheel, the freighter, named *Jana,* failed to radio a standard message to traffic control on the river. Over the next 16 hours, the mystery freighter raised a spectrum of possible worries when it did not call in its location and sestinations to Canadian or U.S. marine officials. When the *Jana* neared another checkpoint, this one on the Detroit River, both U.S. and Canadian officials were monitoring the ship. Only this time, the ship identified itself. As the *Jana* entered U.S. waters, it was escorted by the Coast Guard to a dock in Detroit. The *Jana* was then boarded by U.S. Homeland Security agents, customs and border-protection officials, and a border-patrol dog trained to sniff out drugs and people.

21 JULY 2005 London, United Kingdom.
Terrorists failed in a new spate of suicide-bomb attacks similar to those of 7 July 2005. Three explosive devices were found on the Underground and one on a bus. All devices failed to explode, and CCTV images of all the men involved were obtained.

17 JULY 2005 Kusadasi, Turkey.
A bomb on a minibus leaving a popular Turkish resort killed five people. Authorities believed the PKK Kurd separatist group might have been responsible.

12 JULY 2005 Warsaw, Poland.
The Polish capital's sole underground train line was closed and evacuated following a bomb threat, but a search yielded no devices.

7 JULY 2005 London, United Kingdom.
Terrorists, most likely linked to al-Qaeda, struck London's transport system at the end of the morning rush hour with four bombs on the Underground trains and one bus. Fifty-two people were killed and up to 700 injured.

2 JULY 2005 Ankara, Turkey.
Kurd rebels detonated a bomb on a train in eastern Turkey. Six security guards were killed and 15 people were wounded.

30 JUNE 2005 Mogadishu, Somalia.
A UN aid ship carrying supplies for tsunami victims in Somalia was hijacked by pirates off the Horn of Africa. The pirate demanded a ransom sum of $500,000 and threatened to sink the ship.

10 JUNE 2005 Moscow, Russia.
A bomb derailed a train en route from Chechnya to Moscow. Five casualties were reported.

9 JUNE 2005 Zaragoza, Spain.
An explosion near this Spanish city's airport followed a warning by ETA.

7 JUNE 2005 London, United Kingdom.
An unclaimed mobile phone on a jet from Portugal to Birmingham resulted in the aircraft being diverted under military escort to Stansted airport.

6 JUNE 2005 Kathmandu, Nepal.
Maoist rebels detonated a mine in southern Nepal, which blew up a bus as it passed through Madi. At least 36 passengers were killed. An army convoy had been the planned target.

27 APRIL 2005 Seattle, WA, United States.
The would-be millennium bomber who crossed the border from Canada with a trunkload of explosive materials to blow up LAX provided information on more than 100 suspected terrorists, helped shut down clandestine al Qaeda cells, and exposed valuable organizational secrets of the global terrorist network.

27 APRIL 2005 Kuala Lumpur, Malaysia.
Armed pirates hijacked an Indonesian-owned ship bound for Singapore, ferried it to port in Malaysia, and unloaded its cargo of tin ingots before escaping to Indonesia.

27 APRIL 2005 Newark, NJ, United States.
A British man accused of trying to smuggle shoulder-held missiles into the U.S. and offering to obtain a radioactive "dirty bomb" for terrorists was found guilty in U.S. District Court on all five of the counts he faced. The London man was arrested in a sting operation involving an 18-month collaboration among officials in the U.S., Russia, and Britain. Two other men indicted in the same case for their roles as financial middlemen pleaded guilty to transferring money illegally.

22 APRIL 2005 London, United Kingdom.
British "shoe bomber" accomplice jailed for 13 years. A British man who admitted to conspiring with "shoe bomber" Richard Reid to blow up airliners simultaneously over the Atlantic—but had a change of heart before boarding his flight—was jailed for 13 years. Saajid Badat, 25,

entered a surprise guilty plea. Reid failed in his bid to blow up an American Airlines plane flying from Paris to Miami on December 22, 2001, after passengers and crew overpowered him as he tried to ignite explosives in his shoe. Badat confessed to an identical plot, planned for a flight from Amsterdam to the United States. Reid was sentenced to life imprisonment by a U.S. court in January 2003.

22 APRIL 2005 Moscow, Russia.

Russia blames Chechen sisters for suicide bombings. Two female suicide bombers behind two of the most deadly militant attacks to hit Russia—the Beslan school massacre and a passenger-jet bombing a week earlier—were members of the same family and probably sisters, Russian prosecutors revealed. Roza Nagayeva, 30, was identified by DNA tests as one of two suicide bombers among up to 50 militants who took 1,200 people hostage at a school in Beslan, North Ossetia, in September 2004. She detonated explosives strapped to her chest at the start of the standoff, which ended when special forces stormed the school. In all, 330 people died, half of them schoolchildren, in the fighting that followed. Prosecutors established that she was related to Amnat Nagayeva, one of two suicide bombers behind near-simultaneous explosions on board two passenger planes on August 24, which killed 95 people. She was aboard a Tupolev 134 heading from Moscow's Domodedovo airport to Volgograd with 35 passengers and eight crew members. The second plane crashed six minutes later, killing its 38 passengers and eight crewmembers. Officials have said that Satsita Dzherbikhanova, 37, from Chechen detonated a bomb on board. DNA established their identities with a probability of "over 90 percent" but held out the possibility that they could be cousins. The two Nagayevas and Dzherbikhanova disappeared at the same time in August 2004. Caucasus families often avenge their relatives' deaths, and the fact that two of the women were sisters may help to explain their motivation. Nagayeva's brother, Uvais, 32, was killed by Russian troops in May 2001. A human-rights group recorded that he was dragged from his family home with a friend. They were later forced to lie down on gravestones in a nearby cemetery and shot. The friend was killed outright, but Nagayev feigned death and returned home. Six days later he was again taken from his home by Russian troops but did not return.

21 APRIL 2005 United States and Canada.

Each year the American economy loses $4.13 billion in economic potential because of clogged border crossings into Canada, with the greatest losses suffered by the Great Lakes states. The economic loss to the U.S. is conservatively measured at $471,461 per hour—each and every hour of the day. There is also a corresponding loss of potential employment—particularly in the manufacturing and transportation industries. The study found that Michigan and Ohio bear the brunt of the economic loss, closely followed by New York and Pennsylvania. There are over 5.2 million jobs in the United States that are dependent on trade with Canada.

19 MARCH 2005 United States.

Keeping the nation's ferries safe. While there have been no reported threats to a ferry in the United States, the Federal Bureau of Investigation reported at least seven incidents last year involving surveillance of ships in Washington state. Coast Guard officials say nearly 400 passengers would be likely to die if a large ferry were attacked, more than twice the number of deaths expected from an airplane crash. Officials worry that ferries may be attacked because they often carry cars and large trucks that could hide bombs, they run on a schedule, and they are screened less intensely than airplanes. There have been attacks on ferries elsewhere: a 1,050-passenger ferry sank in the Philippines in February 2004 after a bomb, consisting of eight pounds of TNT packed into a television, killed more than 115 people. More than 700 ferries operate nationally, carrying 175 million passengers a year.

14 MARCH 2005 Herat, Afghanistan.

Land mines killed three and wounded eight traveling on a bus in Afghan's Western Province. It is unknown whether the mine was Taliban-inspired or a "leftover."

26 JANUARY 2005 Glendale, CA, United States.

A commuter-train collision was caused by a suicidal man in Glendale, CA. The man had parked his SUV on the tracks with the intent of taking his own life, but got out of the vehicle before impact and watched the collision, causing the death of eleven people and injuring at least 120 others.

25 JANUARY 2005 Texas, United States.

U.S. officials intercepted a small plane and arrested four illegal Chinese immigrants it was transporting from Mexico to San Antonio, TX. Law-enforcement agents took the passengers, two men and two women, and the Mexican pilot into custody when the plane landed at San Antonio's Stinson Field. The four had valid Chinese passports, and none of the names appeared on the government's database of terrorist suspects.

24 JANUARY 2005 New York City, NY, United States.

Two of New York City's subway lines were crippled after a fire in a lower Manhattan transit-control room that was started by a homeless person trying to keep warm. The lines may not return to normal capacity for three to five years, officials said. The blaze, at the Chambers Street station used by the A and C lines, was described as doing the worst damage to subway infrastructure since the terrorist attack of September 11, 2001. It gutted a locked room that is no larger than a kitchen but that contains some 600 relays, switches, and circuits that transmit vital information about train locations. Long waits and erratic service are likely to be the norm on the two lines, which have a combined ridership of 580,000 each weekday. The fire underscored the fragility of the antiquated equipment that keeps the subways moving and of the sensitive nodes where that equipment is stored. Officials said they believed that there were only two companies in the world that were able to repair the signals. One is based in Pittsburgh, and the other in Paris.

20 JANUARY 2005 Canary Islands.
A boat being towed between the Canaries and Morocco carrying illegal immigrants bound for Spain was cast adrift and lost. At least 10 Africans died.

11 JANUARY 2005 United States.
The Federal Railroad Administration [FRA] issued a safety advisory to all of the nation's railroads to strengthen procedures for monitoring track-switching operations. The advisory was in response to recent incidents involving trains that derailed because switches that divert them from one track to another were left in the wrong position. The safety advisory stated that railroads should document when a manually operated switch in nonsignaled territory is changed from the main track to a siding and returned back to the normal position for main track movements.

11 JANUARY 2005 South Carolina, United States.
Electronic signals could have warned the Norfolk Southern train— carrying toxic chlorine in Graniteville, SC, on January 6—in time to prevent it from slamming into a parked train, according to several railroad experts. The crash site—like about 40 percent of the nation's main tracks—is in a "dark territory" without signals. Without signals, the two-man crew of the 42-car train involved in the wreck that left nine dead likely had no time to stop. The only warning for the crew of the moving train probably was a reflector disk at the intersection of the main and side tracks, where the manual switch is located. Most of the nation's freight, as measured in volume, moves through areas with signals, and "dark territory" areas aren't necessarily more dangerous if other regulations are followed, an FRA spokesman said.

8 JANUARY 2005 Bieber, CA, United States.
The Federal Railroad Administration said that an error similar to the one suspected as the cause of a chlorine tank disaster in Graniteville, SC, may have caused a derailment days later in California. In the second accident, seven locomotives and 14 cars were derailed, two workers were injured, and damages exceeded $970,000.

29 DECEMBER 2004 Los Angeles, CA, United States.
The Port of Los Angeles became the top U.S. international freight gateway by shipment value in 2003, according to a report from the U.S. Department of Transportation's Bureau of Transportation Statistics. Los Angeles' water port handled $17 billion in export trade and $105 billion in imports, totaling $122 billion in business. Los Angeles handled $10 billion more than the $112 billion in freight that moved through JFK International Airport, now the second-ranked international freight gateway in 2003.

24 DECEMBER 2004 Honduras.
Officials reported 28 bus commuters were massacred in the industrial hub of San Pedro Sula.

12 DECEMBER 2004 Dubai.
Dubai Ports Customs and Free Zone Corporation, the sixth largest port operator in the world, committed to target, prescreen, and secure cargo destined for the U.S. Dubai Ports Customs and Free Zone Corporation joined the U.S. Customs and Border Protection's [CBP] Container Security Initiative [CSI], making it the first Middle Eastern port to participate.

7 OCTOBER 2004 Tunis, Tunisia.
A boat carrying over 70 illegal immigrants to Italy sank with the known loss of 22 people, but many more remained missing.

5 OCTOBER 2004 Larnaca, Cyprus.
A Lufthansa 747 flight was diverted to Cyprus en route to Israel from Frankfurt after receiving a bomb threat. Passengers and crew were evacuated, but no devices were found in the subsequent search.

28 SEPTEMBER 2004 Oslo, Norway.
An Algerian asylum seeker on a local Norwegian commuter flight attacked the pilot and passengers with an axe as the aircraft was about to land. Passengers overpowered the man. The pilot was slightly injured.

26 SEPTEMBER 2004 Stansted, United Kingdom.
A Greek Olympic Airways Airbus 340 made an emergency landing at London's designated emergency airport after a Greek newspaper received three calls saying aircraft would be destroyed in revenge for events in Iraq. A full search of the aircraft yielded nothing.

31 AUGUST 2004 Moscow, Russia.
A suicide bomber attacked a metro station, killing 10 and injuring 51. An Islamic group supporting the Chechen government claimed responsibility.

27 AUGUST 2004 New York City, NY, United States.
The FBI arrested two in a plot to blow up the subway station at Herald Square, a block from where the Republican National Convention will be held; the Verrazano Bridge, which connects Brooklyn and Staten Island; and two other New York City subway stations.

26 AUGUST 2004 Istanbul, Turkey.
A Turk who claimed to have explosives in his mobile phone caused an Istanbul airplane to make an emergency landing in Munich.

24 AUGUST 2004 Moscow, Russia.
Terrorists blew up two commercial airlines within minutes of each other in an attempt to disrupt country elections. Eighty-nine persons were killed. One aircraft departed Moscow for Volgograd in south Russia and the other for Sochi on the Black Sea. Both acts were the result of "black widow" suicide bombers who had bypassed or bribed security to get on board.

24 AUGUST 2004 Athens, Greece.
Police arrested a Greek man as he boarded an international flight carrying a loaded gun in his luggage.

22 AUGUST 2004 Dhaka, Bangladesh.
A train was set on fire east of the Bangladeshi capital, injuring 20. The act was apparently in retaliation for the deaths of 20 in an explosion at an opposition rally.

19 AUGUST 2004 New York City, NY, United States.
A government contractor carried highly explosive Soviet munitions from Afghanistan that were not detected until he arrived at New York's John F. Kennedy airport. Shaun Marshall, a medic for defense contractor

DynCorp, arrived from the United Arab Emirates. He was trying to board a United Airlines flight home to California when he was pulled aside for a routine security check. A search of his bags by federal screeners found what police bomb technicians described as a Soviet "projectile point detonating fuse" and a "surface-to-air and air-to-air cartridge." Marshall also had five .50-caliber bullets and four small arms cartridges, which he did not declare to United as required by law. Marshall, 26, told officers he was importing the munitions, which he believed to be inert, for use in DynCorp training exercises, and was released. The FBI arrested Marshall at his Riverside, CA, home after the bomb squad analyzed the munitions and DynCorp officials said Marshall had no involvement in its training operations.

16 AUGUST 2004 New York City, NY, United States.
A Pakistani gave American officials what they regard as credible and specific information indicating that al Qaeda is considering using tourist helicopters in terror attacks in New York City. As a result, security measures for helicopter operators in New York City are being stepped up in a new directive. Among the new measures under review is a requirement for operators to conduct airport-style screenings of passengers for suspicious items.

5 AUGUST 2004 Washington, DC, United States.
Port and border drayage-truck drivers could pose a security risk that is difficult to manage, a Department of Homeland Security [DHS] official told reporters and transportation officials. A spokesperson said that while the DHS was looking at possible vulnerabilities in the entire supply chain, independent drayage drivers posed a unique risk.

4 AUGUST 2004 United States.
Few measures exist to avert truck bombs, experts say. The next terror attacks in the United States, experts say, could very well involve a mixture of two ingredients that are ubiquitous and hard to control: big vehicles and explosive materials as mundane as gasoline or fertilizer. In the chemical and trucking industries, the offices of law-enforcement agencies, and the ranks of antiterrorism experts inside and outside government, there is a widespread sense that little can be done to keep potential terrorists away from trucks, buses, and the materials that can turn them into weapons. Explosive substances like ammonium-nitrate fertilizers are so widely used and so easily stolen and stockpiled that any restrictions imposed now would have little effect. In July 2004, the fertilizer industry, in concert with the Bureau of Alcohol, Tobacco, Firearms and

Explosives, began urging sellers of ammonium nitrate—which has been the main ingredient of the bombs used in at least half a dozen major terror attacks here and abroad—to track sales and require buyers to show identification. Only Nevada and South Carolina have passed laws requiring such tracking. Congress has never pursued similar requirements at the federal level, partly because lawmakers from agricultural states have said it would inconvenience farmers without deterring terrorists.

26 JULY 2004 California, United States.

Severe cargo congestion and labor shortages at American seaports are creating long delays in delivering goods and potential threats to national security, dockworkers and security experts say. The problems are particularly acute at the ports of Los Angeles and Long Beach, the nation's busiest, handling roughly a third of the nine million cargo containers that arrive in the United States each year. A union official said facilities and work crews there could not keep up with the volume of incoming freight, and that as a result some port regulations from the Department of Homeland Security were not being followed. Security and intelligence experts have identified the nation's 361 seaports and the 60,000 mostly foreign-flagged ships that sail in and out of them each year as prime targets for a potential terrorist attack. But ships and seaports have received only a small fraction of the attention given the aviation system since the 9/11 terrorist attacks.

21 July 2004 Atlanta, GA, United States.

An airline passenger wearing only a pair of pajama bottoms stole a baggage tractor at the city's main airport and drove it onto an active runway. Robert W. Buzzell, 31, walked out an exit door that had an alarm at Hartsfield-Jackson Atlanta International Airport. Flights were not affected by the incident, which took place before 6 A.M., airport officials said. The man was stopped by mechanics, who asked him for an employee-identification card. When he could not provide one, they escorted him to an office and called police. Authorities said the man appeared mentally unstable. Buzzell, who had a ticket for a Delta flight, was jailed on charges of unlawful interference with security, theft by taking, and reckless conduct.

20 JULY 2004 Boston, MA, United States.

The U.S. Coast Guard, state police, and other agencies stepped up security patrols in Boston Harbor to prepare for the Democratic National Convention. The Coast Guard is coordinating the overall effort that stretches from the Fleet Center to the federal courthouse to Logan

International Airport on the other side of the harbor. The goal was to keep most of the harbor open for the public but not to be business as usual.

20 JULY 2004 Athens, Greece.

Athens' main port was sealed to allow divers to install an underwater monitoring system as part of an Olympic security network that protesters contended is a privacy invasion. The fiber-optic cables were a key element of an electronic web of cameras, sensors, and other intelligence-gathering devices designed to help safeguard the August 13–29 Olympics. Greece spent $1.24 billion on Olympic security. Police officials reassured Athenians that the electronic monitoring system, which includes a 200-foot blimp [with chemical "sniffers" and high-resolution cameras] and thousands of infrared and high-resolution cameras, would not violate privacy.

19 JULY 2004 Britain, United Kingdom.

British authorities found secret documents abandoned at a petrol station outlining counterterrorist attacks against aircraft and Heathrow International Airport. Police said the documents, drawn up by the Metropolitan Police's air-security section, were discovered by a motorist and handed into the *Sun* newspaper. The incident occurred a day after the biggest anti-terrorism exercise Britain had ever seen was branded a failure when it took three hours to begin decontaminating people supposedly affected by toxic gas.

7 JULY 2004 Yemen.

Six Yemenis were charged in the planning of the October 2000 bombing of the *U.S.S. Cole* and said they belonged to Osama bin Laden's terrorist network. Among the six charged was mastermind Abd al-Rahim al-Nashiri, accused of planning and funding the attack and training the cell members who carried it out. U.S. officials believed the Saudi-born al-Nashiri is a close associate of bin Laden. In addition to the *Cole* attack, he is suspected of helping direct the 1998 bombings of U.S. embassies in Kenya and Tanzania. Court charges included forming an armed gang to carry out criminal acts against the interests of the state, resisting authorities, and forging documents.

6 JULY 2004 Alaska, United States.

Alaska Marine Highway System [AMHS] incorporated National Maritime Security Plan, requiring increased port, harbor, and vessel security. The U.S. Department of Homeland Security required some 3200 port facilities, 9500 vessels, and 40 offshore oil and natural-gas rigs to comply with new requirements under the Maritime Transportation

Security Act and the International Ship and Port Facility Security Code by the July 1, 2004 deadline. New security measures will be physical improvements such as better lighting in terminals and parking lots, closed-circuit cameras on car decks, and checking in firearms.

6 JULY 2004 United States.

The United States denied entry to 42 foreign ships and detained 38 in port since July 1, under tough new United Nations security rules designed to thwart terrorist attacks. The foreign vessels detained or denied entry failed to comply with the requirements of the International Ship and Port Facility Security Code [ISPS], U.N. regulations that came into force on July 1. Washington, fearing an attack or infiltration by al Qaeda from the sea, has vowed to police the new rules strictly by turning away ships that are not security-certified or delaying ones that have called at "contaminated ports." The regulations, signed by 147 governments, require ports, stevedoring companies, and owners of ships larger than 500 tons to draw up plans for responding to a terror threat, implement tighter security around facilities, and train staff.

2 JULY 2004 United States.

U.S. ports remain vulnerable to the kind of speedboat attack that crippled the *USS Cole* and killed 17 sailors in Yemen, Coast Guard officials said. U.S. officials have warned al Qaeda favors such attacks, and that though port security has improved vastly since the attacks of September 11, 2001, it would be very difficult to intercept a small boat loaded with explosives and on a suicide mission. Among the Coast Guard's new capabilities is intelligence coordination. It checks every visiting foreign-ship crewmember's background against multiple national-security databases. It has an intelligence coordination center at the Office of Navy Intelligence and a center that tracks when oceangoing vessels enter U.S. ports.

4 JULY 2004 California, United States.

After two years of intense focus on airline security, the government turned to mass-transit systems and Amtrak passenger trains. Americans take more than 11 million trips a day by bus, train, and subway, compared with 1.8 million by air. Yet Washington has spent only about half a cent for each rider on ground-transit security since the September 11, 2001, attacks, compared with more than $9 for each airline passenger, according to Congressional estimates. Current standards call for each transit system to develop its own security plan in consultation with federal authorities. Many systems, including the California Bay Area's BART system, have launched publicity campaigns to encourage commuters to report abandoned or suspicious bags and parcels. Metal trash cans, which could serve as hiding places for bombs, have been removed or replaced with hardened containers that can direct a blast upwards and away from patrons, and many systems now have dog teams that can detect explosives.

1 JULY 2004 Hong Kong and Singapore.

New rules aimed at protecting shipping from terrorists went into effect on July 1, 2004, and the world's two busiest ports in Hong Kong and Singapore reported no early snags. Under the rules—backed by the United Nations and International Maritime Organization—port facilities, owners of ships larger than 500 tons and the companies that unload them must make detailed plans to prepare for terrorist threats. Ships must also have a security officer, alarm system, automatic identification system and a method of checking IDs of people who board. They must also restrict access to the engine room and bridge. According to Coast Guard statistics, implementing the security plan in the United States will cost $7.3 billion.

1 JULY 2004 United States.

U.S. Coast Guard began boarding all foreign-flagged vessels sailing into U.S. ports to enforce new International Maritime Organization certification standards.

1 JULY 2004 United States.

The Department of Homeland Security found 29 truck drivers licensed to carry hazardous materials with possible ties to terrorist organizations while conducting routine background investigations.

29 JUNE 2004 Iowa, United States.

Iowa to check for radioactive cargo. Drive-through radiation-detection equipment will be installed at five weigh stations on Interstates 35 and 80 later this year to look for radioactive cargo in heavy trucks. Iowa will become one of the first states in the country to routinely check trucks for the illegal cargo as part of an effort to stop terrorists from smuggling bomb-making materials or stolen nuclear weapons. Law-enforcement officers who work at the weigh stations will be given hand-held devices that can check for explosives.

29 JUNE 2004 Singapore.

Indonesia, Malaysia, and Singapore formed a joint antipirate naval task force to patrol the notorious Malacca Straits. Each country will provide seven vessels with crews of 100. In 2003, pirate attacks were estimated at in excess of 440.

29 June 2004 International Ports.
The International Council of Cruise Lines [ICCL] announced that all 118 vessels of its member lines are 100 percent compliant with the International Ship and Port Facility Security [ISPS] Code. The announcement came two days before the July 1 deadline for all ships and port facilities worldwide. The ISPS Code is a set of measures that enhance security of ships and port facilities globally, adopted by the International Maritime Organization [IMO] as an amendment to Safety of Life at Sea [SOLAS] regulations. Key security elements contained in the ISPS Code include security plans, screening measures, access control, waterside security, and communications between ships and ports.

29 June 2004 California, United States.
Unlike the rest of the nation, train-related injuries and deaths in California went up 15 percent in the last three years. Eighty-five deaths and 50 injuries statewide were reported in 2003. Nationally, the Federal Railroad Administration reported that 324 people died in crashes at crossings, and more than 500 died after being hit by trains in 2003. The Federal Railroad Administration also estimated that a person is hit by a train every three hours on one of the nation's 252,000 highway-rail crossings. Although trains are massive and highly visible, people still take chances at the tracks. A freight train traveling 60 mph can take as long as a mile and a half to stop.

28 June 2004 Istanbul, Turkey.
Citing security at Istanbul airport, Israel's El Al suspended flights to Turkey during the NATO summit. El Al Israel Airlines halted the flights due to changes in security arrangements at Ataturk Airport. It gave no details but said the flights were suspended because of Turkey's refusal to uphold agreed security arrangements. Turkey is a top destination for Israeli tourists and businesspeople. Hundreds of passengers were stranded in Istanbul. El Al officials denied the move was linked to a recent bus bombing in Istanbul, in which four people were killed ahead of the NATO summit on June 28–29. Israeli media reported Turkey aimed to limit the number of El Al security guards posted at the airport, contrary to what El Al said was previously agreed. Some El Al officials called for the halting of Turkish Air flights to Israel.

26 June 2004 California, United States.
State's ports seek more security cash. California's Bay Area ports and commercial ship owners have largely complied with a mandate to create new waterway security measures in the wake of 9/11, though funding has

arrived at an extremely slow speed. The U.S. Coast Guard said all 87 waterway facilities, along with all 109 Bay Area registered vessels, met the deadline, though some are operating under alternative plans.

25 JUNE 2004 Piraeus, Greece.

Greece signed a deal with the U.S allowing American customs officers to carry out security scans on commercial shipping containers leaving the country's largest seaport for the U.S. The statement said that under the "declaration of principles" signed in Brussels, with which Greece joined the Container Security Initiative [CSI], Greek and U.S. customs officials will identify and screen cargo containers destined for the U.S. from the port of Piraeus. Greece has the biggest merchant fleet in the EU, and one of the biggest in the world

25 JUNE 2004 Arizona, United States.

The U.S. Department of Homeland Security [DHS] announced the first sustained civilian use of unmanned aerial vehicles [UAVs] to curb illegal activities along Arizona's southern border. Two Hermes 450 UAVs will be used as part of the Arizona Border Control [ABC] Initiative to assist with border surveillance activities and augments manned aircraft, helicopters, and ground sensors already in place. The UAV flights will be controlled and monitored by U.S. Customs and Border Protection's [CBP] border patrol and are scheduled through the summer of 2004. The UAVs are equipped with electro-optic sensors and communications payloads, which provide around-the-clock images to CBP border-patrol agents. These aerial vehicles permit greater border coverage and quicker response times in the rugged, desolate areas of the Southwest border.

25 JUNE 2004 United States.

Concern is growing that world trade may be seriously disrupted July 1 when new ship security rules come into force well before many ships and most world ports have complied. When the IMO published figures on compliance levels this week, only 37 percent of the world's large merchant vessels and 16 percent of ports met the standards, agreed in December 2002. The U.S. Coast Guard refused to soften its insistence on applying the rules in full from the start.

24 JUNE 2004 Istanbul, Turkey.

Three people died and 20 were injured when a bomb exploded on a bus in the Turkish city ahead of a major NATO weekend conference.

23 June 2004 Tampa, FL, United States.

Coast Guard to implement Operation Port Shield. On July 1, 2004, under the International Ship and Port Facility Security [ISPS] Code and the U.S. Maritime Transportation Security Act [MTSA], a suite of international and U.S. maritime-security requirements will take effect. Operation Port Shield is a focused effort to validate U.S. vessel and port-facility compliance with MTSA and ISPS as well as foreign vessels calling on U.S. ports. In the Tampa Captain of the Port area of responsibility, 75 U.S. vessels and 80 port facilities were required to submit security plans for review no later than December 31, 2003. All foreign vessels subject to the ISPS requirements will also be checked for their compliance. Those failing to comply will not be allowed to enter U.S. ports. Marine Inspectors from Marine Safety Office Tampa verify approximately 7 to 10 foreign vessel arrivals per week.

23 June 2004 Medford, OR, United States.

Federal Transportation Security Administration [TSA] officials placed the Medford, OR, airport on a probationary status after the recent firing of two security employees who slept on the job and of a third who was arrested for the theft of a laptop computer from the lost and found, where it was stored after a passenger misplaced it. There was also a security lapse the weekend before; while passengers were boarding a plane, a passenger reportedly walked out of the line heading for a plane, went over to the fence, and may have been handed something from the other side. With fines for security breaches at some airports amounting to as much as $250,000, the airport notified TSA officials immediately about the problem.

10 June 2004 Kitsap Peninsula, WA, United States.

Potential incendiary device removed from near a Washington state ferry terminal. The device was a motorcycle-type battery about half the size of a car battery with wires leading into a brown bottle about eight ounces in size. The bottle was less than half full of an unknown liquid. A bomb squad rendered the device safe, and it was removed for further investigation, including identification of the liquid. A passenger getting off the ferry on the Kitsap Peninsula noticed a 12-foot aluminum skiff tied to the dock at the terminal about 9:45 P.M. and alerted a crewmember, who took a closer look and spotted the device. A crewmember untied the small boat from the dock, and it drifted onto a nearby beach and was tied to a piece of driftwood by the time the first state troopers arrived. Ownership of the skiff was under investigation. No written threats or warning notes were found, nor did investigators know of any earlier verbal threats, security problems, or other sign of trouble at the ferry terminal.

2 JUNE 2004 Philadelphia, PA, and Boston, MA, United States.

Airliners about to take off from Philadelphia and Boston were searched after authorities received a telephoned bomb threat against the two American Airlines flights. Philadelphia International Airport officials evacuated 19 passengers who were rescreened while the plane was searched. Nothing was found, but the flight to Boston was canceled anyway. Later, in Boston, a connecting flight to London's Heathrow Airport was searched at Logan International Airport as it was about to take off. Nothing was found, and the flight was allowed to take off with 159 passengers.

27 MAY 2004 United States.

High oil prices threaten the recovery of the global airline industry, which is emerging from the crisis caused by 2003's SARS epidemic, the Iraq war, and the events of September 11, 2001, said the head of the International Air Transport Association.

24 MAY 2004 San Francisco, CA, United States.

Security experts warn San Francisco's Bay Area Rapid Transit [BART] and other mass-transit systems need more than vigilance to defend against bombings. Since the attack on the railroad in Spain, BART has come under fire for slow responses to unattended bags. In one case, a bag was reported in San Francisco but traveled through the critical underwater Tran-bay Tube and into downtown Oakland before BART police inspected it. In another case, BART called the Alameda County bomb squad to a suspicious case in San Leandro station two hours after discovering it.

3 APRIL 2004 Madrid, Spain.

Six suspected militants who were being hunted in connection with the 11 March Madrid train bombings blew themselves up in their apartment block as police prepared to close in on the building. One policeman also died.

2 APRIL 2004 Madrid, Spain.

A terrorist plan to blow up a high-speed train on the Madrid-Seville line during the Easter holidays was foiled by chance when a rail worker found a 12-kilogram bomb under the track, 60 kilometers south of Madrid.

24 MARCH 2004 Troyes, France.
Police found a second rail-side bomb, probably planted by the AZF group, which had made threats to blow up the French railways if demands for money were not met. The device was found on the Paris-Basel [Suisse] railway line 100 miles south of Pars.

17 MARCH 2004 Ashdod Port, Israel.
Authorities discovered an empty container that originated from Gaza with a false side. Inside fragmentation grenades, clothes, food, and bedding were found. Authorities theorized suicide terrorists who had killed 10 Israelis within the port may have been smuggled from Gaza in the container.

14 MARCH 2004 Ashdod, Israel.
Two suicide bombers blew themselves up at an Israeli seaport. The bombings halted a peace meeting between Israeli and Palestinian leaders. Hamas and the al Aqsa Martyrs Brigades, militants with links to Palestinian leader Yasser Arafat's Fatah party, claimed joint responsibility for the attacks.

11 MARCH 2004 Madrid, Spain.
Several explosions tore through Atocha railway station and two smaller stations at rush hour, killing 199 people. Mobile telephones were almost certainly used to achieve remote synchronization. While Basque separatist group ETA was initially blamed, accumulated evidence pointed to a Moroccan Islamic fundamentalist group. Days after the incident the cell leader and three cell members blew themselves up as police stormed their apartment. One special-forces agent and 15 police officers were killed. The rail attacks occurred three days before the general elections when Spaniards voted the pro-Iraq-war government out of office. A short time later, Spain withdrew its military forces from Iraq.

3 MARCH 2004 Paris, France.
A gang called AZF [which may take its name from a Toulouse chemical factory, Azote de France, that was destroyed in an explosion in 2001] threatened to blow up sections of track on France's SNCF state railway network unless the government handed over millions of Euros. Thousands of miles of French railway were checked. Officials ruled out links with Islamic or Chechen extremists.

FEBRUARY 2004 Manila Bay, Philippines.
A bomb exploded on a ferry, killing 27 and injuring dozens; 100 passengers went missing. Authorities believed Abu Sayyaf, who had been threatening ferry operators since October 2003, was responsible.

FEBRUARY 2004 Florida, United States.
A U.S. airline employee had seven uniforms stolen from her car. Other items of value in the car were untouched.

22 FEBRUARY 2004 Jerusalem, Israel.
A Palestinian suicide bomber killed eight commuters on an Israeli city bus. The incident happened on the eve of an international court hearing about the legality of the security barrier in the West Bank that Israel sees as a key to halting such attacks.

20 FEBRUARY 2004 New Delhi, India.
Police of Jammu and Indian-controlled Kashmir prevented a suicide attack on New Delhi's International Airport as they arrested three Pakistani militants. Before this, Indian troops shot dead six Islamic rebels in the first clash in the province since India and Pakistan agreed to start a peace process.

6 FEBRUARY 2004 Moscow, Russia.
A suicide bomber blew himself up on the Russian Metro during rush hour, killing at least 39 and wounding scores of others. Chechen activists were suspected.

JANUARY 2004 Oklahoma City, OK, United States.
Authorities reported the credentials of an FAA employee were stolen from the individual's car. Credentials could allow the bearer access to airplane cockpits.

JANUARY 2004 Jakarta and Sumatra, Indonesia.
Gunmen believed to be affiliated with GAM hijacked a tanker off Aceh Province. The tanker was carrying 1,000 tons of palm oil from South Africa to Indonesia. A few weeks later, the gunmen shot and killed four crewmembers when their $10,000 ransom demand was not fully met by the ship owner.

30 JANUARY 2004 London, United Kingdom.
Further BA flights to the U.S. and Riyadh were delayed or cancelled after persisting U.S. intelligence reports that al Qaeda-inspired attacks on transatlantic flights were imminent.

14 JANUARY 2004 London, United Kingdom.
A Sudanese man boarded a Virgin flight from Washington to Dubai with five rounds of live ammunition. He was arrested while checking in on a transfer flight at Heathrow.

13 JANUARY 2004 Tashkent, Uzbekistan.
A domestic airliner crashed on approach to the airport in Uzbekistan's capital, killing all 37 aboard.

2003 Guatemala.

Five men armed with automatic weapons fired on a bus transporting U.S. tourists from Utah's Church of Jesus Christ of Latter Day Saints on their way from an archaeological site back to the Mexican border. One passenger was killed. The gunmen led the other tourists off into the forest, where they were forced to the ground and then robbed.

28 DECEMBER 2003 London, United Kingdom.

As security-threat levels were raised during the Christmas holiday in the U.S. and U.K., the British government considered the introduction of air marshals to aircraft flying to the U.S. Airlines and pilots were hostile to the proposals.

26 DECEMBER 2003 Benin, Lebanon.

A Lebanese chartered aircraft en route to Beirut, with mainly Lebanese on board, crashed into the Atlantic shortly after takeoff from West Africa. At least 140 died.

24 DECEMBER 2003 Madrid, Spain.

Spanish police foiled an ETA attempt to blow up the city's main railway station. A man arrested at the scene was found to be carrying 28 kilograms of explosives.

5 DECEMBER 2003 Chechnya, Russia.

At least 44 people were killed when a bomb tore apart a busy commuter train near Russia's rebel Chechnya region. More than 150 people were injured in the blast.

NOVEMBER 2003 Istanbul, Turkey.

Terrorists packed a pickup truck with 30 bags containing 50 kilograms of ammonium-nitrate-fuel-oil slurry each and 10 five-kilogram boxes of explosives connected to a detonator controlled by the suicide driver. The truck was sent from Istanbul to the Turkish Mediterranean port of Antalya, some 500 miles away, where the driver waited for eight days for an Israeli cruise ship to dock. When the cruise ship failed to call at the port of Antalya due to inclement weather, the truck and its suicide driver returned to Istanbul, where the driver attacked the British Consulate General with his lethal 1.5 ton cargo, killing the British Consul General and 17 other people, while injuring hundreds of others and demolishing part of the consulate building.

NOVEMBER 2003, Instanbul, Turkey.

Terrorists affiliated with the Kurdish Hezbollah, who launched suicide truck-bomb attacks against two Turkish synagogues and British targets in

Istanbul in November 2003, killing more than 60 people, had originally cased the southern Turkish port of Antalya, where an Israeli cruise ship was expected to dock. The terrorists had packed a pickup truck with 30 bags containing 50 kilo-grams of ammonium-nitrate-fuel-oil slurry each and 10 five-kilogram boxes of explosives connected to a detonator controlled by the suicide driver. In November 2003, the truck was sent from Istanbul to the Turkish Mediterranean port of Antalya, some 500 miles away, where the driver waited for eight days for an Israeli cruise ship to dock. When the cruise ship failed to call at the port of Antalya due to inclement weather, the truck and its suicide driver returned to Istanbul, where the driver attacked the British Consulate General with his lethal 1.5-ton cargo, killing the British Consul General and 17 other people while injuring hundreds of others and demolishing part of the consulate building.

OCTOBER 2003 New York City, NY, United States.
Authorities reported an airline employee's NY/NJ Port Authority ID cards stolen from her car.

26 OCTOBER 2003 Samba, Kashmir.
A bomb exploded in the toilet of a coach car, causing no injuries but derailing five cars.

SEPTEMBER 2003 Illinois, United States.
A TSA employee reported a uniform, identification card, and badge stolen from his truck.

AUGUST 2003, Malacca Straits.
In the Malacca Straits, gunmen took control of the Malaysian-flagged tanker, *Penrider,* carrying 1,000 tons of fuel oil from Singapore to Penang. The attackers, whom Malaysian police believe were affiliated with GAM, demanded a ransom of $100,000 against the ship's crew of nine but later released them after the ship owner paid $50,000.

19 AUGUST 2003 Jerusalem, Israel.
A suicide bomber riding a bus detonated his explosives, killing 20 persons—five of them U.S. citizens—and injuring 140 others. Hamas claimed responsibility.

JULY 2003 Oklahoma, OK, United States.
Authorities reported a U.S. Air Force member stationed at Tinker Air Force Base had several uniforms and military equipment stolen from his off-base residence.

11 JUNE 2003 Jerusalem, Israel.
Terrorists killed two Americans in a bus bombing near Klal Center on Jaffa Road. Hamas was responsible.

7 JUNE 2003 Kabul, Afghanistan.
In Kabul, a taxi rigged with explosives rammed into a bus carrying German peacekeepers of the International Security Assistance Force, killing five German peacekeepers and injuring 29 others. The U.S.-funded police school, located about 300 feet from the explosion, lost 13 windows. Authorities suspected al Qaeda.

19 MAY 2003 Jeddah, Saudi Arabia.
Officials arrested three Moroccans as members of al Qaeda at Jeddah International Airport. Found in their baggage were knives and documents that resembled a last will and testament. Officials believed they were trying to hijack the plane and fly it into a building in Jeddah.

19 MAY 2003 Srinagar, Kashmir.
Two bombs exploded at Kashmir's busiest bus terminal, injuring 14 persons.

12 MAY 2003 Nashville, TN, United States.
At Nashville International Airport a male passenger was taken into custody after screeners discovered a .357 revolver in his carry-on bag. Five of the six bullets in the handgun were coated with Teflon, which can penetrate many bullet-resistant vests worn by law-enforcement officers. The man was later identified as a convicted murder.

8 MAY 2003 Hawaii, United States.
A man with a handgun entered Lihue Airport through the Hawaiian Airlines baggage claim area, making his way past the security checkpoint and into the secure area, where he fired two shots and then continued down the concourse, ordering people out. He then sat down with the gun pointed at his head before police convinced him to surrender.

5 MAY 2003 Doda, Kashmir.
A bomb exploded at a bus stand, killing one person and injuring 25 others.

24 APRIL 2003 Kefar Saba, Israel.
A security guard of dual Israeli-Russian citizenship was killed and 11 others were wounded when a lone suicide bomber blew himself up at the entrance to a busy train station. The al-Aqsa Martyrs Brigade claimed responsibility.

12 APRIL 2003 Qazigund-Anantnag, Kashmir.
Terrorists threw a hand grenade into a bus station, killing one person and injuring 20 others.

2 APRIL 2003 Davoa, Philippines.
A bomb exploded on a crowded passenger wharf, killing 16 persons and injuring 55 others. The attack may have been carried out by two Indonesian members of JI, a regional terrorist group with links to al Qaeda. The two individuals arrested told investigators they also were involved in the Davao Airport bombing, and JI provided funds. Several Indonesian members of JI have been spotted in terrorist-training camps on the southern island of Mindanao.

MARCH 2003 Italy.
Terrorists clashed with Italian Railway Police, resulting in the death of one terrorist and the death of an Italian security officer. BR/PCC was responsible.

31 MARCH 2003 Havana, Cuba.
A man armed with two hand grenades hijacked a domestic airliner with 46 passengers and crew onboard in an attempt to reach the U.S. After an emergency landing at Havana airport due to insufficient fuel, the plane remained on the runway all night. The next day, more than 20 passengers left the aircraft apparently unharmed. With at least 25 passengers on board, the hijacked plane departed Havana airport and safely landed in Key West, Florida.

13 MARCH 2003 Rajouri, Kashmir.
A bomb exploded on a bus parked at a terminal, killing four persons.

12 MARCH 2003 Miami, FL, United States.
Miguel Rodriquez and his brother Roberto plead guilty to the 1980 hijacking of a Delta Air Lines flight to Havana.

4 MARCH 2003 Davao, Philippines.
A bomb hidden in a backpack exploded in a crowded airline terminal, killing 21 persons [including one U.S. citizen] and injuring 146 others [including three U.S. citizens].

5 MARCH 2003 Haifa, Israel.
A suicide bomber boarded a bus and detonated an explosive device, killing 15 persons—including one U.S. citizen—and wounding 40 others. Hamas claimed responsibility.

2 MARCH 2003 London, United Kingdom.
Three men armed with hammers and crowbars hijacked a truck loaded with computer microchips from the United Airlines warehouse. The truck, flown in from Miami, was valued at $11 million.

1 MARCH 2003 Limerick City, Ireland.
Nearly 300 antiwar protestors demonstrated outside Shannon Airport. Ten attempted to breach the perimeter fence and were arrested by police.

24 FEBRUARY 2003 Limerick City, Ireland.
Officials arrested three antiwar protesters on a roadway beside Shannon Airport in possession of camera equipment and maps detailing key positions inside the perimeter fence.

22 FEBRUARY 2003 Istanbul, Turkey.
Two unidentified persons threw a bomb into a British Airways office, shattering windows. No casualties were reported. KGK was suspected.

20 FEBRUARY 2003 Davao Airport, Philippines.
A car bomb exploded outside a restaurant about 100 yards from the Davao Airport on Mindanao, killing 21 people and injuring 117. Officials believed three Arab men were behind the car bomb, apparently timed for the arrival of Defense Secretary Angelo T. Reyes.

19 FEBRUARY 2003 Rockhampton, Australia.
A man hijacked a Cessna 210 from Hedlow Airfield and forced the pilot to fly to Mackay Airport.

14 FEBRUARY 2003 Miami, FL, United States.
Officials detained a Japanese tourist at Miami International Airport when they discovered stamps on his passport from Somalia, Yemen, Saudi Arabia, Ethiopia, Zimbabwe, and Pakistan. Officials found an 11-ounce metal canister containing gasoline, two boxes of matches, and a barbecue grill.

13 FEBRUARY 2003 London, United Kingdom.
A Venezuelan national arriving on a British Airways flight from Caracas to Gatwick Airport presented a false passport and was discovered to have a live grenade in his baggage. The man, of Bangladeshi origin, was arrested under the Terrorism Act. The discovery caused the closure of the North Terminal for approximately two hours.

13 FEBRUARY 2003 Bogota, Colombia.

A Southern Command-owned airplane carrying five crew and five passengers [four U.S. citizens and one Colombian] crashed in the jungle. All five passengers survived the crash; two of the crewmembers were injured. Terrorists later killed a Colombian army officer and a U.S. citizen, while three other U.S. citizens were missing. FARC claimed to be holding the three missing persons. There were upward of 4,500 individuals involved in nonstop, U.S.-financed search efforts. On 22 April, press reports identified the three missing as civilians doing drug surveillance for the Department of Defense.

11 FEBRUARY 2003 Tribuvan Airport, Nepal.

A U.S. citizen was found in possession of $494,000 and large sums of other foreign currencies while attempting to leave the country.

7 FEBRUARY 2003 Honolulu, HI, United States.

Officials detected traces of explosives on luggage at Honolulu International Airport, but the bag and owner were allowed to leave the checkpoint before TSA agents were aware of the suspicious contents. Several thousand passengers were evacuated from the main and inter-island terminals and allowed to return after bomb-detecting dogs finished their search. Neither the person nor the bag was found. A large number of flights were also delayed, and several were cancelled.

7 FEBRUARY 2003 Istanbul, Turkey.

A Turkish male took two flight attendants hostage at Istanbul International Airport after other passengers exited the aircraft. Police stormed the plane and arrested the hijacker.

7 FEBRUARY 2003 Budapest, Hungary.

Border guards detained a man from Lebanon with a false passport at Ferihegy Airport. While being processed, the man set fire to furniture in an effort to escape and avoid deportation.

5 FEBRUARY 2003 Pakistan.

Officials offloaded two Pakistani residents believed to be al Qaeda members from a Qatar airlines flight bound for Doha out of Lahore International Airport.

2 FEBRUARY 2003 Fuzhou, China.

A male passenger on an Air China flight from Beijing to Fuzhou Airport pulled out a soda can filled with an unidentifiable type of fuel and began pouring the liquid down the aisle, setting several seats on fire.

29 JANUARY 2003 Limerick City, Ireland.
Women used hatchet-type hammers to significantly damage a U.S. Navy C-40 at Shannon Airport.

24 JANUARY 2003 Chengdu Airport, China.
A passenger on board a Sichuan Airlines flight ordered pilots to alter flight course. When the pilots refused, the man detonated a self-made bomb. The hijacker sustained minor injuries. No other injuries or damage to the plane were reported.

16 JANUARY 2003 London, United Kingdom.
Immigration officials arrested three men in transit at Gatwick Airport under antiterrorism legislation.

5 JANUARY 2003 Kulgam, Kashmir.
A hand grenade exploded at a bus station injuring 40 persons, 36 private citizens and four security personnel. No one claimed responsibility.

5 JANUARY 2003 Tel Aviv, Israel.
Two suicide bombers attacked simultaneously, killing 23 persons including 15 Israelis, two Romanians, one Ghanaian, one Bulgarian, three Chinese, and one Ukrainian and wounding 107 other nationalities not specified. The attack took place near the old central bus station where foreign national workers live. The detonations took place within seconds of each other and were approximately 600 feet apart, in a pedestrian mall and in front of a bus stop. The al-Aqsa Martyrs Brigade was responsible.

3 JANUARY 2003 Sinai Peninsula, Egypt.
An Egyptian Boeing 737 crashed into the Red Sea shortly after take-off from Sharm el-Sheikh, killing 148, mainly French, holidaymakers. Terrorism was not suspected.

28 NOVEMBER 2002 Mombasa, Kenya.
Two shoulder-fired missiles fired at an Israeli chartered Arkia Airlines Boeing 757 near Mombasa's airport. Al Qaeda, Government of Universal Palestine in Exile, and the Army of Palestine were responsible.

16 NOVEMBER 2002 Tel Aviv, Israel.
A man attempted to hijack an El Al Israel Airlines flight from Tel Aviv to Istanbul, Turkey, and crashed it into a building in Tel Aviv.

OCTOBER 2002 Yemen.
An al Qaeda affiliate, the Aden-Abyan Islamic Group in Yemen, claimed responsibility for ramming a French super-tanker, the *M/V Limburg,* using

a small boat loaded with explosives in an attack similar to that used against the *USS Cole*. The organizer of the *Limburg* attack was Abd al Rahman al Nashiri.

15 OCTOBER 2002 Khartoum, Sudan.
A Saudi man attempted to hijack a plan from Khartoum to Jeddah, Saudi Arabia, armed with a pistol.

AUGUST 2002 Miami, FL, United States.
A man reporting to the Immigration and Naturalization Service under an alias was discovered to be responsible for the hijacking in 1980 of a Delta Air Lines flight from New Orleans to Havana.

JUNE 2002 Piraeus, Greece.
Authorities uncovered a plot to bomb the Port of Piraeus and arrested members of 17 November.

MAY 2002 Karachi, Pakistan.
A French shuttle bus exploded in two suicide car bombings. LJ members were involved.

JANUARY 2002 Rivolto Air Base, Italy.
Italian police thwarted an attempt by four NTA members to enter the Rivolto Military Air Base.

JANUARY 2002 Horn of Africa.
The Puntland Coast Guard seized a Lebanese-flagged cargo carrier named *Princess Sarah*, along with its crew of 18, and demanded $200,000 in ransom. They held the *Princess Sarah* for two weeks before releasing the ship and its crew after being paid $50,000.

22 DECEMBER 2001 Miami, FL, United States.
An airline flight attendant spotted a passenger, Richard Reid, attempting to ignite a shoe bomb on a transatlantic flight from Paris to Miami. Al Qaeda was responsible.

2 DECEMBER 2001 Haifa, Israel.
A Hamas suicide bomber blew up a bus to avenge the death of a Hamas member, killing 15 and injuring 40.

11 SEPTEMBER 2001 New York City, NY, Pennsylvania, Washington, DC, United States.
World Trade Center and Pentagon attacks. A hijacked commercial airliner struck the North Tower of World Trade Center at 8:45 A.M. Shortly after 9:00 A.M. another hijacked aircraft crashed into the South Tower.

Both 110-story towers collapsed; the South Tower at about 10:00 and the North Tower at about 10:30. Other buildings in the 16-acre site were seriously damaged, including building Seven of the World Trade Center complex, which collapsed later in the evening. Approximately 3,000 people were killed in the incident and about 7,000 were injured. When the towers collapsed, hundreds of responders, including the top leadership of the Fire Department of New York City, were inside. About 450 responders were killed, including 23 from the New York City Police Department, 343 from the Fire Department, and 74 from the Port Authority of New York and New Jersey. Approximately 320 emergency responders were treated for injuries or illnesses at five nearby hospitals; others were threatened at temporary triage stations. At about 9:40 a hijacked airliner crashed into the western side of the Pentagon, killing 125 people on the ground as well as 64 people aboard the plane. Area hospitals treated 88 injured people. The crash damaged or destroyed three of the five interior concentric rings of the Pentagon building. The section where the plane hit had been recently renovated, and many offices were empty or being used for storage at the time. At about 10:03 United Airlines Flight 93 crashed in southwestern Pennsylvania, brought down by passengers trying to retake the hijacked plane. The plane was believed to be heading for the Capitol. Al Qaeda was responsible for all attacks.

JANUARY 2001 India.
Terrorists attacked Srinagar Airport, killing five Indians along with six militants, and a police station, killing at least eight officers and wounding several others. LT claimed responsibility

2001 United States.
Planning for 2001 World Trade Center attacks. "Muscle" hijackers arrived in the U.S. Pilots took cross-country surveillance flights on jets like the ones they later flew. In addition, 19 tickets were purchased between August 25 and September 5.

1999 Los Angeles, CA, United States.
A plot was uncovered to set off a bomb at Los Angeles International Airport during the millennial celebrations. Al Qaeda was responsible.

24 DECEMBER 1999 India.
An Indian airliner with 155 aboard is hijacked, resulting in the release of Masood Azhar, an important leader in the former Harakat ul-Ansar, imprisoned by the Indians in 1994, and Ahmed Omar Sheik, who was convicted of the abduction and murder in 2002 of U.S. journalist Daniel Pearl. HUM and HUA were responsible.

19 DECEMBER 1999 Washington, United States and Canadian border.
An Algerian national was arrested at the Canadian border in Washington as he attempted to smuggle sophisticated explosive materials into the U.S. He was connected to an Algerian terrorist-cell group located in Montreal, believed to have ties to Osama bin Laden. Two other members of the group were arrested as they attempted to illegally enter Vermont from Canada.

14 DECEMBER 1999 Port Angeles, WA, United States.
U.S customs agents found explosives in the trunk of a vehicle driven by Ahmed Ressam as he was attempting to drive across the Canadian-U.S. border. The plot to blow up Los Angeles Airport was later thwarted.

1999 Planning for 2001 World Trade Center Attacks.
Authorities reported Osama bin Laden gave Khalid Shaikh Mohammed go-ahead for "planes operation." Mohammad began collecting information and training materials for the operation.

26 DECEMBER 1998 Angola.
United Nations officials reported that a transport plane carrying 10 U.N. officials and four crewmembers was shot down over an area of intense fighting between UNITA rebels and government troops. National Radio Services stated that UNITA shot down the plane. A U.N. rescue team arrived at the crash site on 8 January 1999, reporting that no one survived the crash and that the bodies of all 14 persons aboard the plane were accounted for.

9 DECEMBER 1998 Sanaa, Yemen.
Yemeni passengers on a chartered Egyptian airliner demanded to be flown to Libya. The Egyptian pilot landed the plane in Tunisia and told the 150 passengers he could not fly the plane to Libya due to the U.N. sanctions. The plane and passengers remained on the ground for 15 hours before returning to Yemen.

9 DECEMBER 1998 Bandipura, Kashmir.
Muslim militants threw a grenade at a group near a bus station, killing three persons and injuring 20 others.

3 DECEMBER 1998 Cauca, Colombia.
Guerrillas kidnapped one German citizen and two Colombians from a bus at a false roadblock in Cauca. The guerrillas set the bus on fire and dynamited a toll booth after stealing the money. Authorities suspected the FARC or ELN was responsible. On 8 January, the ELN released the German citizen unharmed.

17 NOVEMBER 1998 Anantnag, India.
Muslim militants detonated a grenade, killing three persons and injuring 35 others.

21 SEPTEMBER 1998 Sukhumi, Georgia.
Unidentified assailants opened fire on a bus, wounding three U.N. military observers and one other U.N. mission employee. The injured include two Bangladeshis and one Nigerian.

14 AUGUST 1998 Sri Lanka.
The Liberation Tigers of Tamil Eelam [LTTE] seized a Dubai-owned cargo ship and abducted 21 crewmembers, including 17 Indian nationals. The LTTE evacuated the crew before the Sri Lankan Air Force bombed and destroyed the ship, on the suspicion that the vessel was transporting supplies to the LTTE. The 17 Indian hostages were released to the International Committee of the Red Cross on 19 August, but the LTTE continued to hold four Sri Lankans hostage.

10 AUGUST 1998 Anantnag Kashmir, India.
Unidentified assailants threw a grenade and fired automatic weapons into a crowded bus, killing four persons and injuring seven others. Authorities suspected Pakistani-backed separatists.

26 JULY 1998 New Delhi, India.
A bomb exploded on an empty bus parked at the interstate bus terminal in New Delhi, killing two persons and injuring at least eight others. The bomb destroyed the bus and caused major damage to six others.

24 JULY 1998 India.
A bomb exploded near the railroad tracks moments after the Shalimar Express passed by in Jammu and Kashmir, killing one soldier and injuring two civilians. Indian officials believed that Muslim militants were responsible.

JUNE 1998 United States.
Mohammed Rashid was turned over to U.S. authorities overseas and brought to the United States to stand trial on charges of planting a bomb in 1982 on a Pan Am flight from Tokyo to Honolulu that detonated, killing one passenger and wounding 15 others. Rashid had served part of a prison term in Greece in connection with the bombing until that country released him from prison early and expelled him in December 1996. The nine-count U.S. indictment against Rashid charged him with murder, sabotage, bombing, and other crimes in connection with the Pan Am explosion.

23 JUNE 1998 Kashmir, India.

A remote-controlled bomb exploded under the Delhi-bound Shalimar Express in Kashmir, injuring at least 35 of the 2,000 passengers and derailing seven cars, according to press reports. Muslim militants were suspected.

7 JUNE 1998 Pakistan.

A bomb ripped through an 18-car passenger train en route from Karachi to Peshawar, killing 23 persons and wounding at least 32 others and destroying one railcar. Pakistan blamed India's Research and Analysis Wing for the bombing. Indian officials denied the accusation.

1 MAY 1998 Shupiyan, India.

A bomb exploded under a crowded bus in Shupiyan, injuring six persons. Muslim militants were suspected.

15 APRIL 1998 Mogadishu, Somalia.

Militiamen abducted nine Red Cross and Red Crescent workers at an airstrip north of Mogadishu. The hostages included a U.S. citizen, a German, a Belgian, a French, a Norwegian, two Swiss, and one Somali. The gunmen were members of a subclan loyal to Ali Mahdi Mohammed, who controlled the northern section of the capital. On 24 April, the hostages were released unharmed, and no ransom was paid.

1997 New York City, NY, United States.

A plot was foiled involving a terrorist suicide bombing of New York subways.

1996 Planning for 2001 World Trade Center attacks.

Authorities report **Khalid Shaikh Mohammed** first suggested to **Osama bin Laden** training pilots to fly planes into buildings within the U.S.

JANUARY 1995 Manila, Philippines.

"Bojinka" operation plot foiled. In 1994, the veteran jihadists Khalid Sheikh Mohammed [KSM]—the uncle of Ramzi Ahmed Yousef, the mastermind of the 1993 World Trade Center bombing—joined Yousef in the Philippines to plan what has become known as the "Bojinka" operation. The Bojinka plot involved creating undetectable bombs to be smuggled onto U.S. aircraft with the goal of blowing up the commercial jets over the Pacific in a two-day period. The plot was uncovered when Philippine authorities discovered Yousef's bomb-making equipment in Manila.

1995 Philippines.
Authorities uncovered scheme to coordinate training of Islamic pilots at U.S. schools and then fly airliners into buildings in the U.S. Known sites included the CIA, the Pentagon, the World Trade Center, the Sears Tower, the Transamerica Tower, and a nuclear-power plant. Al Qaeda was responsible.

24 DECEMBER 1994 Paris, France.
Terrorists' attempt to fly hijacked Air France flight into the Eiffel Tower failed. Algerian Islamic terrorists with ties to Osama bin Laden carried out the hijacking.

11 DECEMBER 1994 Japan.
A PAL flight en route to Japan is bombed, killing 1 passenger. The bomb was constructed of diluted nitroglycerine in a contact-lens-cleaner bottle with a wristwatch as a timer. It was designed by Yousef, who placed it on the first leg of a two-leg flight. This attack was considered a practice run for the Bojinka plot.

1993 New York City, NY, United States.
A plot was uncovered involving terrorists linked to al Qaeda, who planned to detonate truck bombs in the city's commuter tunnels and bridges.

3 FEBRUARY 1993 London, United Kingdom.
Bombs exploded in South Kensington underground station, following a warning and evacuation.

18 FEBRUARY 1991 London, United Kingdom.
A bomb exploded in Paddington Station. Three hours later, a second bomb exploded at Victoria Station, killing one person and injuring 54.

1990s Paris, France.
Algerian extremists from the Armed Islamic Group [GIA] set off bombs in the Paris subway.

21 DECEMBER 1988 Lockerbie, Scotland.
Terrorists planted bomb on Pan Am Flight 103, which exploded and crashed in Lockerbie. All 259 passengers were killed, including U.S. students and military personnel. Eleven others were killed on the ground. The Popular Front for the Liberation of Palestine-General Command and the Libyan government were responsible.

AUGUST 1988 Bahawalpur, Pakistan.
A bomb planted in a C-130 Hercules aircraft exploded just after takeoff. Pakistani President General Zia Al Haq, a U.S. ambassador, and 37 others were killed.

JULY 1988 Mediterranean Sea.
Three terrorists boarded the cruise ferry, *City of Poros,* in the Mediterranean Sea, which was carrying 500 passengers. The terrorists opened fire with automatic weapons and tossed hand grenades around the deck, killing nine passengers and injuring 100. ANO claimed responsibility.

SEPTEMBER 1986 Karachi, Pakistan.
Terrorists hijacked PAN AM Flight 73. ANO claimed responsibility.

2 APRIL 1986 Athens, Greece.
A bomb planted on TSW Flight 840 exploded as the plane approached Athens Airport. Four U.S. citizens were killed, and 9 injured. The Palestinian splinter group Ezzedine Kassam was responsible.

DECEMBER 1985 Italy and Austria.
Terrorists simultaneously attacked U.S. and Israel airport check-in counters at El Al Airlines in Rome and Vienna airports, killing 20. ANO claimed responsibility.

23 NOVEMBER 1985 Valetta, Malta.
Terrorists hijacked Egypt Flight 648, followed by a 30-hour standoff between the Egyptian commandos and the hijackers. Abu Nidal was responsible.

7 OCTOBER 1985 Port Said, Egypt.
Four members of the Pro-PLO led by Muhammad Abbas attacked the Italian cruise liner *Achille Lauro,* kidnapping 700 hostages. A disabled American, Leon Klinghoffer, was shot and thrown overboard. Hijackers demanded the release of all Palestinians imprisoned worldwide. Egypt offered the terrorists safe haven, and the hostages and ship were released.

23 JUNE 1985 Cork, Ireland.
Air India Flight 182, en route from Montreal to London, exploded about 150 miles southwest of Cork, Ireland killing all 329 people aboard. Most of the wreckage and remains were never recovered, hidden beneath thousands of feet of ocean.

23 JUNE 1985 Tokyo, Japan.
Bombs in passenger luggage killed two baggage handlers during transfer from one Air India flight to another at Tokyo Narita Airport. The incident occurred about one hour before the Flight 182 explosion. Both bombs were made and planted by Inderjit Singh Reyat.

13–14 JUNE 1985 Athens, Greece.
Terrorists hijacked TWA Flight 847 at takeoff and forced it to fly to Beirut, Lebanon, holding 8 crewmembers and 145 passengers hostage. The aircraft was flown twice to Algiers, and finally returned to Beirut after Israel releases 435 Lebanese and Palestinian Shiite prisoners. During the hijacking, a U.S. Navy diver was murdered and dumped on the runway. Hezbollah was responsible.

13 DECEMBER 1984 Kuwait.
A truck bomb exploded, killing 6 and injuring dozens at U.S. and French embassies. The Islamic Jihad is responsible.

AUGUST 1982 Honolulu, HI, United States.
Terrorists planted a bomb on a Pan Am flight and exploded the aircraft over Hawaii. 1 passenger was killed, and several were injured. Mohammad Rashid was responsible.

1980s Spain.
The Basque separatist group ETA attempted to blow up a Spanish patrol ship using a four-four remote-control boat packed with explosives.

27 JUNE 1976 France.
An Air France airliner was hijacked and forced to land in Uganda, taking 258 passengers hostage. The Baader-Meinhof Group and the PFLP were responsible.

17 DECEMBER 1973 Rome, Italy.
Terrorists attacked an airport terminal, including Pan Am Flight 202, killing 29 and taking 5 hostages. The terrorists demanded the release of two Arab terrorists.

1972 Israel.
The JRA hijacked two Japanese airliners, killing 24.

MAY–JULY 1972 Israel and Italy.
A Belgian Sabena flight en route from Vienna, Austria, to Tel Aviv was hijacked. A Tel Aviv bus terminal was bombed, injuring 11, and an oil refinery in Trieste, Italy, was attacked. Black September was responsible.

1970s Middle East.

During the 1970s, the JRA carried out a series of attacks around the world, including the massacre in 1972 at Lod Airport in Israel, two Japanese airliner hijackings, and an attempted takeover of the U.S. Embassy in Kuala Lumpur.

15 SEPTEMBER 1970 Jordan.

In retaliation for plane hijackings, Jordan's army attacked Palestinian positions and expelled PLO officials and commandos from the country. Up to 20,000 died when Palestinian areas and refugee camps were bombed. The PLO moved its base of operations to Beirut, Lebanon.

7–9 SEPTEMBER 1970 Jordan.

Terrorists hijacked two American and one Swiss and British commercial airliner, taking passengers and crews hostage. Palestinian guerrillas were responsible.

6 SEPTEMBER 1970 The Netherlands, Switzerland, Germany, Jordan, and Egypt.

Terrorists hijacked TWA flights carrying a total of 400 passengers, and the aircraft are forced to land in Zerqa, Jordan, and Cairo, Egypt, where each plane was blown up on the ground. PFLP was responsible.

10 FEBRUARY 1970 Munich, Germany.

Terrorists attacked a bus at the airport, killing 1 passenger and injuring 11 others. The Action Organization for the Liberation of Palestine and the Popular Democratic Front for the Liberation of Palestine were responsible.

SEPTEMBER–DECEMBER 1969 Europe.

Palestinian terrorists used hand grenades and bombs to attack El Al offices in Brussels, Belgium; Athens, Greece; and Berlin, Germany.

29 AUGUST 1969 United States and Syria.

A TWA flight out of Los Angeles, CA, was hijacked and forced to land in Damascus, Syria. Six passengers were held hostage. PFLP was responsible.

18 FEBRUARY 1969 Zurich, Switzerland.

Palestinian terrorists attacked an El Al airliner, killing one pilot and wounding another.

23 JULY 1968 Italy, Israel, and Algeria.

Terrorists hijacked an El Al flight en route from Rome to Tel Aviv and forced the plane to land in Algiers, taking 42 hostages. PFLP was responsible.

1956

Large groups of terrorists entered Israel, threatening the security of Israel ports.

APPENDIX B: TRANSPORTATION-SECTOR INITIATIVES

Several industry and government initiatives and programs have been introduced post-9/11 to deal with a variety of strategic and programmatic issues. Some examples for the transportation sector are listed below.

General Initiatives

Following are some general programs that affect the entire transportation sector:

- C-TPAT [Customs-Trade Partnership against Terrorism] Program. C-TPAT is a joint government-industry initiative to build cooperative relationships that strengthen the overall supply chain and border security. It is designed to secure the entire supply chain to protect shipments so that materials of terrorism [or terrorists themselves] do not enter the U.S. through ports, terminals, or across borders. Participants conduct comprehensive self-assessments of their supply-chain security using the C-TPAT security guidelines. These guidelines encompass access control, conveyance security, education and training, manifest procedures, personnel security, physical security, and procedural security. Once they submit their profiles to customs, they must develop and implement programs to enhance security throughout the supply chain in accordance with C-TPAT guidelines. [*Security Management*, December 2003]
- TAPA [Technology Asset Protection Association]. TAPA was established in 1997 by leading technology companies Intel, Compaq, and Sun Microsystems, which banded together to find solutions to the security threats that are common to the technology industry. The non-profit association now comprises 65 high-tech logistics and freight-forwarder members.
- SST [Smart and Secure Tradelanes]. SST is leveraging a system pioneered by the DOD called the total-asset visibility [TAV] network, which was designed to track military shipments via truck, train, and ship from the manufacturer to the battlefield. It is built on existing U.S. and international standards and on the universal data appliance protocol, which allows open, plug-and-play integration of automatic data-collection devices such as RFID and GPS, along with sensors, scanning, and biometric systems. Three port-operating companies— which together account for 70 percent of the world's container-port operations—have spearheaded and funded the effort with the support

of several technology and services companies. Phase I of the project includes deploying baseline infrastructure, hardware [including electronic seals, sensor devices, and sophisticated scanners], and web-based software to secure and track containers in real time.

Aviation Initiatives

Airport-security failures on 9/11 have placed the aviation industry under intense public scrutiny. To regain the public's confidence in air travel, public and private organizations have made substantial investments to increase airport security. Much work remains. The DHS, as the lead federal department for the transportation sector, has agreed to work with the DOT, industry, and state and local governments to organize, plan, and implement needed protection activities. Additional security initiatives include efforts to:

- Identify vulnerabilities, interdependencies, and remediation requirements. The DHS and the DOT have agreed to work with representatives from state and local governments and industry to implement or facilitate risk assessments to identify vulnerabilities, interdependencies, and remediation requirements for operations and coordination-center facilities and systems, such as the need for redundant telecommunications for air-traffic command and control centers.
- Identify potential threats to passengers. The DHS and the DOT have agreed to work with airline and airport-security executives to develop or facilitate new methods for identifying likely human threats while respecting constitutional freedoms and privacy.
- Improve security at key points of access. The DHS and the DOT have agreed to work with airline and airport security executives to tighten security or facilitate increased security at restricted-access points within airport terminal areas as well as the perimeter of airports and associated facilities, including operations and coordination centers.
- Increase cargo-screening capabilities. The DHS and the DOT have agreed to work with airline and airport security officials to identify and implement or facilitate technologies and processes to enhance airport baggage-screening capacities.
- Identify and improve detection technologies. The DHS and the DOT have agreed to work with airline and airport security executives to implement or facilitate enhanced technologies for detecting explosives. Such devices will mitigate the impact of increased security on

passenger check-in efficiency and convenience and also provide a more effective and efficient means of assuring vital aviation security.

Rail and Public-Transportation Initiatives

As with the aviation sector, the rail industry also faces the additional costs of sustaining increased security during periods of heightened alert. Since the events of 9/11, railroads across the country have in effect been working at surge capacity to meet the security requirements of the increased threat environment, which entails assigning overtime and hiring temporary security personnel. Such reservoirs of capacity are costly to maintain. Nevertheless, the rail sector has had to adopt these heightened security levels and the mandatory requirements of the Department of Homeland Security as the new "normal" state. Some cash-strapped operators now face tradeoffs between providing increased levels of security and going out of business.

Homeland Security's mandatory measures cover a broad range of security issues and provide flexibility to meet the specific needs of rail operators. They substantiate existing best practices in the rail industry and will ensure enhanced security across the nation's passenger-rail systems. A sampling of measures required by rail operators is described below:

- Rail owners/operators must designate coordinators to enhance security-related communications with the TSA.
- Passengers and employees will be asked to report unattended property or suspicious behavior.
- At certain locations operators will be required to remove trash receptacles, except clear plastic or bomb-resistant trash containers.
- When needed, canine explosive teams may be utilized to screen passenger baggage, terminals, and trains.
- Facility inspections will be conducted by rail operators for suspicious or unattended items.
- Rail operators will ensure that security is at appropriate levels consistent with the established threat levels.

Other protective measures that have been put in place to strengthen rail security include:

- Conducting comprehensive vulnerability assessments of rail and transit networks that operate in high-density urban areas.

- Training for rail personnel in preventing and responding to potential terrorist events.
- Developing new technologies, including chemical and biological countermeasures.

The rail and public-transportation sectors have been working actively with the DOT to establish a surface-transportation ISAC to facilitate the exchange of information related to both cyber and physical threats specific to the railroads.

Additional rail-protection initiatives include efforts to:

- Develop improved decision-making criteria regarding the shipment of hazardous materials. The DHS and the DOT, coordinating with other federal agencies, state and local governments, and industry, have agreed to facilitate the development of an improved process to assure informed decision-making with respect to hazardous-materials shipment.
- Develop technologies and procedures to screen intermodal containers and passenger baggage. The DHS and the DOT have agreed to work with sector counterparts to identify and explore technologies and processes to enable efficient and expeditious screening of rail passengers and baggage, especially at intermodal stations.
- Improve security of intermodal transportation. The DHS and the DOT have agreed to work with sector counterparts to identify and facilitate the development of technologies and procedures to secure intermodel containers and detect threatening content. They will work with the rail industry to devise or enable a hazardous-materials-identification system that supports the needs of first responders yet avoids providing terrorists with easy identification of a potential weapon.
- Clearly delineate roles and responsibilities regarding surge requirements. The DHS and the DOT have agreed to work with industry to delineate infrastructure protection roles and responsibilities to enable the rail industry to address surge requirements for resources in the case of catastrophic events. This includes convening a group consisting of government and industry representatives to identify options for the implementation of surge capabilities, including access to federal facilities and capabilities in extreme emergencies.

Trucking and Busing Initiatives

As with the other major transportation modes, the trucking and busing industry has assessed its own security programs in light of the 9/11 attacks. However, the sector's vast heterogeneous nature requires further expanded coordination among stakeholder organizations to assure a more consistent, integrated national approach. Additionally, a better understanding of the overall system would lead to more adaptable, less intrusive, and more cost-effective security processes. Trucking and busing protection initiatives include efforts to:

- Facilitate comprehensive risk, threat, and vulnerability assessments. The DHS, working closely with DOT and other key sector stakeholders, has agreed to facilitate comprehensive risk, threat, and vulnerability assessment for this mode.
- Develop guidelines and standard criteria for identifying and mitigating chokepoints. The DHS, working with the DOT and other key sector stakeholders, has agreed to develop guidelines and standard criteria for identifying and mitigating choke points both nationally and regionally.
- Harden industry infrastructure against terrorism through technology. The DHS has agreed to work jointly with industry and state and local government to explore and identify potential technology solutions and standards that will support analysis and afford better and more cost-effective protection against terrorism.
- Create national transportation operator-security education and awareness programs. The DHS and the DOT have agreed to work with industry to create national operator-security education and awareness programs to provide the foundation for greater cooperation and coordination within this highly diverse sector.

Maritime Initiatives

Following the 9/11 attacks, initial risk assessments were conducted for all ports. These assessments have helped refine critical infrastructure and key-asset designations, assess vulnerabilities, guide the development of mitigation strategies, and illuminate best practices. Most port authorities and private facility owners have also reexamined their security practices. Based on these preliminary risk assessments, the DOT increased vessel-notification requirements to shift limited resources in order to maintain positive control of movement of high-risk vessels carrying high-consequence cargoes and large numbers of passengers. The DOT and the U.S. Coast Guard have also established a sea-marshal program and deployable maritime safety and security teams to implement these activities. Other maritime initiatives are to:

- Identify vulnerabilities, interdependencies, best practices, and remediation requirements. The DHS and the DOT have agreed to undertake or facilitate additional security assessments to identify vulnerabilities and interdependencies, enable the sharing of best practices, and issue guidance or recommendations on appropriate mitigation strategies.
- Develop a plan for implementing security measures corresponding to varying threat levels. The DHS and the DOT have agreed to work closely with other appropriate federal departments and agencies, port-security committees, and private-sector owners and operators to develop or facilitate the establishment of security plans to minimize security risks to ports, vessels, and other critical maritime facilities.
- Develop processes to enhance maritime-domain awareness and gain international cooperation. The DHS and the DOT have agreed to work closely with other appropriate federal departments and agencies, port-security committees, port owners and operators, foreign governments, international organizations, and commercial firms to establish a means for identifying potential threats at ports of embarkation in addition to monitoring identified vessels, cargo, and passengers en route to the U.S.
- Develop a template for improving physical and operational port security. The DOT has agreed to collaborate with appropriate federal departments, agencies, and port owners and operators to develop a template for improving physical and operational port security. A list of possible guidelines will include workforce-identification measures, enhanced port-facility designs, vessel-hardening plans, standards for international container seals, guidance for the research and development of noninvasive security and monitoring systems for cargo and ships, real-time and trace-back capability information for containers, prescreening processes for high-risk containers, and recovery plans. Activities will include reviewing the best practices of other countries
- Develop security and protection guidelines and technologies for cargo and passenger ships. The DHS and the DOT have agreed to work with international maritime organizations and industry to study and develop appropriate guidelines and technology requirements for the security of cargo and passenger ships.
- Improve waterway security. The DHS and the DOT, working with state and local governments and owners and operators, have agreed to develop guidelines and identify needed support for improving security of waterways, including developing electronic monitoring

systems for waterway traffic, modeling shipping systems to identify and protect critical components, and identifying requirements and procedures for periodic waterway patrols.

Mass-Transit Initiatives

Mass transit is localized and varies significantly in size and design from system to system. Identifying critical guidelines and standards for planning are keys to unifying security activities. Panels in the transit cooperative-research program have recommended and are overseeing 10 research projects in the areas of prevention, mitigation, preparedness, and response. Their recommendations can provide additional input to the development of these planning areas. Additional mass-transit protection initiatives include efforts to:

- Identify critical planning areas and develop appropriate guidelines and standards. The DHS, working closely with the DOT and other federal, state, and local mass-transit officials, has agreed to identify critical planning areas and develop appropriate guidelines and standards to protect mass-transit systems. Such critical planning areas and guidelines include design and engineering standards for: facilities, and rail and bus vehicles, emergency guidance for operations staff, screening methods and training programs for operators, security-planning-oversight standards, mutual-aid policies, and continuity-of-operations planning.
- Identify protective impediments and implement security enhancements. The DHS, working closely with the DOT and sector representatives, has agreed to review legal, legislative, and statutory regimes to develop an overall protective architecture for mass-transit systems and to identify impediments to implementing needed security enhancements.
- Work with other sectors to manage unique risks resulting from interdependencies. The DHS, in collaboration with the DOT, has agreed to convene cross-sector working groups to develop models for integrating priorities and emergency-response plans in the context of interdependencies between mass-transit and other critical infrastructures.

Chapter 14

THE CHEMICAL AND HAZARDOUS-MATERIALS SECTOR

This chapter outlines the:

- Criticality of the chemical sector to national interests
- Chemical-sector attractiveness to terrorist and criminal elements
- Chemical-sector vulnerabilities
- Tailoring the S^3E Security Methodology to the chemical sector
- Chemical-sector challenges facing the security-assessment team
- Applying the security-assessment methodology to the chemical sector
- Preparing the chemical security-assessment report
- Historical overview of selected chemical incidents
- U.S. government chemical initiatives

CHEMICAL AND HAZARDOUS-MATERIALS CRITICALITY TO NATIONAL INTERESTS

The chemical sector provides products that are essential to the United States' standard of living; its products are fundamental elements of every other economic sector. For example, it produces fertilizer for agriculture, chlorine for water purification, and polymers for innumerable household and industrial products. Additionally, more than $97 billion of the sector's products go to healthcare alone. Currently, **the chemical sector is the nation's top exporter,** accounting for 10 cents out of every dollar. The industry is also one of our country's most innovative. It earns one out of every seven patents issued in the U.S., a fact that enables our country to remain competitive in the international chemical market. The sector itself is highly diverse in terms of company size and geographic dispersion. Its product- and service-delivery system depends on raw materials, manufacturing plants and processes, and distribution systems, as well as research facilities and supporting infrastructure services, such as transportation and electricity. Across the nation, there are some 60,000 facilities that manufacture, use, or store hazardous chemicals in quantities that could potentially put large numbers of Americans at risk of injury or death in the event of a chemical release. [1]

Chemical facilities manufacture a host of products including basic organic chemicals, plastic materials and resins, petrochemicals, and industrial gases, to name a few. Other facilities, such as fertilizer and pesticide factories, pulp and paper manufacturers, water facilities, and refineries, also house large quantities of chemicals. According to the EPA, over 15,000 facilities in various industries produce, use, or store one or more of the identified 140 toxic and flammable chemicals that pose the greatest risk to human health and the environment. Exhibit 14.1 outlines the number and percent of processes in different industry sectors that maintain more than threshold amounts of these hazardous chemicals. [2]

Public confidence is important to the continued economic robustness and operation of the chemical industry. Uncertainty regarding safety impacts both the producers and the commercial users of the product. With respect to process safety, numerous federal laws and regulations exist to reduce the likelihood of accidents that could result in harm to human health or the environment. However, there is currently no clear, unambiguous legal or regulatory authority at the federal level to help ensure comprehensive, uniform security standards for chemical facilities. [3]

Industry Sector	Number of Processes	Percent of Processes
Agriculture and farming, farm supply, fertilizer production, pesticides	6,317	31%
Water supply and wastewater treatment	3,753	18%
Chemical manufacturing	3,803	18%
Energy production, transmission, transport, and sale	3,038	15%
Food and beverage manufacturing and storage [including refrigerated warehousing]	2,366	11%
Chemical warehousing [not including refrigerated warehousing]	318	2%
Other [1]	1,075	5%
Total [2]	20,670	100%

Source: The Environmental Protection Agency

[1] This represents a large variety of industry sectors including pulp mills, iron and steel mills, cement manufacturing, and computer manufacturing.

[2] The total number of covered processes is not equal to the 15,000 facilities, because some facilities have more than one covered process [i.e. a process containing more than a threshold amount of covered hazardous chemical].

Exhibit 14.1 Number and Percent of Risk-Management-Plan-Covered Processes by Industry Sector

As in most other industries, the chemical industry relies on the availability, continuity, and quality of services and supplies from other critical infrastructures. For example, the chemical industry is the nation's third largest consumer of electricity. An assured supply of natural gas at competitive prices is another crucial resource for the sector. [4]

Target for Terrorism

Experts agree that chemical facilities present an attractive target for terrorists intent on causing massive damage, because many facilities house toxic chemicals that could become airborne and drift to surrounding areas if released. Alternatively, terrorists could steal chemicals, which could be used to create weapons capable of causing harm.

Loosely affiliated terrorist groups have demonstrated a growing interest in chemical weapons, but explosives are still the most frequently used method of deploying such weapons. During the 1990s, both international and domestic terrorists attempted to use explosives to release chemicals from manufacturing and storage facilities. Most of these attempts were abroad in war zones such as Croatia and included attacks on a plant producing fertilizer, carbon-black, and light-fraction petroleum products; other plants producing pesticides; and a pharmaceutical factory using ammonia, chlorine, and other hazardous chemicals. All these facilities were close to population centers. In the United States there were at least two instances during the late 1990s when criminals attempted to cause chemical releases from facilities. One involved a large propane-storage facility and the other a gas refinery. In addition, one of the 1993 WTC bombers, Unidal Ayyad, used company stationery to order chemical ingredients to make a bomb. Testimony at the trial of the bombers indicated that they had successfully stolen cyanide from a chemical facility and were training to introduce it into the ventilation systems of office buildings. More recently, chemical trade publications reportedly were found in al Qaeda hideaways [5]. A historical overview of selected terrorist attacks, criminal incidents, industry mishaps, and government actions within the chemical sector are displayed in Appendix A.

According to the EPA, 123 chemical facilities located throughout the nation have toxic "worst-case" scenarios in which more than one million people could be at risk of exposure to a cloud of toxic gas. About 700 facilities could each potentially threaten between 100,000 and one million people, and about 2,300 facilities could each potentially threaten between 10,000 and 100,000 people within their vulnerability zones. Such zones are determined by drawing a circle around a facility whose radius is equal to the distance a toxic gas cloud would travel before dissipating to relatively harmless levels. However, the EPA requirements for reporting "worst-case" release analyses tend to result in consequence estimates that are significantly higher than what is likely to actually occur, as release analyses do not take into account active mitigation measures facilities often employ to reduce the consequences of release. Also, in an actual event, the toxic cloud would most likely only cover a fraction of the vulnerability zone, and its reach would depend on weather conditions and other factors, rendering it unlikely to result in exposure of the entire population estimated in the "worst-case" scenario. [6]

Vulnerabilities

Potential terrorist acts against chemical facilities might include direct attacks on facilities or chemicals on site or efforts to use business contacts, facilities, and materials to gain access to potentially harmful materials. In either case, terrorists may be employees [saboteurs] or outsiders, acting alone or in collaboration with others. In the case of a direct attack, traditional or nontraditional weapons may be employed, including explosives, incendiary devices, firearms, airplanes, computer programs, or weapons of mass destruction [nuclear, radiological, chemical, or biological].

The potential impact on public health and the environment from a sudden release of hazardous chemicals has long concerned security and safety experts. More than 41 million Americans live within a range of a potential toxic cloud that could result from a chemical release at a facility located in their home zip code. The risk of an attack varies among facilities depending upon several factors, including their location and the types of chemicals they use, store, or manufacture. Many facilities are located in populated areas where a chemical release could result in injuries or death as well as economic harm. [7]

Facilities that contain large amounts of toxic chemicals and are located near population centers may be at higher risk of a terrorist attack if one assumes that the objective is a catastrophic release. Attacks on such facilities could harm large numbers of people, with health effects ranging from mild irritation to death, cause large-scale evacuations, and disrupt the local regional economy. Therefore, the need to reduce the sector's vulnerability to acts of terrorism is important to safeguard the economy and protect our citizens and the environment.

TAILORING THE S^3E SECURITY ASSESSMENT PROCESS FOR CHEMICAL AND HAZARDOUS-MATERIALS FACILITIES

While **the federal government currently does not require chemical facilities to take security measures to protect against terrorist attacks,** it does require certain facilities to take preventive measures to protect against accidental release and security precautions against trespassing and theft. These requirements, however, do not cover a wide range of chemical facilities or the full range and level of all reasonable threats.

In December 2006, the Department of Homeland Security introduced for public review an aggressive and comprehensive set of proposed regulations that require chemical facilities fitting certain profiles to complete a secure online risk assessment to assist in determining their overall level of risk [8]. High-risk facilities will then be required to conduct vulnerability assessments and submit site security plans that meet the DHS performance standards. The DHS will validate submissions through audits and site inspections and will provide technical assistance to facility owners and operators as needed. Performance standards will be designed to achieve specific outcomes, such as securing the perimeter and critical targets, controlling access, deterring theft of potentially dangerous chemicals, and preventing internal sabotage. Security strategies necessary to satisfy these standards will depend upon the level of risk at each facility. Other federal requirements include the following:

- **The Conservation and Recovery Act of 1976,** which requires facilities that house hazardous materials to take certain security measures such as posting warning signs, using 24-hour surveillance, or surrounding the active portion of the facility with a barrier. It is only applicable to 21 percent of the 15,000 chemical facilities. While these measures may help impede a terrorist's access to a facility, they are aimed at keeping out trespassers and wanderers, not intentional and dedicated intruders.

- **The Drug Enforcement Agency [DEA],** which requires any chemical facility that manufactures one of the 32 chemicals that can be used as a precursor to illegal drugs or controlled substances to securely store, restrict access, and monitor inventories of these chemicals. According to the EPA and the DEA, these requirements only apply to 62 of the 15,000 facilities that have these chemicals.

In addition to these federal requirements, some states and localities have imposed their own safety requirements on chemical facilities and in some instances have addressed security from terrorism. For example, Contra Costa County, California, requires chemical facilities to incorporate inherently safer technologies. New Jersey has imposed criminal penalties for any toxic-chemical manufacturer who recklessly allows an unauthorized individual to obtain access to a chemical. Baltimore, Maryland, passed a city ordinance addressing the threat of terrorism that requires chemical manufacturers to follow a set of safety and security regulations devised by its fire and police commissions. Companies that fail to comply with the ordinance may face penalties such as the withholding or suspension of facility operating permits.

Applying the investigative principles outlined in Part I, the security-assessment team can use the tailored version of the S^3E **Security Assessment Methodology Model** shown below as a road map to conducting security assessments for the chemical and hazardous-materials sector.

The exhibit on the next page highlights in grey shaded areas those specific elements that are tailored for application to the chemical sector.

Challenges Facing the Security-Assessment Team

The security-assessment team must understand the criticality of assurance of supply to downstream users of chemical products. Many large municipal waterworks maintain only a few days' supply of chlorine for disinfecting their water. Agricultural chemicals, particularly fertilizers, must be applied in large volumes during very short time periods. Some products cannot be transferred between transportation modes. Facilities with just-in-time delivery systems maintain fewer and smaller chemical stockpiles.

Another challenge is that the contamination of chemicals vital to many applications could impact a wide range of other industries, thereby affecting public health and the economy. For instance, in addition to the risk of contamination at product storage facilities, many chemicals are also inherently hazardous and therefore represent potential risks to public health and safety in a malicious context.

A third hurdle is that many current statutes related to the handling of highly toxic substances were created decades ago and may no longer be effective for monitoring and controlling access to dangerous substances. For example, although licensed distributors of pesticides can only sell them to licensed purchasers, license requests, which are granted at the state level by county agents, are fairly easy to obtain. In addition, the basis for licensing varies from state to state.

The greatest challenge for the security-assessment team rests in risk profiles of chemical plants, which differ tremendously because of differences in technologies, product mix, design, and processes. No single security solution fits all situations.

APPLYING THE S^3E SECURITY-ASSESSMENT METHODOLOGY TO THE CHEMICAL AND HAZARDOUS-MATERIALS SECTOR

The security-assessment team should use its collective expertise to:

- Identify the important mission and functions of the chemical facility [**Task 1—Operational Environment**] and incorporate the

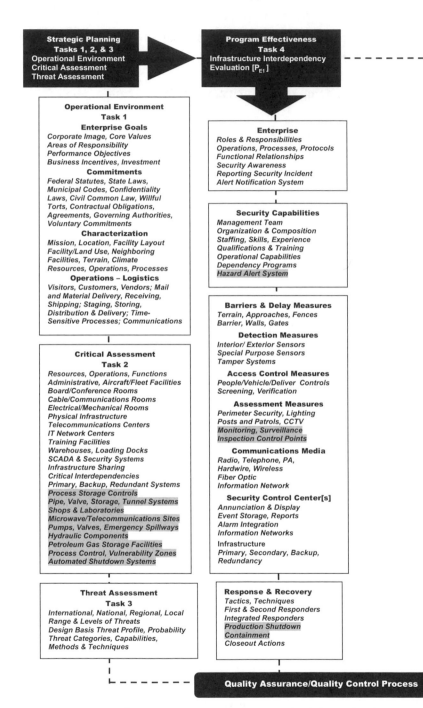

Exhibit 14.2 The S^3E Security-Assessment Methodology for the Chemical and Hazardous-Materials Sector

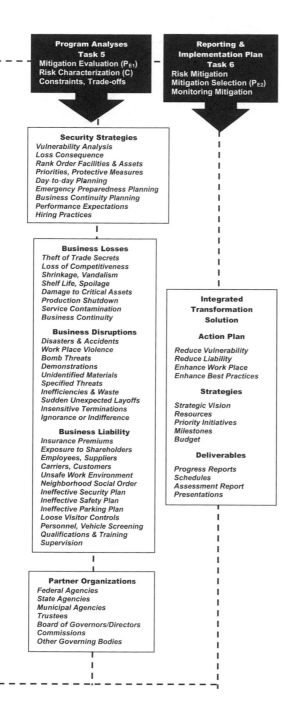

Exhibit 14.2 *(continued)*

Areas of Responsibility	Performance Objectives
Business Incentives & Investment	Power Production & Distribution
Corporate Image & Core Values	Power Supply & Demand
Mission & Services	Public Safety & Public Confidence

Exhibit 14.3 Identify Chemical and Hazardous-Materials Enterprise Business Goals and Objectives

development of all business operations and continuity expectations. Examination focuses on services and the customer base to determine the business contributions to the community, its culture, and business objectives. These include but are not necessarily limited to the identification of business goals and objectives, identification of the customer base, identification of enterprise commitments, and characterization of facilities and boundaries, as shown in Exhibits 14.3 through 14.6.

• Describe the utility configuration, operations, and other elements [**Task 2—Critical Assessment**] by examining conditions, circumstances, and situations relative to safeguarding public health and safety and reducing the potential for disruption of chemical services. Focus is on an integrated evaluation process to determine which assets need protection in order to minimize the impact of threat events. The security-assessment team identifies and prioritizes facilities, processes, and assets critical to chemical service objectives that might be subject to malevolent acts or disasters that could result in a potential series of undesired consequences. The process takes into account the impacts that could substantially disrupt the ability of the utility to provide safe chemical services and strives to reduce security risks associated with the consequences of significant events. This includes but is not necessarily limited to the assessment of facilities, assets, and operations and prioritization of assets, as shown in Exhibits 14.7 and 14.8.

General Public, Critical Customers	Local, Regional, National, International
Government, Military	Medical, Firefighting
Industrial, Business, Defense Industrial Base	Size of Population Served

Exhibit 14.4 Identify Chemical and Hazardous-Materials Enterprise Customer Base

29 CFR 1910.36 - Means of Egress	Family Education Rights and Privacy Act 1974 [As Amended]
29 CFR 1910.151 - Medical Services and First Aid	Family and Medical Leave Act 1993 [As Amended]
29 CFR 1910.155 - Fire Protection	Federal Polygraph Protection Act [As Amended]
18 U.S.C. 2511 - Technical Surveillance	
21 U.S.C. 801 - Drug Abuse and Control	Federal Privacy Act 1974 [As Amended]
26 U.S.C. 7201 - Tax Matters	Freedom of Information Act [U.S.C. 552] [As Amended]
29 U.S.C. 2601 - Family and Medical Leave Act	
31 U.S.C. 5322 - Financial Transaction Reports	National Labor Relations Act [29 U.S.C. 151] [As Amended]
41 U.S.C. 51 - U.S. Contractor Kickbacks	Interagency Agreements
41 U.S.C. 701 - Drug Free Workplace Act	Internal Revenue Service
46 U.S.C. 1903 - Maritime Drug Enforcement	Labor Management Relations Act
50 U.S.C. 1801 - Foreign Intelligence Surveillance Act	Memorandum of Agreement
Americans with Disabilities Act [42 U.S.C. 12101]	Memorandum of Understanding
	Oil Pollution Act 1990 [As Amended]
Architectural Barrier Act 1968 [As Amended]	Omnibus Crime Control and Safe Streets Act, 1968 [As Amended]
Bioterrorism Preparedness and Response Act, 2002 [As Amended]	Policy Coordination Committee [PCC]
Carriers, Suppliers, Vendors, Insurance	Privacy Act [5 U.S.C. 522e] [As Amended]
Civil Rights Act 1964 [U.S.C. 200e] [As Amended]	Rehabilitation Act, 1973 [As Amended]
Commerce Commission	State and Local Industry Administrators and Authorities
Comprehensive Drug Abuse Prevention Act 1970 [As Amended]	State and Local Emergency Planning Officials and Committees
Comprehensive Environmental Response, Compensation and Liability Act 1980 [As Amended]	Uniform Trade Secrets Act
	U.S. Department of Commerce
Controlled Substance Act [As Amended]	U.S. Department of Homeland Security [DHS]
Drug Free Workplace Act 1988 [As Amended]	U.S. Department of Justice [DOS]
Emergency Planning & Right-to-Know Act 1986 [As Amended]	U.S. Department of Transportation [DOT]
	U.S. Environmental Protection Agency [EPA]
Employee Polygraph Protection Act 1988 [29 U.S.C. 2001] [As Amended]	U.S. Patriot Act, 2002 [As Amended]
	U.S. Postal System
Fair Credit Reporting Act [15 U.S.C. 1681] [As Amended]	U.S. Occupational Safety and Health Administration [OSHA]

Exhibit 14.5 Identify Chemical and Hazardous-Materials Enterprise Commitments

Commercially Leased Facilities	Infrastructure Sharing
Dedicated site owned and operated by the enterprise.	Neighboring Facilities
	Operated and maintained through reciprocal agreements with internal and external entities.
Dispersed Locations	
Facility Land Use	
Facility Layout and Location	Terrain & Climate
	Types of assets at specific locations

Exhibit 14.6 Characterise Configuration of Chemical and Hazardous-Materials Enterprise Facilities and Boundaries

Administrative, Aircraft/Fleet Facilities	Pipes, Control Valves, Tunnel Systems
Automated Shutdown Systems	Population & Population Throughputs
Cable/Communications Rooms & Centers	Primary & Backup Power Systems
Chemical Handling, Hazards	Primary, Backup, Redundancy Systems
Crisis Management Centers	Process Control, Vulnerability Zones
Critical Interdependencies	Pumps, Isolation Valves, Emergency Spillways
Electrical/Mechanical Rooms	
Hazard Alert System	Resources, Operations, Functions
Hours of Operation	SCADA & Security C3 Centers
Hydraulic Components	Shipping & Receiving
Infrastructure Sharing	Shops & Laboratories
IT Network Computer Centers	Telecommunications Centers
Microwave/Telecommunications Sites	Time-Sensitive Process, Storage, Controls
Mission, Service Capabilities	Training Facilities
Operational Centers and Control Rooms	Visitors, Mail and Material Delivery
Petroleum Gas Storage Facilities	Warehousing, Loading Docks
Physical Infrastructure	

Exhibit 14.7 Critical Assessment of Chemical and Hazardous-Materials Enterprise Facilities, Assets, Operations Processes, and Logistics

Least critical assets based on value system	Most critical assets based on value system
Least critical facilities based on value system	Most critical facilities based on value system

Exhibit 14.8 Prioritize Critical Chemical and Hazardous-Materials Enterprise Assets in Relative Importance to Business Operations

- Review the existing design-basis threat profile [**Task 3—Threat Assessment**] and update it as applicable to the client. If a threat profile has not been established, then the security-assessment team needs to develop one. Both the security-assessment team and the client must have a clear understanding of undesired events and consequences that may impact chemical services to the community and business-continuance-planning goals. Emphasis is on prioritizing threats through the process of identifying methods of attack, past or recent events, and types of disasters that might result in significant consequences; the analysis of trends; and the assessment of the likelihood of an attack occurring. Focus is on threat characteristics, capabilities, and target attractiveness. Social-order demographics to develop trend analysis in order to support framework decisions are also considered. Under this task the security-assessment team should also validate previous chemical-assessment studies and update conclusions and recommendations into a broader framework of strategic planning. Potential undesirable events include but are not necessarily limited to malevolent acts, other disruptions, types of perpetrators, impacts of loss, and threat attractiveness and likelihood of occurrence, as shown in Exhibits 14.9 through 14.13.
- Examine and measure the effectiveness of the security system [**Task 4—Evaluate Program Effectiveness**]. The security assessment evaluates business practices, processes, methods, and existing protective measures to assess their level of effectiveness against vulnerability and risk and to identify the nature and criticality of business areas and assets of greatest concern to chemical operations. Focus is on vulnerability and vulnerable access points where penetration may be accomplished and how. Emphasis is placed on the physical characteristics of vulnerable points, accessibility to the location, and the effectiveness of protocol obstacles an adversary would likely have to overcome to reach a critical chemical asset. These include but are not necessarily limited to evaluation of security operations security organization, interface with partners, SCADA and security-system performance, adversary plan and path analysis, and effectiveness of response and recovery, as shown in Exhibits 14.14 through 14.19.
- Recognize the importance of bringing together the right balance of people, information, facilities, operations, processes, and systems [**Task 5—Program Analyses**] to deliver a cost-effective solution. The security-assessment team develops an integrated approach to measurably reduce security risk by reducing vulnerability and consequences

Armed attack	Intentional shutdown of electrical power
Arson	Loss of key personnel
Assassination, kidnapping, assault	Loss of proprietary information
Barricade/hostage incident	Mail fraud
Bomb threats and bomb incidents	Misuse/damage of control systems
Chemical, Biological, Radiological [CBR] contamination of assets, supplies, equipment, and areas	Misuse of chemical supply chain
	Misuse of wire and electronic communications
	Phishing, hacking, computer crime
Computer-based extortion	Power failure
Computer system failures and viruses	Rape and sodomy
Computer theft	Sabotage
Cyber attack on SCADA or other systems	Social engineering
Cyber stalking, software piracy	Software piracy
Damage/destruction of interdependency systems resulting in production shutdown	Storage of criminal information on the system
Damage/destruction of disinfection capability	Unlawful access to stored wire and electronic communications and transactional records access
Damage/destruction of control/isolation valve systems	
Damage/destruction of pump stations and motors	Supplier, courier disruptions
	Telecommunications failure
Damage/destruction of service capability	Terrorism
Damage/destruction of pipelines and cabling	Theft or destruction of chemicals
Delay in processing transactions	Torture
Disable manufacturing, production, distribution and delivery processes	Treason, sedition, and subversive activities
Email attacks	Unauthorized transfer of American technology to rogue foreign states
Encryption breakdown	Unidentified materials or wrong deliveries
Explosion	Unlawful access to stored communications
FAX security	Unlawful wire and electronic communications interception and interception of oral communications
Hacker attacks	
Hardware failure, data corruption	
HAZMAT spill	Use of cellular/cordless telephones to disrupt network communications or in support of other crimes
Inability to access system or database	
Inability to perform routine transactions	
Intentional opening/closing of control/isolation values and gates	Voice mail system
	Wide-scale disruption and disaster
Intentional release of dangerous chemicals	Wide-scale evacuation

Exhibit 14.9 Determine Types of Chemical and Hazardous-Materials Enterprise Malevolent Acts That Could Reasonably Cause Undesirable Events

Bribery, graft, conflicts of interest	Natural disasters
Civil disorders, civil rights violations	Obscenity
Conspiracy	Obstruction of justice
Eavesdropping	Online pornography and pedophilia
Economic espionage censorship	Planned production shutdown or outage
Embezzlement, extortion and threats	Racketeering
Historical rate increases	Racketeer influenced and corrupt
Homicide	organizations
Human error	Release and misuse of personal
Industrial accidents and mishaps	information
Inefficiencies and waste	Resource constraints and competing
Ineffective hiring practices	priorities
Ineffective personnel and vehicle controls	Robbery and burglary
Ineffective safety and security plans	Sexual abuse, sexual exploitation
Ineffective supervision and training	Shrinkage, shelf-life, spoilage
Ineffective use of resources and time	Sudden unexpected layoffs
Ineffective visitor controls	Tracking effects of outage overtime
Ignorance and indifference	Tracking effects of outage across related
Insensitive terminations	resources and dependent systems
Institutional corruption, fraud, waste, abuse	Theft of trade secrets, intellectual property, confidential information, patents, and copyright infringements
Loss of competitiveness	Trespassing and vandalism
Maintenance activity	

Exhibit 14.10 Assess Other Disruptions Impacting Chemical and Hazardous-Materials Enterprise Operations

Activist groups or cults	State-sponsored or independent terrorist groups
Disenfranchised individuals	
Insiders	Vandals
Lone wolves	

Exhibit 14.11 Identify Chemical and Hazardous-Materials Enterprise Category of Perpetrators

Compromise of public confidence	Inability to store or transfer chemical
Cost to repair	Local civil unrest
Dealing with unanticipated crisis	Loss of energy sources or ability to generate energy
Disabling of or loss of key personnel	
Duration of loss of emergency services	Loss of life [customers and employees]
Duration of loss of energy source	Loss or disruption of communications
Economic loss	Loss or disruption of information
Illness [customers and employees]	Number of critical customers impacted
Impact on regional economic base	Number of users impacted
Impact on utility ratepayers	Production shutdown
Inability of customers to conduct business	Slowdown and interruptions of delivery services
Inability to control chemical processes	
Inability to distribute chemical	Temporary closure of financial institutions

Exhibit 14.12   Assess Initial Chemical and Hazardous-Materials Enterprise Impact of Loss Consequence

Consequence analysis	Social demographics
Economic conditions	Social/political status

Exhibit 14.13   Assess Initial Likelihood of Chemical and Hazardous-Materials Enterprise Threat Attractiveness and Likelihood of Malevolent Acts of Occurrence

Alert Notification System	Roles & Responsibilities
Emergency Preparedness Planning	Reporting Security Incidents
Functional Interfaces	Security Awareness
Operations, Processes, Protocols	Security Plans, Policies & Procedures

Exhibit 14.14   Evaluate Existing Chemical and Hazardous-Materials Security Operations and Protocols $[P_{E1}]$

Dependency Programs	Qualifications & Training
Management Team	Staffing, Skills, Experience
Operational Capabilities	Supervision
Organization & Composition	

Exhibit 14.15   Evaluate Existing Chemical and Hazardous-Materials Security Organization $[P_{E1}]$

Agency for Toxic Substances and Disease Registry	International Electronics Supply Group [IESG]
	International Parking Institute
American Chemical Association	International Program on Chemical Safety
American Chemistry Council	International Organization for Standardization [ISO]
American Electronics Association [AEA]	
American Institute of Certified Public Accountants	Interstate Commerce Commission [ICC]
	Joint Terrorism Task Force [JTTF]
American National Standards Institute	Law Enforcement Support Center [LESC]
American Psychological Association	Library of Congress, Congressional Research Service
American Society for Industrial Security	
American Society for Testing and Materials	Medical Practitioners and Medical Support Personnel
American Society for Testing Materials, Vaults	
Center for Chemical Process Safety [CCPS]	Multi-lateral Expert Groups
	Mutual Aid Associations
Centers for Disease Control and Prevention [CDC]	National Advisory Commission on Civil Disorder
Chemical Information Sharing and Analysis Centers [ISAC]	National Crime Information Center [NCIC]
Commerce Commission	National and State Public Health Associations
Computer Security Institute	National Electrical Manufacturing Association [NEMA]
Critical Infrastructure Protection Board [PCIPB]	National Fire Protection Association [NFPA]
Defense Central Investigation Index [DCII]	National Information Infrastructure [NII]
Drug Enforcement Agency	National Infrastructure Protection Center
EPA National Security Research Center	National Institute of Law Enforcement and Criminal Justice
EPA Office of Air and Radiation	
EPA Office of Prevention, Pesticides and Toxic Substances	National Institute of Standards and Technology [NIST]
El Paso Intelligence Center [EPIC]	National Insurance Crime Bureau [NICB] Online
Factory Mutual Research Corporation	National Labor Relations Board
Federal, state, and local agency responders	National Parking Association
Illuminating Engineers Society of North America	National Safety Council
	National White Collar Crime Center [NWCCC]
Institute for a Drug Free Workplace	Occupational Safety and Health Review Commission [OSHRC]
Institute of Electrical and Electronic Engineers [IEEE]	
	Office of National Drug Control Policy
International Association of Chiefs of Police	Policy Coordination Committee [PCC]
International Association of Professional Security Consultants	Private Security Advisory Council
	Regional Information Sharing System [RISS]:
International Computer Security Association	Middle Atlantic-Great Lakes Organized Crime Law Enforcement Network
International Criminal Police Organization [INTERPOL]	[MAGLOCLEN] *(continued on next page)*

Exhibit 14.16 Evaluate Chemical and Hazardous-Materials Enterprise Interface and Relationship with Partner Organizations $[P_{E1}]$

Mid-States Organized Crime Information Center [MOCIC]	U.S. Department of Commerce [DOC]
	U.S. Department of Homeland Security [DHS]
New England State Police Information Network [NESPIN]	U.S. Department of Justice [DOJ]
	U.S. Environmental Protection Agency [EPA]
Regional Organized Crime Information Center [ROCIC]	U.S. Federal Bureau of Investigation [FBI]
	U.S. Drug Enforcement Agency [DEA]
Rocky Mountain Information Network [RMIN]	U.S. Food and Drug Administration [FDA]
	U.S. Government Accounting Office [GAO]
Western States Information Network [WSN]	U.S. Nuclear Regulatory Commission [NRC]
Safe Manufactures National Association	U.S. Postal System
The Terrorism Research Center	U.S. Secret Service
U.S. Central Intelligence Agency [CIA]	U.S. Treasury Department
U.S. Coast Guard	Underwriters Laboratories [UL]
U.S. Customs	World Health Organization [WHO]
U.S. Department of Agriculture	

Exhibit 14.16 *(continued)*

Assessment Capability	Modem & Internet Access
Badge Controls	Performance Standards & Vulnerabilities
Barrier/Delay Systems	Personnel & Vehicle Access Control Points
Chemical & Other Vendor Deliveries	Physical Protection System Features
Communications Capability	Post Orders & Security Procedures
Cyber Intrusions, Firewalls, Other Protection Features	Response Capability
Display & Annunciation Capability	Security Awareness
Evacuation Response Plans	Security Training/Exercises/Drills
Intrusion Detection Capability	System Capabilities & Expansion Options
Lock & Key Controls, Fencing & Lighting	System Configuration & Operating Data

Exhibit 14.17 Evaluate Existing Chemical and Hazardous-Materials SCADA and Security-System Performance Levels [P_{E1}]

through a combination of workable solutions that bring real tangible enterprise value to the table. Emphasis is placed on assessing the current status and level of protection provided against desired protective measures to reasonably counter the threat. Operational, technical, and financial constraints are examined, as well as future expansion plans and constraints. These include but are not necessarily limited to analysis of security strategies and operations, revised analysis of undesirable consequences of an event, revised analysis of threat attractiveness and likelihood, analysis of risk-reduction and -mitigation plans, analysis of residual vulnerability, and cost estimation, as shown in Exhibits 14.20 through 14.26.

Likely approaches and escape routes	Delay-penetration obstacles
Deception actions	Target selection and alternative targets
Interception analysis	Time-delay response analysis

Exhibit 14.18 Define the Chemical and Hazardous-Materials Adversary Plan, Distractions, Sequence of Interruptions, and Path Analysis

Closeout Actions	Integrated Responders
Containment	Production Shutdown
Divert Production	Tactics & Techniques
First & Second Responders	

Exhibit 14.19 Assess Effectiveness of Chemical and Hazardous-Materials Enterprise Response and Recovery [P_{E1}]

Business Continuity Planning	Loss Consequence Approach
Day-to-Day Planning	Performance Expectations
Emergency Response Planning	Rank Order Facilities & Assets
Hiring Practices	Vulnerability Analysis

Exhibit 14.20 Analyze Effectiveness of Chemical and Hazardous-Materials Enterprise Security Strategies and Operations [P_{E1}]

Compromise of public confidence	Inability to store or transfer chemical
Cost to repair	Local civil unrest
Dealing with unanticipated crisis	Loss of energy sources or ability to generate energy
Disabling of or loss of key personnel	
Duration of loss of emergency services	Loss of life [customers and employees]
Duration of loss of energy source	Loss or disruption of communications
Economic loss	Loss or disruption of information
Illness [customers and employees]	Number of critical customers impacted
Impact on regional economic base	Number of users impacted
Impact on utility ratepayers	Production shutdown
Inability of customers to conduct business	Slowdown and interruptions of delivery services
Inability to control chemical processes	
Inability to distribute chemical	Temporary closure of financial institutions

Exhibit 14.21 Refine Previous Analysis of Undesirable Consequences That Can Affect Chemical and Hazardous-Materials Enterprise Functions

Consequence analysis	Social demographics
Economic conditions	Social/political status

Exhibit 14.22 Refine Previous Analysis of Likelihood of Chemical and Hazardous Materials Enterprise Threat Attractiveness and Malevolent Acts of Occurrence

Creation and/or revision or modification of sound business practices and security policies, plans, protocols, and procedures.	Development, revision, or modification of interdependency requirements.
Creation and/or revision or modification of emergency operation plans including dependency support requirements.	Physical infrastructure, SCADA/security system upgrades to improve deter, delay, detection and assessment capabilities.

Exhibit 14.23 Analyze Selection of Specific Risk-Reduction Actions Against Current Risks and Develop Prioritized Plan for Chemical and Hazardous-Materials Enterprise Security-Mitigation Solutions [P_{E1}]

Reasonable and prudent mitigation options	Mirrors business culture image

Exhibit 14.24 Develop Short- and Long-Term Chemical and Hazardous-Materials Enterprise Mitigation Solutions

Performance effectiveness calculations	Residual vulnerability consequences

Exhibit 14.25 Evaluate Effectiveness of Developed Chemical and Hazardous-Materials Enterprise Mitigation Solutions and Residual Vulnerability [P_{E2}]

Practical, cost-effective recommendations	Reasonable return on investments

Exhibit 14.26 Develop Cost Estimate for Chemical and Hazardous-Materials Enterprise Short- and Long-Term Mitigation Solutions

PREPARING THE CHEMICAL SECURITY-ASSESSMENT REPORT

Chapter 10 provides a recommended approach to reporting the security-assessment results.

CHEMICAL AND HAZARDOUS-MATERIALS SECTOR INITIATIVES

Efforts to enhance chemical safety include a variety of initiatives. Appendix B summarizes some significant actions taken or currently underway by the U.S. Department of Homeland Security and industry.

REFERENCES

1. "The National Strategy for the Physical Protection of Critical Infrastructure and Key Assets." (February 2003).
2. U.S. Environmental Protection Agency Annual Report (2005)
3. "The National Strategy for the Physical Protection of Critical Infrastructure and Key Assets." (February 2003).
4. Ibid.
5. U.S. Congressional Record (2002).
6. "The National Strategy for the Physical Protection of Critical Infrastructure and Key Assets." (February 2003).
7. Laplante, Allison. "Too Close to Home: A Report on Chemical Accident Risks in the United States," GAO Report 03-439 Homeland Security: Voluntary Initiatives Are Under Way at Chemical Facilities But The Extent of Preparedness is Unknown.
8. U.S. Department of Homeland Security (Draft Regulation for Public Review: Chemical Facility Vulnerability Assessments, December 2006).

APPENDIX A: A HISTORICAL OVERVIEW OF SELECTED TERRORIST ATTACKS, CRIMINAL INCIDENTS, AND INDUSTRY MISHAPS WITHIN THE CHEMICAL AND HAZARDOUS-MATERIALS SECTOR

The author has compiled the information in this appendix in a chronological sequence to share the historical perspective of incidents with the reader interested in conducting further research. The listing, along with the main text of Chapter 2, presents a capsule review of terrorist activity and other criminal acts perpetrated around the world and offers security practitioners specializing in particular regions of the world a quick reference to range and level of threat.

Source: *The information is a consolidated listing of events, activities, and news stories as reported by the U.S. Department of Homeland Security, U.S. Department of State, and various other government agencies. Contributions from newspaper articles and news media reports are also included. They are provided for a better understanding of the scope of terrorism and other criminal activity and further research for the interested reader.*

10 OCTOBER 2006 Arkansas, United States.
Hazardous-materials specialists were called to the governor's office after a letter containing a white powder led authorities to briefly evacuate the state capitol building. The powder was quickly determined to be harmless.

26 MAY 2006 Tennessee, United States.
White supremacist Demetrius Van Crocker was convicted of attempting to acquire chemical weapons and explosives to use in terrorist attacks against the American government and African American neighborhoods.

6 MARCH 2006 Pennsylvania, United States.
At the Wharton School, University of Pennsylvania, the Center for Risk Management and Decision Processes reported that of 14,500 chemical facilities, 1,145 reported 1,913 accidents between June 21, 1994 and June 20, 1999. Of the 1,145 facilities reporting accidents, 346 had multiple incidents. Half of the chemicals were involved in accidents. Half of the accidents resulted in injuries to workers. Accidents caused 1,897 injuries and 33 deaths to employees, 141 injuries and no deaths to nonemployees. Over 200,000 residents were involved in evacuations and shelter-in-place incidents.

02 MARCH 2006 Kiev, Ukraine.
Ukraine's SBU security service arrested a man at Kiev's Boryspil Airport who had a case containing radioactive uranium-238 in his car. Depleted uranium can theoretically be used to make nuclear "dirty bombs" but is often used in gun ammunition and armor because of its high density.

6 FEBRUARY 2006 Washington, DC, United States.
The Director of the CIA warns of the potential for attack by al Qaeda on chemical facilities.

15 JUNE 2005 United States.
The U.S. government must impose tighter regulation of chemical facilities to help prevent terrorist attacks, a top official of the Department of Homeland Security [DHS] said. The American Chemistry Council, which represents 132 companies, says its members already have adopted "extraordinary measures" to secure some 2,040 facilities at a cost exceeding $2 billion. Still, such companies represent only part of the chemical industry. High-risk facilities representing 20 percent of U.S. chemical operating capacity aren't governed by any kind of voluntary practice or voluntary security code.

27 APRIL 2005 Washington, DC, United States.
U.S. chemical plants are not adequately protected against terror-ists, critics said, pushing for federal regulation. Investigators described spotty results in how well the chemical industry is prepared to respond in the event of an attack. About one-fifth of the nation's 15,000 chemical facilities are close to population centers. At issue is whether the government should regulate security at privately operated chemical plants. The Department of Homeland Security [DHS] has identified 297 chemical facilities where a toxic release could affect 50,000 or more people. The DHS is working with plant owners and operators and local and state authorities "to put in place security measures, surveillance equipment and effective response plans." Such measures, however, are voluntary. In recent accidental releases at plants, mass exposures to dangerous toxins were avoided only because of rainstorms, shifting winds that scrubbed chemicals from the air, and other factors. Industry officials who oppose regulation said they have installed voluntary security measures at chemi-cal plants.

14 APRIL 2005 London, United Kingdom.
Ricin may still be at large in the UK. An al Qaeda suspect told offi-cials that he and another man, Kamal Bourgass, made two batches of ricin

from castor beans. The cell planned to put ricin on door handles of cars and shops and on open packs of toothbrushes in supermarkets. The highly toxic substance was never found. There was also evidence of a plot for a cyanide attack on the subway. Ricin is 6,000 times more poisonous than cyanide: an amount equivalent to a grain of salt is enough to kill an adult. Bourgass, 31, was jailed for 13 years for a plot to kill civilians with home-made poisons and explosives. The Algerian was previously jailed for life for murdering a Special Branch detective. A raid on the cell uncovered ingredients and equipment to manufacture a range of poisons and gases, recipes, and plans for explosives.

18 MARCH 2005 Honolulu, HI, United States.
Hazmat crews remove chemical vials from Hawaii home. More than 50 police officers, firefighters, Hawaii National Guard members who specialize in weapons of mass destruction, city paramedics, and state Department of Health environmental emergency-response personnel and hazardous-materials crews descended on a home in a quiet neighborhood to remove 78 glass vials thought to be from a military test kit for mustard gas. None of the vials were broken, and four were empty. They were transported from the site inside a triple-sealed container system to Wheeler Army Airfield under police escort. A team from the Army's Aberdeen Proving Ground in Maryland was expected to examine the vials. Authorities closed a portion of a local street and told neighbors within a quarter-mile radius to stay indoors or leave the area entirely. [Outcome was not disclosed or reported to the public.]

7 MARCH 2005 Ohio, United States.
A tanker truck crashed, spilling a liquid plastic additive considered a fire hazard and snarling commuter traffic on U.S. 23 near Kingston and Chillicothe, OH.

6 MARCH 2005 Salt Lake City, UT, United States.
A railroad tanker car leaking a mixture of chemicals sent a plume of orange fumes above Salt Lake City, UT, causing the evacuation of as many as 6,000 residents and the closing of Interstate 15. Workers found acid bubbling from three holes in the tanker. Special equipment was brought in from Las Vegas, NV, to pierce the tanker and drain the liquid into other containers. More than 100 emergency crews responded to the chemical spill, which they were initially told was composed of sulfuric, nitric, hydrofluoric, and hydrochloric acids. The chemicals are dangerous on a number of levels—any one of them could burn the skin on contact and if inhaled could damage the lungs and esophagus and cause difficulty

breathing, nausea, and vomiting. Residents were allowed to return home later in the day.

2 MARCH 2005 Boryspil Airport, Ukraine.

Ukraine's SBU security service arrested a man at Kiev's airport who had a case containing radioactive uranium-238 in his car. The man was detained at the main international gateway with 1.28 pounds of uranium. Depleted uranium, where uranium-238 is normally found, can theoretically be used to make nuclear "dirty bombs" but is often used in gun ammunition and armor because of its high density.

21 FEBRUARY 2005 Melbourne, Australia.

One of Australia's main airport terminals was shut down for eight hours as emergency crews hunted in vain for the cause of a mystery illness that struck down nearly 60 staff and passengers in the building. Paramedics, firefighters, and hazardous-materials crews in full protective clothing rushed to Melbourne Airport after staff in a domestic terminal began suffering nausea and vomiting, dizziness, headaches, and shortness of breath. Up to 2,000 staff, passengers, and their friends were evacuated and the terminal shut down while emergency crews tested air-conditioning units and other facilities in an unsuccessful search for what was causing the illness. Those affected were mostly security and airline staff working in the departure area for domestic carrier Virgin Blue. The shutdown of the terminal caused chaos around the airport, forcing the cancellation or postponement of scores of Virgin Blue and other domestic flights. The terminal reopened and flights resumed about two hours later. While some kind of toxic gas was the main suspect in the incident, the exact cause of the contamination remained a mystery.

19 FEBRUARY 2005 United States.

Since the 2001 anthrax attacks, research has focused on developing improved sensors to detect potential chemical or biological terror agents. But these devices themselves cannot head off terrorist attacks, and while they should be part of an overall protection strategy, reliance on such technology can create a false sense of security, warned the Georgia Institute of Technology. A systems approach would include central command centers, response strategies tailored to the facility, protection of water and air-circulation systems—and neutralizing and sterilizing chambers built into air-circulation systems to limit the spread of terror agents. Almost every public building in the United States has a heating and air-conditioning system that circulates the air, which also provides a vehicle for introducing both chemical and biological agents.

The concept would be to insert into that system a sterilization chamber that would disable the biological agents and decompose the chemical agents. A chamber exposing the air to ultraviolet light could inactivate most biological agents. And because of their reactive nature, most chemical agents could be neutralized with a small number of chemical processes built into filtering systems.

7 FEBRUARY 2005 United States.
Concerned about terrorism, federal officials are considering stripping the large toxic-chemical warning labels from cargo trains, which would remove the only indicator available to the general public. At the same time, officials are considering plans to reroute trains carrying hazardous cargo around metropolitan areas. The cost, however, could run to billions of dollars. A January derailment in South Carolina that killed nine people and prompted a mile-wide evacuation, as well as two derailments that killed four people last year in San Antonio, have revived a national discussion about rail safety. Chemicals accounted for about 10 percent of train cargo, or 160 million tons, in 2003, according to the Association of American Railroads. Trains carry a wide range of chemicals, including chlorine and ammonia, which can be deadly when released into the air. Federal laws require hazardous cargo to be carried in fortified container cars, with the strictest requirements for radioactive waste. Engineers must also carry lists, or manifests, of what is on each train.

12 JANUARY 2005 Ocala, FL, United States.
The FBI arrested a man after they found the biotoxin ricin in his possession in the home he shares with his mother. The former waiter had at least 83 castor beans and other byproducts consistent with the manufacture of ricin in addition to several weapons, including an AK-47 and an Uzi. He was charged with possession of a biological weapon.

11 JANUARY 2005 United States and the Bahamas.
The United States and the Bahamas are to cooperate on detecting illicit shipments of nuclear material. The U.S. Energy Department's National Nuclear Security Administration [NNSA] and the Commonwealth of the Bahamas signed an agreement to install special equipment at one of the Bahamas' busiest seaports to detect hidden shipments of nuclear and other radioactive material. The Bahamas will be the first country in the Caribbean to deploy this type of detection system. The NNSA goal is to detect, deter, and interdict illicit shipments of nuclear and other radioactive materials. NNSA works with foreign partners to equip seaports with radiation-detection equipment and to provide training to

appropriate law-enforcement officials. This is the sixth cooperative agreement and joins efforts currently underway in the Netherlands, Greece, Sri Lanka, Belgium, and Spain.

6 JANUARY 2005 South Carolina, United States.

A train collision in South Carolina caused by human error raised new concerns about the safety of transporting hazardous materials. Two were dead and 180 treated after the chemical spill [chlorine] that followed the crash, which derailed 13 cars. A switch used to park locomotives was left open, allowing the oncoming freight train to steer onto a side track and collide with a parked train. The Governor declared a state of emergency for the county, and the FAA—at the request of the South Carolina Law Enforcement Division—imposed temporary flight restrictions in the area. More than 90,000 shipments of chlorine alone are transported across the country every year. According to a study by the Naval Research Lab, 100,000 people could die in only 30 minutes if there was a major breach of a single chlorine tanker in a populated area. Deaths from hazardous spills or explosions are relatively rare, but the number of hazardous rail shipments has doubled in the last 20 years, so the potential for major accidents is growing. The government says more than 52,000 people were evacuated between 1985 and 1995 due to accidents involving the transportation of hazardous materials.

5 JANUARY 2005 Vero Beach, FL, United States.

A law firm threatened with anthrax in previous weeks received an envelope of white powder and was evacuated and quarantined for a brief time.

30 DECEMBER 2004 United States.

Terrorists who might try to manufacture biological weapons face technical problems that would confound even skilled scientists who tried to help them, biological-warfare experts say. But specialists also say it is all but inevitable that a terrorist group will gain the expertise to launch small-scale biological attacks and eventually inflict mass casualties. Information on the mechanics of creating bioweapons is easily accessible on the Internet and in technical manuals, and the equipment to do the job is readily found. Advances in bioscience and the rapid dissemination of this knowledge are making it easier for even undergraduates to create dangerous pathogens. A biowarfare consultant to the Pentagon said that while there are 1,000 to 10,000 "weaponeers" worldwide with experience working on biological arms, there are more than one million and perhaps many millions of "broadly skilled" scientists who, while lacking training in that narrow field, could construct bioweapons. Toxins such as botulinum are the easiest to turn into weapons. Bacterial agents such as anthrax are more difficult to manufacture. Viruses are tougher still.

26 NOVEMBER 2004 Iraq.
Iraqi forces found chemical materials in a Fallujah lab.

24 NOVEMBER 2004 United States.
Backed by the Pentagon's Defense Advanced Research Projects Agency [DARPA], scientists are recruiting insects, shellfish, bacteria, and even weeds to act as "bio-sentinels," which give early warning of biological and chemical attacks, detect explosives, or monitor the spread of contamination. Biologists at Virginia Commonwealth University believe bugs can check their habitats for noxious materials from anthrax to chemicals more thoroughly, cheaply, and reliably than man-made sensors. "It's more than bugs fighting terrorism. It's developing a new kind of technology to detect and map biological and chemical contaminants in the environment." Other research involves genetically modifying common weeds like the ones found in sidewalk cracks to make them change color if exposed to a biochemical attack.

27 OCTOBER 2004 United States.
Facilities identified as potentially attractive targets for a terrorist attack due to the presence of large volumes of hazardous materials have not done an adequate job of preventing and preparing for such an event. These are the findings of a survey conducted by the Paper, Allied-Industrial, Chemical, and Energy Workers [PACE] International Union in the chemical, paper, oil-refining, and other industries. There are 15,000 facilities in the United States that produce or store large quantities of 140 highly hazardous chemicals that are regulated by the U.S. Environmental Protection Agency under the Risk Management Program [RMP]. Tens of millions of people live in the areas surrounding these RMP sites. PACE represents approximately 50,000 workers at 189 RMP sites in 38 states.

25 OCTOBER 2004 United Kingdom.
A report by the British Medical Association [BMA] paints a bleak picture of the global community's ability to cope with advances in biological and genetic weapons technology. The report, "Biotechnology, Weapons, and Humanity II," warns that the "window of opportunity" to take action on this issue is shrinking fast and analyzes whether terrorist attacks such as 9/11, anthrax attacks in the U.S. in 2001, and the Moscow

Theatre siege in 2002 have impacted the development of biological weapons. If the development of biological and genetic weapons is not curtailed, a future scenario could see weapons that target specific ethnic groups, imitation viruses, bioregulators, genetically engineered anthrax, and synthetic polio virus. The problem is that the same technology being used to develop new vaccines and find cures for debilitating diseases could also be used for malign purposes.

19 OCTOBER 2004 Bratislava, Slovakia.

A post office finds two more envelopes containing suspicious white powder in addition to the five that turned up there the previous week. The previously received suspect envelopes were addressed to politicians, including the prime minister and deputy prime minister.

30 SEPTEMBER 2004 North Dakota, United States.

Three barrels of sodium cyanide fell off the back of a truck somewhere between Devils Lake and Cavalier, ND, and attracted the attention of the Federal Bureau of Investigation and the Department of Homeland Security. Two barrels were located soon after, but officials were still searching for the missing third barrel a week later. Beekeepers were going to utilize the chemical to sterilize equipment and to kill bees at the season's end. Sodium cyanide is registered for use in the commercial chrome-plating business and in mining for extracting gold and silver from ore. An official said that it isn't illegal to possess the compound, but it also isn't registered as a pesticide anywhere in the United States. As a result of the incident, officials discovered that sodium cyanide is exempt from the U.S. Department of Transportation's placard rules regarding the display of hazardous warning signs by vehicles carrying the chemical. It was also discovered that the use of sodium cyanide as a pesticide is more common than anticipated.

29 SEPTEMBER 2004 Maryland, United States.

A Maryland man convicted in an extortion scheme was charged with possessing two weaponized toxic substances, ricin and nicotine sulfate, and at least eight hand grenades. Agents investigating a $17 million extortion plot that Myron Tereshchuk, 42, directed against a business competitor discovered the prohibited substances during a search of his home. The indictment alleged that he was not properly registered to possess ricin or hand grenades.

29 SEPTEMBER 2004 Kyrgyzstan.

The Kyrgyz security agency arrested a man for trying to sell plutonium amid rising worries of a growing black-market trade in radioactive materials. Kyrgyz National Security Service agents posing as buyers arrested the man after confirming that he was in possession of pluto-nium-239. Authorities did not say how much of the radioactive material —which can be used in atomic weapons and as a reactor fuel—was con-fiscated. The plutonium was held in 60 small containers. The suspect's identity was not released. Plutonium-239 is not used in Kyrgyzstan, a former Soviet republic in central Asia, and authorities do not know where it was obtained.

22 SEPTEMBER 2004 London, UK.
Police arrested four men for terrorist offences, including the buying of radioactive material for a "dirty bomb" to detonate in the UK capital.

27 AUGUST 2004 Watertown, MA, United States.
A pipe bomb ripped through Amaranth Bio Inc., a biotech-research laboratory. Amaranth Bio is involved in innovative cell research focused on organ regeneration. No one claimed responsibility.

18 SEPTEMBER 2004 Russia.
Russian officials admitted they can not fully rule out the possibility that fissile materials, including highly enriched uranium and pluto-nium as well as technologies suitable for manufacturing nuclear weapons, may fall into the hands of international terrorists. An inspec-tion operation carried out by the International Atomic Energy Agency uncovered a diversified and highly organized clandestine network engaged in illegally selling nuclear materials and technologies.

1 SEPTEMBER 2004 Russia.
Russia deployed extra troops to guard dozens of nuclear facilities across the country after militants seized a school in the south and a suicide bomb attack in Moscow. Russia, the world's number-two atomic power after the United States, has come under international pressure to do more to protect its Soviet-era nuclear facilities against attack. Russia runs dozens of atomic reactors, uranium-enrichment facilities, and nuclear-research reactors—some in the far-flung corners of Siberia that are poorly guarded. Reactors are also attractive to militants because atomic fuel stored at many sites can be used in nuclear bombs.

14 AUGUST 2004 Malaysia.
An envelope containing suspicious white powder and threats against Americans was delivered to the U.S. embassy in Malaysia, prompting health checks on three staff at the embassy. The staff—includ-ing the second secretary—was briefly quarantined for detoxification. The embassy remained open. Malaysian police said the envelope was delivered

by post to the embassy and carried the home address of one of the staff. A leaflet was enclosed with the powder from a previously unknown group, threatening to blow up the embassy and kill Americans in Malaysia. The leaflet, from a group calling itself Jemaaah Muhahidin Mohamad, warned Washington to "take your army out of Iraq and remove the sanctions on Sudan or face the consequences."

25 JULY 2004 McCook, NE, United States.
The McCook, NE, Police Department reported that an anhydrous-ammonia leak had occurred at the Frenchman Valley Coop and that area residents were encouraged to stay inside, close doors and windows, bring pets inside, and turn off air conditioners that might draw in outside air. Anhydrous ammonia is a hydroscopic compound, which means that it seeks water from the nearest source, including the human body. This attraction places the eyes, lungs, and skin at greatest risk because of their high moisture content. Caustic burns result when the anhydrous ammonia dissolves into body tissue. Most deaths from anhydrous ammonia are caused by severe damage to the throat and lungs from a direct blast to the face. When large amounts are inhaled, the throat swells shut and victims suffocate. Exposure to vapors or liquid also can cause blindness. An additional concern is its low boiling point. The chemical freezes on contact at room temperature. It will cause burns similar to but more severe than those caused by dry ice.

22 JULY 2004 Kentucky, United States.
A chemical spill at a Dow Corning plant released a cloud of hydrochloric-acid vapor, disrupting traffic along the area's highway, water, and rail lines. Two employees were taken to a hospital, treated, and released the same day. Details of the injuries were not released. Officials said the vapor was contained on site and there was no danger to the public. Hydrochloric-acid fumes can irritate the eyes, skin, mouth, and respiratory system. Authorities closed nearby rail lines and a four-mile section of U.S. 42 and stopped water traffic on the Ohio River for about four hours. Residents within two miles were told to stay indoors.

22 JULY 2004 California, United States.
A stretch of Interstate 15 reopened in California's North County after hazmat crews finished clearing a caustic chemical that spilled from a tanker truck. A corrosive mix of hydrochloric, nitric, chromic, and hydrofluoric acids spilled onto Interstate 15 in the Hidden Meadows area, prompting authorities to close several miles of the freeway in both directions and evacuate 10 homes in the area. A man driving north on

Interstate 15 saw thick green smoke pouring out of the side of the 8,000-gallon tanker truck and motioned to the truck driver to pull over. No one was injured and the driver was not arrested.

7 JULY 2004 Kyrgyzstan.

Kyrgyzstan is preparing to respond to possible chemical attacks. The country's National Security Service mentioned possible terrorist attacks against U.S. citizens deployed at military bases in Kyrgyzstan and neighboring Uzbekistan. There are 49 waste dumps and 20 chemically dangerous sites in the country. Rescue workers from Kyrgyzstan joined their counterparts from Central Asian neighbors Kazakhstan, Tajikistan, and Uzbekistan in a five-day chemical and radioactive rescue-training session sponsored by the Organization for the Prohibition of Chemical Weapons.

5 JULY 2004 United States.

Experts warn that al Qaeda-linked terrorist groups could try to launch biological or chemical attacks against U.S. allies and secular Muslim governments in Asia using widely available materials. Authorities suspect Jemaah Islamiyah [JI] to be exploiting anti-western sentiment over Iraq to recruit new members and raise funds that could be used to obtain or develop such weapons. Warning signs include the discovery in October 2003 of manuals on bioterrorism at a JI hideout in the southern Philippines and the arrest in June 2003 of a man who tried to sell cesium 137, a radioactive material that could be used to make dirty bombs.

2 JULY 2004 Iraq.

Terrorists may have been close to obtaining munitions containing the deadly nerve agent cyclosarin, which Polish soldiers recovered last month in Iraq. Polish troops had been searching for munitions as part of their regular mission in south-central Iraq when they were told by an informant in May 2004 that terrorists had made a bid to buy the chemical weapons, which date back to Saddam Hussein's war with Iran in the 1980s.

1 JULY 2004 United Nations.

The Organization for the Prohibition of Chemical Weapons [OPCW] hopes to have all U.N.-recognized countries pledge their opposition to chemical weaponry by 2007. Between the group and its goal, however, are some of the most impoverished, secretive, and strife-ridden nations in the world. There are 164 countries party to the Chemical Weapons Convention, each agreeing not to develop or use chemical agents such as mustard gas and sarin and to destroy any existing stocks. Another 30 nations have signed but not ratified the treaty, said a OPCW spokesperson. Holdouts include North Korea, Israel, Egypt, and

Syria—all of which are believed to have had chemical-weapons programs. Other nonmembers are grouped in Africa, the Caribbean, and the Pacific Islands—developing regions whose leaders might simply not see a reason to endure the cost and work involved in joining the treaty, experts said.

1 JULY 2004 Williamsburg, KY, United States.

For nearly nine hours, hazmat, police, and fire crews were on the scene of a synthetic-drug lab in Williamsburg, KY. Police say some new tenants were cleaning out the basement of 544 North 6th Street when they noticed a strange smell coming from more than a dozen containers. Hazmat officials say this wasn't their typical drug discovery. The chemicals hidden in the basement were so toxic they can burn away flesh in seconds. It was a synthetic-drug lab—capable of producing drugs like PCP and heroin—chemicals that when mixed are so dangerous precautions like encapsulated suits and decontamination are required. The chemicals will be taken to a lab for processing. Police say they have two suspects and arrests are pending.

1 JULY 2004 United States.

The Department of Homeland Security found 29 truck drivers licensed to carry hazardous materials with possible ties to terrorist organizations. In addition, a man whom authorities believe may have been part of an al Qaeda "sleeper cell" obtained a license to haul hazmat materials months after he was identified as a suspected terrorist by the FBI. The FBI identified Mohamad Kamal Elzahabi as a suspected terrorist before the September 11, 2001, attacks, according to unidentified law-enforcement officials. However, Minnesota Department of Public Safety officials said they did not know that Elzahabi was suspected of having al Qaeda connections when he applied in early 2002 for a CDL to drive a school bus and to haul hazmat materials.

29 JUNE 2004 Atlanta, GA, United States.

The discovery of a chemical that had solidified with age into a potential explosive forced an evacuation at Clark Atlanta University for about two hours until the substance could be neutralized. About 200 students and faculty members were affected by the evacuation after a cleaning crew found several 2 1/2- to 5-pound containers of dried picric acid at the school's science center. When picric acid solidifies, it can be equivalent to dynamite. School officials did not know how long the acid had been stored there.

29 JUNE 2004 Preston County, WV, United States.

Residents within a one-mile radius of an old mine site in Preston County, WV, were asked to leave their homes after an ammonia leak. No injuries were reported, but emergency officials ordered evacuations because vapors from anhydrous ammonia can irritate the eyes and respiratory tract. The chemical, which is used to treat acid mine drainage, can be fatal if inhaled in large doses. The leak occurred at the Squire's Creek Mine when a valve on a storage tank failed. Officials with the state Department of Environmental Protection said about 6,000 gallons of ammonia turned to gas and produced a chemical cloud. Patriot Mining Company, Inc., a subsidiary of Anker Coal Co. owns the inactive mine. Though the ammonia is stored in a 12,000-gallon tank, Environmental Protection Agency records available online indicate the tank is never filled more than 85 percent of capacity. EPA records indicate there have been no accidents at the site in the past five years. Patriot employees are trained in the safe handling, storage, and use of anhydrous ammonia as well as in what to do if there is an accidental release.

29 JUNE 2004 San Antonio, TX, United States.

Federal investigators began their on-the-scene investigation of a freight-train wreck near San Antonio, TX, that killed from one to three people. The collision between a Union Pacific and a Burlington Northern-Santa Fe freight train unleashed a cloud of chlorine gas and ammonium nitrate. A 23-year-old railroad worker was killed. Bexar County medical examiners also investigated whether the deaths of two women [an 84 year-old and her 59 year-old daughter] found in a home a mile away were a result of the chemical leak. The wreck derailed 40 cars in a rural area southwest of San Antonio. Fire officials say as many as 50 people suffered minor respiratory irritation.

28 JUNE 2004 Hamburg, Germany.

A German ship carrying nearly 1000 tons of sulfuric acid sunk in the German port of Hamburg and raised fears of a major environmental accident. The vessel, the ENA-2, went down after hitting a container vessel. Nine port employees and two police officers were taken to the hospital with breathing difficulties after inhaling gas from the ship's ventilation system. Experts said a disaster would have been inevitable had the double-hull structure of the ship been damaged when it struck the container ship during docking procedures. Reports that the captain of the German-registered chemical carrier had been drunk during docking were investigated. The ENA-2, owned by Norddeutsche Affinerie [NA], had been maneuvering into the Hamburg oil terminal when it collided with the container vessel. The tanker managed to dock but then sank at its moorings.

26 JUNE 2004 Mansfield, OH, United States.

A chemical spill shut down Ohio Highway 13 just north of Mansfield for most of the day. A CCX semi pulling two trailers turned onto the highway and the rear trailer tipped over, spilling the chemicals. The truck was carrying soaps, cleaning solutions, paint, and paint-related materials to a site in the city. Mansfield police believed the truck driver might have been driving too fast when he made the turn. Among the chemicals was a drum of sodium metasilicate, a corrosive cleaning solution used on metals. Businesses in the area were asked to shut down their air-conditioning units and close all windows.

25 JUNE 2004 Shaanxi, China.

At least 3,500 people were evacuated when liquefied-petroleum gas spilled at a gas station that belongs to a chemical plant in the northwestern Shaanxi province. The gas station of the Shaanxi provincial Chemical Co. is located on the outskirts of Weinan city and is close to four villages of more than 3,500 people. Witnesses said a LPG truck drove into the gas station in the early morning and within 30 minutes a strong smell of gas was noticed in the air. Over 120 policemen and firemen were mobilized to help the locals evacuate. Officials said the gas had leaked from sand holes in the valve of the LPG tank.

23 JUNE 2004 Vandalia, OH, United States.

A chemical cloud covered a Vandalia, OH, field and forced the evacuation of several homes. A motorist passing by the field noticed a cloud hovering over the roadway and called authorities. Upon arrival, firefighters found that the valve of one anhydrous-ammonia tank had been opened, and they immediately closed several roads leading to the area. Officials said the chemical can cause burning eyes and breathing problems. As a precaution, authorities evacuated around 10 homes, but the residents were allowed to return after the tank valve was closed. There were no injuries or health problems as a result of the chemical leak.

21 JUNE 2004 Pennsylvania, United States.

A tanker truck exploded, spilling chemicals on a freeway and injuring one person. Police said a corrosive-chemical mixture of sodium hydrochlorite and sodium hydroxite was mistakenly placed into the truck's aluminum tank. The mixture should have been transported in a stainless-steel tank. State police had stopped the truck around 8 P.M. because it was leaking its load onto Interstate 90 near the Interstate 79 interchange. The truck exploded about 30 minutes later, injuring one bystander. Emergency crews tried to contain the chemical, which leaked into the highway-drainage system. Crews worried that heavy rain showers forecast for the next morning would spread the contamination.

21 JUNE 2004 Clay, KY, United States.
An attempted anhydrous-ammonia theft at a farm store in the western Kentucky town of Clay caused a chemical leak that forced evacuations and left two people in need of treatment. Anhydrous ammonia is a common ingredient in methamphetamine, an illegal drug that has become pervasive in western Kentucky. The county sheriff said one or more people slipped under a chain-link fence and placed a hose on a 1,000-gallon tank of ammonia. They attempted to pump the chemical into a small tank they had brought, but the hose blew off, releasing about 700 pounds of ammonia into the air. As many as 300 people in Webster County were asked to leave their homes, and most took shelter in a local church. It took workers about an hour to cap off the leaking ammonia and another hour for the chemical to dissipate. At press time there were no suspects. They did leave the tank and hose behind, causing officials to believe they may have received a dose of the ammonia.

10 JUNE 2004 Oakland, CA, United States.
At least five workers developing an anthrax vaccine at a children's hospital-research lab in Oakland, CA, were accidentally exposed to the bacterium because of a shipping mistake. Officials said none of the researchers showed symptoms of infection since the first exposure two weeks earlier, but each was treated with precautionary antibiotics. The researchers believed they were working with syringes full of a dead version of anthrax; instead they were shipped live anthrax by a lab of the Southern Research Institute in Frederick, MD. Anthrax produces severe flulike symptoms in most of its victims. If inhaled, ingested, or otherwise introduced into the body, it can kill. Though the five workers were exposed, state health officials and the hospital did not believe anyone was infected because the researchers took proper safety precautions.

1 JUNE 2004 United States.
Concerns rise over chemicals as targets. Homeland Security watchdogs call them "prepositioned weapons of mass destruction" for terrorists: huge tanks of concentrated deadly gases that the chemical industry stores near densely populated areas and that railroads bring through cities en route to somewhere else. The U.S. harbors more than 100 chemical facilities where an accident would put more than a million people at risk,

according to documents filed with the EPA. The presence of these highly toxic chemicals in the midst of cities may be the most vulnerable point in the nation's defenses. But proposals to reduce that risk by requiring the use of alternative chemicals or rerouting hazardous tankers around a city have faltered. Fear of such an attack on a chemical facility prompted bipartisan momentum in Congress after the September 11, 2001, attacks to require the chemical industry to switch to less dangerous processes where possible. Nevertheless, nearly three years later, the laws regulating chemical plants remained the same as before 9/11.

24 MAY 2004 Mihailesti, Romania.

A truck loaded with pesticides overturned and subsequently exploded in southeastern Romania—in the town of Mihailesti, some 70 kilometers [45 miles] northeast of the capital—killing 16 people, including several firefighters, in a powerful blast, the government said.

23 MAY 2004 Argonia, KS, United States.

Eight cars of a Burlington Northern Santa Fe train jumped the tracks in Argonia, KS, forcing the temporary evacuation of an area near a fertilizer business. One of the cars may have come in contact with a tank of anhydrous ammonia at the fertilizer business. Two of the cars contained hazardous materials. Some ammonia chloride leaked from one of the cars.

APRIL 2004 United Kingdom.

British antiterrorist officers uncovered a possible plot to use osmium tetroxide in an attack. The chemical can cause death or blindness if dispersed in an explosion. The alleged plotters were reported to have been in direct contact with extremists in Pakistan. The plot was discovered when their telephone calls were monitored by GCHQ, the UK government electronic-surveillance center. A related plot was uncovered by the French in January 2004.

8 FEBRUARY 2004 Islamabad, Pakistan.

After months of cross-continental pressure and intelligence gathering, Abdul Qadeer Khan—father of Pakistan's nuclear program—admitted passing nuclear and bomb-making information to Iran, North Korea, and Libya.

3 FEBRUARY 2004 Washington, DC, United States.

Ricin powder was found in the office building of the U.S. Republican Senate Majority Leader and sparked fears of a bioattack. Three Senate buildings were closed and personnel were decontaminated.

4 JUNE 2003 Brussels, Belgium.
Letters containing the nerve agent adamsite were sent to the U.S., British, and Saudi embassies; the government of Prime Minister Guy Verhofstadt; the Court of Brussels; a Belgian ministry; the Oostende airport; and the Antwerp port authority. After exposure to the substance, at least two postal workers and five policemen were hospitalized with skin irritation, eye irritation, and breathing difficulty. In Oostende, three persons exposed to the tainted letter were hospitalized. Belgium police suspected a 45-year-old Iraqi political refugee opposed to the U.S. war in Iraq. The police searched his residence and confiscated a document and plastic bag containing some powder. The antiterrorism investigators also suffered skin irritation, eye irritation, and breathing difficulty. The Iraqi was charged with premeditated assault.

MARCH 2003 Syria.
Western intelligence agencies received information that components of Saddam's WMD went to Syria before the war. On the eve of the Iraq war, intelligence agencies monitored the movement of large convoys of high-volume trucks from a presidential palace in Iraq to a presidential palace in Syria. Subsequent intelligence could not confirm the cargo carried by the convoys.

2 MARCH 2003 Karachi, Pakistan.
A plot was uncovered to construct and detonate dirty nukes on United States soil; it was revealed that Osama bin Laden, al-Zawarhiri, and Dr. X planned the scheme. The plan was discovered during questioning of arrested Shaikh Mohammed, al Qaeda's military-operations chief.

31 MAY 2002 Lowell, MA, United States.
A rail car leaked hydrochloric acid, creating a cloud of hazardous vapor.

JANUARY 2002 Minot, ND, United States.
A Canadian Pacific Railway derailment ruptured seven tank cars carrying anhydrous ammonia, creating a vapor plume approximately 5 miles long and 2 1/2 miles wide. The hazardous material released affected 15,000 people, causing one death and more than 300 injuries.

OCTOBER-DECEMBER 2001 United States.
The first deadly use of biological weapons by terrorists was the late-2001 U.S. mailings of anthrax-laced letters by persons still unknown. Between early October and early December 2001, five people died from anthrax infection, and at least 13 others contracted the disease

in Washington, DC; New York City; Trenton, New Jersey; and Boca Raton, Florida. Anthrax spores were found in a number of government buildings and postal facilities in these and other areas. Most of the confirmed anthrax cases were tied to contaminated letters mailed to media personalities and U.S. senators. Thousands of people were potentially exposed to the spores and took preventive antibiotics. Numerous mail facilities and government buildings were shut down for investigation and decontamination. In the wake of these incidents, federal, state, and local emergency-response agencies across the United States had to respond to thousands of calls to investigate suspicious packages, unknown powders, and other suspected exposures. Almost all of these incidents turned out not to involve an actual biohazard.

NOVEMBER 2001 Russia.
Intelligence agencies reported nuclear weapons were available to Osama bin Laden for $10 million to $20 million.

25 NOVEMBER 2001 Middle East.
Mullah Mohammed Omar reported that the nuclear destruction of the United States is underway.

11 OCTOBER 2001 United States.
Intelligence agencies reported two nuclear suitcases may have reached al Qaeda operatives in the United States.

13 SEPTEMBER 2001 Israel.
Authorities prevented a radioactive backpack bomb from entering Israel. Al Qaeda operatives were suspected.

JULY 2001 Baltimore, MD, United States.
The derailment of a CSX transportation train in the Howard Street Tunnel and the ensuing fire, fueled by hazardous materials, disrupted the city for several days.

FEBRUARY 2000
Al Qaeda defector Jamal Ahmed al-Fadl testified that Osama bin Laden had tried to buy uranium on the black market for $1.5 million in a presumed attempt to develop nuclear weapons.

20 MARCH 1995 Japan.
Terrorists simultaneously released the chemical nerve agent sarin on several Tokyo subway trains, killing 12 persons and injuring up to 6,000. Members of Aum Shinrikyo, a Japanese cult, were responsible.

Authorities further report that Aum Shinrikyo had planned to use remote-control helicopters to spray dangerous substances from the air. The plan was abandoned when the helicopters crashed during testing. The group was responsible for other mysterious chemical accidents in Japan in 1994.

1994 Moscow, Russia.
Officials reported 3 kilograms of 90 percent enriched uranium stolen.

NOVEMBER 1992 United States.
Planning for 1993 World Trade Center bombing. Authorities reported Yousef first called chemical companies for raw material. In January 1993, he moved to Jersey City, where he started to build a urea-nitrate bomb.

1992 Podolsk, Russia.
Officials reported 1.5 kilograms of 90 percent enriched uranium stolen from the Luch Production Association.

1988 Iraq.
Saddam Hussein used chemical weapons against the Kurds.

DECEMBER 1984 Bhopal, India.
Accidental release of methyl isocyanate at the Union Carbide plant killed thousands.

APPENDIX B: UNITED STATES GOVERNMENT AND INDUSTRY CHEMICAL AND HAZARDOUS-MATERIALS INITIATIVES

Currently, parts of the chemical industry have taken positive, voluntary steps to protect sector infrastructure. For example, several trade associations have developed or are developing security codes to help their members address the need to reduce vulnerabilities. These commendable efforts will make important contributions to protecting against terrorist attacks, but they are in the early stages of implementation. It should also be noted that a significant percentage of companies that operate major hazardous-chemical facilities do not abide by voluntary security codes developed by other parts of the industry.

Chemical- and hazardous-materials-sector protection initiatives include efforts to:

- **Promote enhanced site security.** The DHS, in concert with the EPA, will work with Congress to enact legislation that would require certain chemical facilities, particularly those that maintain large quantities of hazardous chemicals in close proximity to population centers, to undertake vulnerability assessments and take reasonable steps to reduce the vulnerabilities identified.

- **Review current laws and regulations that pertain to the sale and distribution of pesticides and other highly toxic substances.** The EPA, in consultation with the DHS and other federal, state, and local agencies and other appropriate stakeholders, will review current practices and existing statutory requirements on the distribution and sale of highly toxic pesticides and industrial chemicals. This process will help identify whether additional measures may be necessary to address related security issues.

- **Continue to develop the chemical Information Sharing Analysis Center [ISAC] and recruit sector constituents to participate.** The purpose of the chemical sector's ISAC (in the early stages of development) is to facilitate advanced warnings on security threats and the sharing of other security-related data. The DHS and the EPA, in concert with chemical-industry officials, will promote the ISAC concept within the sector in order to draw increased participation from the industry at large.

Chapter 15

THE AGRICULTURE AND FOOD SECTOR

This chapter outlines the:

- Importance of the agriculture and food sector to the social, economic, and political stability of the nation
- Agriculture-and-food-sector attractiveness to terrorist and criminal elements
- Agriculture-and-food-sector vulnerabilities
- Tailoring the S^3E Security Methodology to the agriculture and food sector
- Agriculture-and-food-sector challenges facing the security-assessment team
- Applying the security assessment methodology to the agriculture and food sector
- Preparing the Energy Security Assessment Report
- Historical overview of selected agriculture & food incidents
- U.S. government agriculture & food initiatives

IMPORTANCE TO THE SOCIAL, ECONOMIC, AND POLITICAL STABILITY OF THE NATION

By any reasonable measure, agriculture is not just a vital component of the national economy but the global economy as well. The agriculture and food chain includes the farmers and ranchers who grow our food products and raise our animals as well as restaurants and retail outlets where food is purchased and consumed and home kitchens.

These industries are a source of essential commodities in the U.S. and account for close to one-fifth of the gross domestic product. Within the United States there are over 1.9 million farms, 87,000 food-processing plants, and 1 million restaurants and food-service outlets. According to the Agriculture Department, farming directly employs about 3 percent of the American population, and one out of eight Americans work in an occupation directly supported by food production. **This makes the food and agricultural sector the nation's largest employer.** Crops make up more than one-half the total value of American farm commodities and comprise the major components of prepared feeds for livestock, poultry, and farm-raised fish. The farm sector alone generates $50 billion and another $54 billion a year through meat and milk sales.

Agricultural exports exceed $50 billion a year [approximately one-quarter of its farm and ranch products], making it the largest positive contributor to the U.S. trade balance. Exports of American agricultural products account for 15 percent of all global agricultural exports. This equates to about $140 billion a year. According to the U.S. Department of Agriculture, in 1998 the U.S. produced nearly half of the world's soybeans, more than 40 percent of its corn, 20 percent of its cotton, 12 percent of its wheat, and 16 percent of its meat. According to government officials, food production in 2001 constituted 9.7 percent of the U.S. gross domestic product, generating sales of $991.5 billion.

An Attractive Target for Terrorists

The United States spends more than $1 billion every year to keep America's food supply safe. However, even without terrorism, food-borne diseases cause about 5,000 deaths and 325,000 hospitalizations each year, according to the Centers for Disease Control and Prevention. **Today, the greatest threats to the food and agricultural systems are disease and contamination.** [1] Due to the sector's many points of entry, detection represents a challenge to assuring their protection.

There have been many instances throughout recorded history in which civilian food supplies have been sabotaged deliberately during military campaigns to terrorize or otherwise intimidate civilian populations [2]. Prior to the 1991 Gulf War, Iraq cultivated foot-and-mouth disease and camelpox [3]. In Afghanistan after the 9/11 attacks, the U.S. military found hundreds of pages of U.S. agricultural documents, including copies of the al Qaeda training manual devoted to agricultural terrorism—the destruction of crops, livestock, and food processing operations. During World War II the British experimented with "cattle cakes" [cow snacks laced with anthrax] to disrupt the German beef industry. The United States field-tested both hog cholera and exotic Newcastle disease (END). The Soviet Union cultivated foot-and-mouth disease, rinderpest, and sheep and goat pox. During the apartheid years, the Republic of South Africa weaponized both foot-and-mouth disease and African swine fever for use in Angola, Namibia, and Zimbabwe. A historical overview of selected terrorist attacks, criminal incidents, and industry mishaps within the agriculture and food sector are displayed in Appendix A.

The U.S. agriculture sector is an obvious target for terrorists: infecting plants or animals with deadly disease is easier, cheaper, and less risky than infecting humans directly; the economic consequences of a widespread attack would be enormous; and the panic and fear such an attack might reap could lead to wide-scale social disruption. Terrorists would use the resulting fear and alarm to their advantage to create an overall atmosphere of anxiety without actually having to carry out indiscriminate civilian-directed attacks. An act of agroterrorism through the release of contagious pathogens against livestock or the contamination of the farm-to-table continuum through the introduction of toxic or bacterial agents could have devastating consequences, not only in terms of human health but also because of economic impact, loss of consumer confidence in the safety of the nation's food supplies, and undermining of public confidence in and support of government. [4]

Complete surveillance of U.S. agricultural holdings is not a realistic or cost-effective option. With more than 1.9 million farms, millions of acres of farmland, 87,000 processing plants, and 1 million restaurants and food-service outlets, no inspection regime could fully guarantee safety and security.

Agriculture and Food Vulnerabilities

Food-production systems range from farms with goods marketed to neighboring communities to large corporations with global production and distribution systems. Many foods, such as fish, meat, poultry, fruits, and vegetables, are consumed with minimal processing. Others, such as cereal products and cooking oils, undergo considerable processing before reaching the consumer. The production system and the steps vulnerable to attack are therefore different for each type of food. **Often it is the interfaces between components of the chain, where food changes hands, that are the most vulnerable to sabotage.** The potential for intentional contamination of products is likely to increase as the point of contamination nears the points of production and distribution. However, the potential for greater individual morbidity or mortality usually increases as the agent is introduced closer to the point of consumption. [5]

The agriculture and food sector is vulnerable to deliberate and accidental disruption because [6]:

- The agriculture and food sector operates in a relatively open manner.
- Security at animal auctions and other sale events lacks surveillance or monitoring.
- Food-processing and -packing plants lack security precautions.
- Reporting unusual occurrences is inconsistent and does not promote early warning and identification of pathogenic outbreaks.
- A short supply of qualified veterinarians exists to recognize and treat exotic livestock diseases.

Government officials and security experts are concerned that food-related terrorism could involve either attempts to introduce poisons into the food supply or attacks that would ruin domestically cultivated crops or livestock. **Many agricultural experts believe the greatest threat to U.S. agriculture would be the deliberate or accidental introduction of foot-and-mouth disease,** the highly contagious viral disease that attacks cloven-footed animals including cattle, swine, sheep, deer, and elk. While humans cannot contract the disease from animals, its effect on animals is so swift and debilitating that milk and meat production could be severely cut nationwide. With thousands of animals being transported across state lines every day, an outbreak could spread within days before animal-health officials would even be able to provide a definitive diagnosis. No cases of foot-and-mouth disease have been diagnosed in the U.S. since 1929, but with more than 100 million head of cattle, 70 million pigs, 10 million sheep, and more than 40 million wild cloven-footed animals, the country remains at great risk for the disease. The U.S. Department of

Agriculture estimates that an attack on livestock with a contagious disease such as foot and mouth could cause $10 to $30 billion in damage to the U.S. economy. [7]

The number and variety of food-borne, crop, and livestock diseases make it hard to distinguish terrorist attacks from natural disasters. It took a year for U.S. officials to conclude that the Oregon religious cult—which poisoned salad bars in various restaurants with salmonella bacteria—was in fact a deliberate attack and not a general food-contamination incident. An attack could occur against crops or livestock raised on American soil or at any point between farm and table by introducing toxins anywhere along the food chain. Imported food also could be tainted with biological or chemical agents before entering the United States. Such attacks could cause death and illness depending on whether toxic substances were introduced into food and how quickly the contamination was detected. **Much of the knowledge required to produce chemical and biological agents that could be used to purposely contaminate food is in the public domain.** This deliberate contamination may be very difficult to recognize, especially if the agent is uncommon and the clinical symptoms are obscure. In some cases, deliberate contamination of food may reveal itself through disease clusters in animals [8].

Even attacks that don't directly affect human health can cause panic, undermine the economy, and erode confidence in the United States government. It is highly unlikely that acts of agroterrorism can be completely prevented. It is even more unlikely, if not impossible, to prevent hoaxes, such as spreading false rumors about unsafe foods via the mass media or the Internet. Food-terrorism hoaxes may overwhelm emergency-response systems and cause economic and political disruption. They often follow actual terrorist incidents or threats against other infrastructure sectors. **Publicity about unsubstantiated threats can be as effective as an actual attack in eroding public confidence in the food supply.** In addition to generating panic, such publicity often propagates further hoaxes such as "copycat" impulses, which can complicate the emergency response.

TAILORING THE S^3E SECURITY ASSESSMENT PROCESS FOR THE AGRICULTURE AND FOOD SECTOR

The exhibit on the next page highlights in grey shaded areas those specific elements that are tailored for special application to the agriculture and food sector.

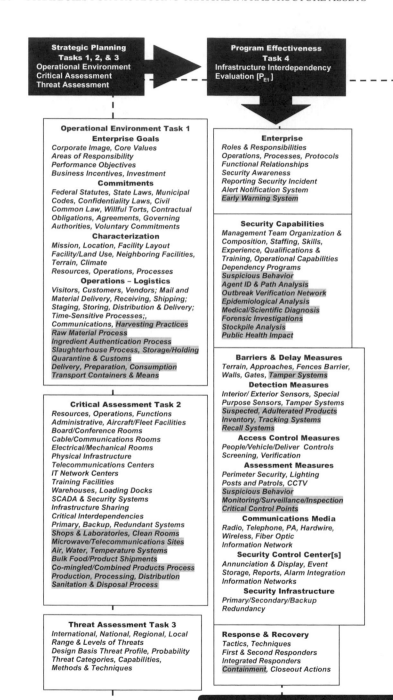

Exhibit 15.1 The S^3E Security-Assessment Methodology for the Agriculture and Food Sector

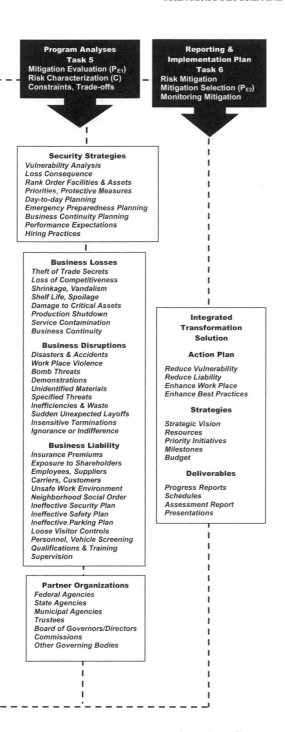

Program Analyses
Task 5
Mitigation Evaluation (P_{E1})
Risk Characterization (C)
Constraints, Trade-offs

Reporting &
Implementation Plan
Task 6
Risk Mitigation
Mitigation Selection (P_{E2})
Monitoring Mitigation

Security Strategies
Vulnerability Analysis
Loss Consequence
Rank Order Facilities & Assets
Priorities, Protective Measures
Day-to-day Planning
Emergency Preparedness Planning
Business Continuity Planning
Performance Expectations
Hiring Practices

Business Losses
Theft of Trade Secrets
Loss of Competitiveness
Shrinkage, Vandalism
Shelf Life, Spoilage
Damage to Critical Assets
Production Shutdown
Service Contamination
Business Continuity

Business Disruptions
Disasters & Accidents
Work Place Violence
Bomb Threats
Demonstrations
Unidentified Materials
Specified Threats
Inefficiencies & Waste
Sudden Unexpected Layoffs
Insensitive Terminations
Ignorance or Indifference

Business Liability
Insurance Premiums
Exposure to Shareholders
Employees, Suppliers
Carriers, Customers
Unsafe Work Environment
Neighborhood Social Order
Ineffective Security Plan
Ineffective Safety Plan
Ineffective Parking Plan
Loose Visitor Controls
Personnel, Vehicle Screening
Qualifications & Training
Supervision

Integrated
Transformation
Solution

Action Plan

Reduce Vulnerability
Reduce Liability
Enhance Work Place
Enhance Best Practices

Strategies

Strategic Vision
Resources
Priority Initiatives
Milestones
Budget

Deliverables

Progress Reports
Schedules
Assessment Report
Presentations

Partner Organizations
Federal Agencies
State Agencies
Municipal Agencies
Trustees
Board of Governors/Directors
Commissions
Other Governing Bodies

Exhibit 15.1 *(continued)*

Agriculture and Food Challenges Facing the Security-Assessment Team

The United States has a strong, well-functioning food-safety system to protect the public against unintentional contamination of food products.

In addition to agriculture and food industry measures, the overall mechanism includes extensive analyses of critical control points in the food-supply chain and federal, state, and local inspections of food-processing and -storage facilities, as well as food-service establishments. However, in an industry as complex and varied as agriculture, security is an elusive concept. From sprawling farms to feed lots, from state fairs to food-processing plants, there are countless points at which terrorists could access the food-supply system with relative ease. [9]

Changes in the ways that foods and crops are produced, distributed, and consumed present new challenges for the security-assessment team in terms of devising workable security solutions. More of our food is grown abroad, many foods are transported long distances, and we eat away from home more frequently. The security-assessment team must be sensitive to public confidence in the safety of agricultural and food-processing and -packaging systems, as these represent a key part of sustaining the economic viability of the sector. The security-assessment team must also understand that America's reputation as a reliable supplier of safe, high-quality foodstuffs is likewise essential to maintaining the confidence of foreign customers, who are important to the national economy as a whole. [10]

APPLYING THE S^3E SECURITY-ASSESSMENT METHODOLOGY TO THE AGRICULTURE AND FOOD SECTOR

The security-assessment team should use its collective expertise to:

- Identify the important mission and functions of the food or agriculture facility [**Task 1—Operational Environment**] and incorporate the development of all business operations and continuity expectations. Examination focuses on services and the customer base to determine the business contributions to the community, its culture, and business objectives. This includes but is not necessarily limited to identification of business goals and objectives, customers, and commitments and characterization of facilities and boundaries, as shown in Exhibits 15.2 through 15.5.

Areas of Responsibility	Performance Objectives
Business Incentives & Investment	Public Safety & Public Confidence
Corporate Image & Core Values	Safe Storage, Transport, and Distribution
Harvesting, Production, Processing	Safe Consumption
Mission & Services	

Exhibit 15.2 Identify Agriculture and Food Enterprise Business Goals and Objectives

Local, Regional	Industrial & Business
General Public, Critical Customers	National, International
Government, Military	Size of Population Served

Exhibit 15.3 Identify Agriculture and Food Enterprise Customer Base

29 CFR 1910.36 - Means of Egress	Architectural Barrier Act 1968 [As Amended]
29 CFR 1910.151 - Medical Services and First Aid	
29 CFR 1910.155 - Fire Protection	Bioterrorism Preparedness and Response Act, 2002 [As Amended]
18 U.S.C. 2511 - Technical Surveillance	Carriers, Suppliers, Vendors, Insurance
21 U.S.C. 801 - Drug Abuse and Control	Civil Rights Act 1964 [U.S.C. 200e] [As Amended]
26 U.S.C 7201 - Tax Matters	Commerce Commission
29 U.S.C. 2601 - Family and Medical Leave Act	Comprehensive Drug Abuse Prevention Act 1970 [As Amended]
31 U.S.C. 5322 - Financial Transaction Reports	Comprehensive Environmental Response, Compensation and Liability Act 1980 [As Amended]
41 U.S.C. 51 - U.S. Contractor Kickbacks	
41 U.S.C. 701 - Drug Free Workplace Act	Controlled Substance Act [As Amended]
46 U.S.C. 1903 - Maritime Drug Enforcement	Drug Free Workplace Act 1988 [As Amended]
50 U.S.C. 1801 - Foreign Intelligence Surveillance Act	Emergency Planning & Right-to-Know Act 1986 [As Amended]
Americans with Disabilities Act [42 U.S.C. 12101]	Employee Polygraph Protection Act 1988 [29 U.S.C. 2001] [As Amended]
Animal and Plant Health Inspection Service	Fair Credit Reporting Act [15 U.S.C. 1681] [As Amended]

(continued on next page)

Exhibit 15.4 Identify Agriculture and Food Enterprise Commitments

Family Education Rights and Privacy Act 1974 [As Amended]	Privacy Act [5 U.S.C. 522e] [As Amended]
	Rehabilitation Act, 1973 [As Amended]
Family and Medical Leave Act 1993 [As Amended]	State and Local Industry Administrators and Authorities
Federal Polygraph Protection Act [As Amended]	State and Local Emergency Planning Officials and Committees
Federal Privacy Act 1974 [As Amended]	Uniform Trade Secrets Act
Food Safety and Inspection Service	U.S. Department of Agriculture [DOA]
Freedom of Information Act [U.S.C. 552] [As Amended]	U.S. Department of Commerce
	U.S. Department of Homeland Security [DHS]
National Labor Relations Act [29 U.S.C. 151] [As Amended]	U.S. Department of Justice [DOS]
Interagency Agreements	U.S. Department of Transportation [DOT]
Internal Revenue Service	U.S. Drug Enforcement Agency [DEA]
International Atomic Emergency Agency	U.S. Environmental Protection Agency [EPA]
International Program on Chemical Safety	
Labor Management Relations Act	U.S. Food and Drug Administration [FDA]
Memorandum of Agreement	U.S. Patriot Act, 2002 [As Amended]
Memorandum of Understanding	U.S. Postal System
Oil Pollution Act 1990 [As Amended]	U.S. Occupational Safety and Health Administration [OSHA]
Omnibus Crime Control and Safe Streets Act, 1968 [As Amended]	World Health Organization [WHO]
Policy Coordination Committee [PCC]	

Exhibit 15.4 *(continued)*

- Describe the utility configuration, operations, and other elements
  [**Task 2—Critical Assessment**] by examining conditions, circumstances, and situations relative to safeguarding public health and
  safety and to reducing the potential for disruption of food and agriculture services. Focus is on an integrated evaluation process to
  determine which assets need protection in order to minimize the
  impact of threat events. The security-assessment team identifies and
  prioritizes facilities, processes, and assets critical to agriculture and
  food-service objectives that might be subject to malevolent acts or
  disasters that could result in a potential series of undesired consequences. The process takes into account the impacts that could substantially disrupt the ability of the utility to provide safe services and

Commercially Leased Facilities	Neighboring Facilities
Dedicated site owned and operated by the enterprise.	Operated and maintained through reciprocal agreements with internal and external entities.
Dispersed Locations	
Facility Layout & Location	Terrain & Climate
Facility Land Use	Types of assets at specific locations.
Infrastructure Sharing	

Exhibit 15.7 Characterize Configuration of Agriculture and Food Enterprise Facilities and Boundaries

strives to reduce security risks associated with the consequences of significant events. These include but are not necessarily limited to assessment of assets, operations, and processes and prioritization of assets and operations, as shown in Exhibits 15.6 and 15.7.

Administrative, Aircraft/Fleet Facilities	Mixing Processes
Air, Water, Temperature Systems	Monitoring, Surveillance, Inspection
Board/Conference Rooms	Operational Command Centers
Bulk Shipments, Storage, and Holding	Packaging, Storage, Holding, Processes
Cable/Communications Rooms & Centers	Physical Infrastructure
Clean Rooms	Population & Population Throughputs
Co-mingled/Combined Products Process	Primary & Redundancy Systems
Crisis Management Centers	Primary and Backup Power Sources
Critical Control Checkpoints	Quarantine & Customs
Crop Harvesting Practices	Raw Material Storage, Transport, Delivery
Distribution & Transmission Systems	Receiving & Shipping
Electrical/Mechanical Rooms	Restaurant & Home Consumption
Generators, Pumps, Valves	Sanitation & Disposal Systems
Hours of Operations	SCADA & Security C3 Centers
Hydraulic Components	Shops & Laboratories
Ingredient Authentication Process	Slaughter House, Storage, Holding, Processes
Institutional Food Services	
Inventory Tracking System	Staging, Sorting, Distribution & Delivery
IT Network Computer Centers	Telecommunications Centers
Microwave/Telecommunications Sites	Temperature/Time-Sensitive Processes
Mission, Service Capabilities	Visitors, Mail and Material Delivery

Exhibit 15.6 Critical Assessment of Agriculture and Food Enterprise Assets, Operations, Processes, and Logistics

• Review the existing design-basis threat profile [**Task 3—Threat Assessment**] and update it as applicable to the client. If a profile has not been established, then the security-assessment team needs to develop one. Both the security-assessment team and the client must have a clear understanding of undesired events and consequences that may impact food and agriculture services to the community and business continuance-planning goals. Emphasis is on prioritizing threats through the process of identifying methods of attack, recent or past events, and types of disasters that might result in significant consequences; the analysis of trends; and the assessment of the likelihood of an attack occurring. Focus is on threat characteristics, capabilities, and target attractiveness. Social-order demographics to develop trend analysis to support framework decisions are also considered. Under this task the security-assessment team should also validate previous agriculture and food studies and update conclusions and recommendations into a broader framework of strategic planning. These include but are not necessarily limited to determination of types of malevolent acts, assessment of other types of disruptions, identification of categories of perpetrators, assessment of loss consequences, and assessment of threat attractiveness and likelihood of occurrence, as shown in Exhibits 15.8 through 15.12.

Least critical facilities based on value system	Most critical facilities based on value system
Least critical assets based on value system	Most critical assets based on value system

Exhibit 15.7 Prioritize Critical Agriculture and Food Enterprise Assets and Operations in Relative Importance to Business Operations

Armed attack	Computer system failures and viruses
Arson	Computer theft
Assassination, kidnapping, assault	Cyber attack on SCADA or other systems
Barricade/hostage incident	Cyber stalking, software piracy
Bomb threats and bomb incidents	Damage/destruction of interdependency systems resulting in production shutdown
Chemical, Biological, Radiological [CBR] contamination of assets, supplies, equipment, and areas	Damage/destruction of disinfection capability
	Damage/destruction of pipelines and cabling
Computer-based extortion	*(continued on next page)*

Exhibit 15.8 Determine Types of Malevolent Acts That Could Reasonably Cause Agriculture and Food Enterprise Undesirable Events

Damage/destruction of control/isolation valve systems	Misuse of wire and electronic communications
	Phishing, hacking, computer crime
Damage/destruction of pump stations and motors	Power failure
	Rape and sodomy
Damage/destruction service capability	Sabotage
Damage/destruction of pipelines and cabling	Sexual abuse, sexual exploitation
Destruction of disease-ridden products	Social engineering
Disable manufacturing, production, distribution and delivery processes	Software piracy
	Storage of criminal information on the system
Disease contamination, quarantine	Supplier, courier disruptions
Delay in processing transactions	Telecommunications failure
Email attacks	Terrorism
Encryption breakdown	Theft of trade secrets, intellectual property, confidential information, patents, and copyright infringements
Explosion	
FAX security	
Hacker attacks	Theft or destruction of chemicals
Hardware failure, data corruption	Theft or tampering of food ingredients
HAZMAT spill	Theft of trade secrets, intellectual property, confidential information, patents, and copyright infringements
Inability to access system or database	
Inability to perform routine transactions	
Industrial accidents and mishaps	Torture
Intentional opening/closing of control/isolation values and gates	Tracking effects of outage overtime
	Tracking effects of outage across related resources and dependent systems
Intentional release of dangerous chemicals	
Intentional release of contagious pathogens against livestock	Treason, sedition, and subversive activities
	Trespassing and vandalism
Intentional release of pathogens designed to kill humans and animals	Unauthorized transfer of American technology to rogue foreign states
Intentional shutdown of electrical power	Unidentified materials or wrong deliveries
Introduction of pathogens designed to undermine public confidence and trigger mass economic destabilization	Unlawful access to stored wire and electronic communications and transactional records access
Introduction of toxic or bacterial agents down the food chain	Use of cellular/cordless telephones to disrupt network communications or in support of other crimes
Loss of key personnel	
Maintenance activity	Unlawful wire and electronic communications interception and interception of oral communications
Misuse/damage of sanitation and disposal systems	
Misuse of food supply chain	Voice mail system
Misuse of the water supply chain	Wide-scale disruption and disaster
Misuse/damage of control systems	Wide-scale evacuation

Exhibit 15.8 *(continued)*

Bribery, graft, conflicts of interest	Loss of competitiveness
Civil disorders, civil rights violations	Loss of proprietary information
Conspiracy	Mail fraud
Eavesdropping	Natural disasters
Economic espionage censorship	Obscenity
Embezzlement, extortion and threats	Obstruction of justice
Historical rate increases	Online pornography and pedophilia
Homicide	Planned production shutdown or outage
Human error	Racketeer influenced and corrupt organizations
Ignorance and indifference	
Ineffective hiring practices	Racketeering
Ineffective personnel and vehicle controls	Release and misuse of personal information
Ineffective safety and security plans	
Ineffective supervision and training	Resource constraints and competing priorities
Ineffective use of resources and time	
Ineffective visitor controls	Robbery and burglary
Inefficiencies and waste	Shrinkage, shelf-life, spoilage
	Sudden unexpected layoffs
Insensitive terminations	
Institutional corruption, fraud, waste, abuse	

Exhibit 15.9 Assess Other Disruptions Impacting Agriculture and Food Enterprise Operations

Activist groups, subversive groups or cults	Terrorist organizations posing as legitimate entities
Disenfranchised individuals, hackers	
Legitimate entities serving as conduits for terrorist financing	Terrorist organizations posing as legitimate entities using the system for criminal purposes
State-sponsored or independent terrorist groups	
	Vandals, lone-wolves, insiders
Organized criminal groups	

Exhibit 15.10 Identify Category of Agriculture and Food Enterprise Perpetrators

Cluster or communicable illness	Local civil unrest
Compromise of public confidence	Loss of life [customers and employees]
Cost to repair	Loss or disruption of communications
Dealing with unanticipated crisis	Loss or disruption of information
Destruction of disease-ridden products	Loss or inability to maintain air/water/temperature, time-sensitive capabilities
Disabling of or loss of key personnel	
Disease contamination and control	
Duration of loss cycle	Number of users impacted
Illness [customers and employees]	Production shutdown
Inability to control production, manufacture, ingredient processes	Slowdown and interruption of delivery services
Inability to store, transfer, distribute commodities	Temporary closure of financial institutions

Exhibit 15.11   Assess Initial Impact of Agriculture and Food Enterprise Loss Consequence

Consequence analysis	Impact of economic base
Economic, social, political impact	Social demographics

Exhibit 15.12   Assess Initial Likelihood of Agriculture and Food Enterprise Threat Attractiveness and Likelihood of Malevolent Acts of Occurrence

- Examine and measure the effectiveness of the security system [**Task 4—Evaluate Program Effectiveness**]. The security assessment evaluates business practices, processes, methods, and existing protective measures to assess their level of effectiveness against vulnerability and risk and to identify the nature and criticality of business areas and assets of greatest concern to food operations. Focus is on vulnerability and vulnerable access points where penetration may be accomplished and how. Emphasis is placed on the physical characteristics of vulnerable points, accessibility to the location, and the effectiveness of protocol obstacles an adversary would likely have to overcome to reach a critical agriculture or food asset. This includes but is not necessarily limited to evaluation of security operations, existing security operation, relationship with partner organizations, and SCADA and security-system performance, definition of adversary plan and path analysis, and assessment of response and recovery, as shown in Exhibits 15.13 through 15.18.

Alert Notification System	Roles & Responsibilities
Early Warning System	Security Plans, Policies and Procedures
Emergency Preparedness Planning	Security Awareness
Functional Interfaces	Reporting Security Incidents
Operations, Processes, and Protocols	

Exhibit 15.13 Evaluate Existing Agriculture and Food Enterprise Security Operations and Protocols [P_{E1}]

Agent ID and Path Analysis	Operational Capabilities
Critical Control Points	Organization & Composition
Dependency Program	Outbreak Verification Network
Epidemiological Analysis	Public Health Impact
Forensic Investigations	Qualifications and Training
Identifying Suspicious Behavior	Recall Systems
Inventory and Tracking Systems	Staffing, Skills, Experience
Management Team	Stockpile Analysis
Medical/Scientific Diagnosis	Supervision
Monitoring, Surveillance and Inspection	Suspected, Adulterated Products

Exhibit 15.14 Evaluate Existing Agriculture and Food Enterprise Security Organization [P_{E1}]

Agency for Toxic Substances and Disease Registry	Centers for Disease Control and Prevention [CDC]
American Electronics Association [AEA]	Commerce Commission
American Institute of Certified Public Accountants	Computer Security Institute
	Critical Infrastructure Protection Board [PCIPB]
American National Standards Institute	Defense Central Investigation Index [DCII]
American Psychological Association	Drug Enforcement Agency
American Society for Industrial Security	El Paso Intelligence Center [EPIC]
American Society for Testing and Materials	Factory Mutual Research Corporation
American Society for Testing Materials, Vaults	Federal, state, and local agency responders
Animal and Plant Health Inspection Service	Food Safety and Inspection Service

(continued on next page)

Exhibit 15.15 Evaluate Agriculture and Food Enterprise Interface and Relationship with Partner Organizations [P_{E1}]

Illuminating Engineers Society of North America

Institute for a Drug Free Workplace

Institute of Electrical and Electronic Engineers [IEEE]

International Atomic Emergency Agency

International Association of Chiefs of Police

International Association of Professional Security Consultants

International Computer Security Association

International Criminal Police Organization [INTERPOL]

International Electronics Supply Group [IESG]

International Parking Institute

International Program on Chemical Safety

International Organization for Standardization [ISO]

Interstate Commerce Commission [ICC]

Joint Terrorism Task Force [JTTF]

Law Enforcement Support Center [LESC]

Library of Congress, Congressional Research Service

Medical Practitioners and Medical Support Personnel

Multi-lateral Expert Groups

Mutual Aid Associations

National Advisory Commission on Civil Disorder

National and State Public Health Associations

National Crime Information Center [NCIC]

National Electrical Manufacturing Association [NEMA]

National Fire Protection Association [NFPA]

National Information Infrastructure [NII]

National Infrastructure Protection Center

National Institute of Law Enforcement and Criminal Justice

National Institute of Standards and Technology [NIST]

National Insurance Crime Bureau [NICB] Online

National Labor Relations Board

National Parking Association

National Safety Council

National White Collar Crime Center [NWCCC]

Occupational Safety and Health Review Commission [OSHRC]

Office of National Drug Control Policy

Policy Coordination Committee [PCC]

Private Security Advisory Council

Protective Trade Embargoes

Regional Information Sharing System [RISS]:

  Middle Atlantic-Great Lakes Organized Crime Law Enforcement Network [MAGLOCLEN]

  Mid-States Organized Crime Information Center [MOCIC]

  New England State Police Information Network [NESPIN]

  Regional Organized Crime Information Center [ROCIC]

  Rocky Mountain Information Network [RMIN]

  Western States Information Network [WSN]

Safe Manufactures National Association

The Terrorism Research Center

U.S. Central Intelligence Agency [CIA]

U.S. Department of Agriculture

U.S. Department of Commerce [DOC]

U.S. Department of Homeland Security [DHS]

U.S. Department of Justice [DOJ]

U.S. Department of Transportation [DOT]

U.S. Department of Transportation Federal Highway Administration

U.S. Drug Enforcement Agency [DEA]

U.S. Food and Drug Administration [FDA]

U.S. Environmental Protection Agency [EPS]

U.S. Federal Bureau of Investigation [FBI]

U.S. Government Accounting Office [GAO]

U.S. Postal System

Underwriters Laboratories [UL]

Water Information Sharing and Analysis Centers [ISAC]

World Health Organization [WHO]

Exhibit 15.15 *(continued)*

Alert Warning Systems	Lock & Key Controls, Fencing & Lighting
Assessment Capability	Modem & Internet Access
Badge Controls	Performance Standards & Vulnerabilities
Barrier/Delay Systems	Personnel & Vehicle Access Control Points
Chemical & Other Vendor Deliveries	Physical Protection System Features
Communications Capability	Post Orders & Security Procedures
Cyber Intrusions, Firewalls, Other Protection Features	Response Capability
Display & Annunciation Capability	Security Awareness
Evacuation Response Plans	Security Training/Exercises/Drills
Intrusion Detection Capability	System Capabilities & Expansion Options
Inventory, Tracking, Recall Systems	System Configuration & Operating Data

Exhibit 15.16 Evaluate Existing Agriculture and Food SCADA and Security System Performance Levels [P_{E1}]

Deception actions	Introduction down the food chain
Delay-penetration obstacles	Introduction in crop harvesting
Interception analysis	Target selection and alternative targets
Introduction in animals and feeding grounds	Time-delay response analysis

Exhibit 15.17 Define the Agriculture and Food Enterprise Adversary Plan, Sequence of Interruptions, and Path Analysis

Agent ID & Path Analysis	Medical, Scientific Diagnosis
Contain Suspected, Adulterated Products	Monitoring, Surveillance, Inspection
Critical Control Points	Outbreak Verification Network
Epidemiological Analysis	Production Shutdown
Forensic investigations	

Exhibit 15.18 Assess Effectiveness of the Agriculture and Food Enterprise Response and Recovery [P_{E1}]

- Recognize the importance of bringing together the right balance of people, information, facilities, operations, processes, and systems [**Task 5—Program Analyses**] to deliver a cost-effective solution. The security-assessment team develops an integrated approach to

measurably reduce security risk by reducing vulnerability and consequences through a combination of workable solutions that bring real,
tangible enterprise value to the table. Emphasis is placed on assessing
the current status and level of protection provided against desired
protective measures to reasonably counter the threat. Operational,
technical, and financial constraints are examined, as well as future
expansion plans and constraints. These include but are not necessarily limited to analysis of security strategies and operations, refinement of consequence analysis and threat attractiveness and likelihood
of occurrence, analysis of risk-reduction and mitigation plans, development of short- and long-term mitigation solutions, evaluation of
solutions and residual vulnerability, and development of cost estimates, as shown in Exhibits 15.19 through 15.25.

Business Continuity Planning	Performance Expectations
Day-to-Day Planning	Priorities, Protective Measures
Emergency Preparedness Planning	Rank Order Facilities & Assets
Hiring Practices	Vulnerability Analysis
Loss Consequence	

Exhibit 15.19 Analyze Effectiveness of the Agriculture and Food
Enterprise Security Strategies and Operations [P_{E1}]

Cluster or communicable illness	Local civil unrest
Compromise of public confidence	Loss of life [customers and employees]
Cost to repair	Loss or disruption of communications
Dealing with unanticipated crisis	Loss or disruption of information
Destruction of disease-ridden products	Loss or inability to maintain air/water/temperature, time-sensitive capabilities
Disabling of or loss of key personnel	
Disease contamination and control	
Duration of loss cycle	Number of users impacted
Illness [customers and employees]	Production shutdown
Inability to control production, manufacture, ingredient processes	Slowdown and interruption of delivery services
Inability to store, transfer, distribute commodities	Temporary closure of financial institutions

Exhibit 15.20 Refine Previous Analysis of Agriculture and Food
Enterprise Undesirable Consequences That Can Affect Functions

Consequence analysis	Impact of economic base
Economic, social, political impact	Social demographics

Exhibit 15.21 Refine Previous Analysis of Agriculture and Food Enterprise Likelihood of Threat Attractiveness and Malevolent Acts of Occurrence

Creation and/or revision or modification of sound business practices and security policies, plans, protocols, and procedures.	Development, revision, or modification of interdependency requirements.
Creation and/or revision or modification of emergency operation plans including dependency support requirements.	SCADA/security system upgrades to improve detection and assessment capabilities.

Exhibit 15.22 Analyze Selection of Specific Risk-Reduction Actions Against Current Risks, and Develop Prioritized Plan for Agriculture and Food Enterprise Mitigation Solutions $[P_{E2}]$

Reasonable and prudent mitigation options	Mirrors business culture image

Exhibit 15.23 Develop Short- and Long-Term Agriculture and Food Enterprise Mitigation Solutions

Effectiveness calculations	Residual vulnerability consequences

Exhibit 15.24 Evaluate Effectiveness of Developed Agriculture and Food Enterprise Mitigation Solutions and Residual Vulnerability $[P_{E2}]$

Practical, cost-effective recommendations	Reasonable return on investments

Exhibit 15.25 Develop Cost Estimate for Short- and Long-Term Agriculture and Food Enterprise Mitigation Solution

PREPARING THE AGRICULTURE AND FOOD SECURITY-ASSESSMENT REPORT

Chapter 10 provides a recommended approach to reporting the security-assessment results.

U.S. GOVERNMENT AGRICULTURE-AND-FOOD-SECTOR INITIATIVES

Efforts to enhance security include a variety of initiatives. Appendix B highlights some significant efforts taken or currently underway by the U.S. Department of Homeland Security and industry.

REFERENCES

1. Chalk, Peter, "Hitting America's Soft Underbelly: The Potential Threat of Deliberate Biological Attacks Against the U.S. Agriculture and Food Industry," Rand National Defense Research Institute (2004).
2. The World Health Organization Annual Report (2004)
3. Cole, L., *The Eleventh Plague: The Politics of Biological and Chemical Warfare* (New York: W.H. Freeman and Company, 1997).
4. Department of Health and Social Security Food and Drug Administration Center for Food Safety and Applied Nutrition–Food Safety and Security: Operational Risk Management Systems Approach (November 2001)
5. The World Health Organization: Terrorist Threats to Food: Guidance for Establishing and Strengthening Prevention and Response Systems Annual Report (2002)
6. "The National Strategy for the Physical Protection of Critical Infrastructures and Key Assets" (February 2003).
7. The World Health Organization: Terrorist Threats to Food: Guidance for Establishing and Strengthening Prevention and Response Systems Annual Report (2002)
8. Ibid
9. The National Strategy for the Physical Protection of Critical Infrastructures and Key Assets
10. Ibid.

APPENDIX A: A HISTORICAL OVERVIEW OF SELECTED TERRORIST ATTACKS, CRIMINAL INCIDENTS, AND INDUSTRY MISHAPS WITHIN THE AGRICULTURE AND FOOD SECTOR

The author has compiled the information in this appendix in a chronological sequence to share the historical perspective of incidents with the reader interested in conducting further research. The listing, along with the main text of Chapter 2, presents a capsule review of terrorist activity and other criminal acts perpetrated around the world and offers security practitioners specializing in particular regions of the world a quick reference to range and level of threat.

Source: *The information is a consolidated listing of events, activities, and news stories as reported by the U.S. Department of Homeland Security, U.S. Department of State, and various other government agencies. Contributions from newspaper articles and news media reports are also included. They are provided for a better understanding of the scope of terrorism and other criminal activity and further research for the interested reader.*

1 OCTOBER 2006 Baghdad, Iraq.
Gunmen kidnapped 26 workers from a Baghdad meat-processing plant. Men in civilian clothes drove into the factory in the southwest Amel district and escaped with their captives in three trucks. Officials claim the motive was partly fiscal and partly sectarian.

12 SEPTEMBER 2006 CALIFORNIA, UNITED STATES.
An outbreak of E. coli linked to fresh spinach hospitalized 92 people and killed one woman and a two-year-old boy. The outbreak was linked to three farms in California and affected 25 states.

1 JANUARY 2005 UNITED STATES.
Federal authorities asked the fertilizer and explosives industries to be on the lookout for any suspicious attempts by buyers to purchase large amounts of ammonium nitrate, a fertilizer chemical that can be used to construct deadly fertilizer bombs. The warning came after a Canadian company reported that a man using a Middle Eastern name and potentially falsified construction documentation attempted to purchase massive amounts of the fertilizer, which was the same ingredient that Oklahoma City bomber Timothy McVeigh used in his terrorist attack that killed 160 people. The man also contacted several other fertilizer vendors via e-mail about purchasing between 500 to 1,000 metric tons of the fertilizer, an amount that is smaller than typically would be used in construction, farming, or explosives work but bigger than the amount that McVeigh used to blow up the federal building in Oklahoma City.

3 DECEMBER 2004 United States.
An estimated 76 million cases of food-borne disease occur each year in the U.S. The U.S. Centers for Disease Control and Prevention estimates that there are 325,000 hospitalizations and 5,000 deaths related to food-borne diseases each year.

12 AUGUST 2004 United States.
Cues from chatter gathered around the world are raising concerns that terrorists might try to attack the domestic food and drug supply, particularly illegally imported prescription drugs.

28 JULY 2004 Irvine, CA, United States.
FDA finds ground castor beans, not ricin, in tampered baby food. Contrary to the impression given by some early reports, the Food and Drug Administration [FDA] did not find purified ricin in two baby-food jars involved in an apparent tampering case in the Irvine, CA, area. The FDA, which conducted the analyses of these products, found what appeared to be the ground-up remnants of castor beans. Although ricin can be purified through chemical-extraction processes from castor beans, the material found in these jars was far less toxic than purified ricin. On June 16, a man called police in Irvine to report he found a note inside a jar of baby food warning that the container had been contaminated. A similar case was reported by an Irvine couple on May 31 involving the same baby food, Gerber Banana Yogurt. Neither baby became ill.

26 JULY 2004 United States.
Experts say a biological attack on a cattle ranch or wheat farm would kill few if any people, but the economic impact of a deliberately introduced disease could prove disastrous, a realization that has prompted a rethinking of how the risk of agroterrorism, should be assessed in the United States. Whether by terrorists or by Mother Nature, scientists say, potentially devastating plant and animal diseases will almost certainly continue to threaten farmers in the coming years, underscoring the urgent need for improved contingency plans. Many researchers agree that the nation's porous borders pose enormous challenges for strategies focused mainly on prevention, especially when only about 2 percent of all imported agricultural products can be inspected. Even within the U.S., security can be problematic on farms that contain tens of thousands of animals or encompass tens of thousands of acres. So far, 19 universities and institutions have been tapped for the National Plant Diagnostic Network and its sister group, the National Animal Health Laboratory Network. A main goal of each is to create a first-line defense system in the event of a deliberate or accidental disease outbreak.

21 JULY 2004 Albuquerque, NM, United States.

Some 200 county agricultural agents and livestock inspectors in New Mexico have learned how to respond to terrorism or biological threats such as mad-cow and foot-and-mouth diseases. New Mexico State University conducted the training in a two-day workshop in Albuquerque. The homeland-security training coordinator for New Mexico State said agricultural agents and livestock inspectors are important because many animal diseases travel very quickly, they must be isolated and contained, and the right people have to be notified so as not to expose others. The group learned the right way to collect animal and soil samples and to detect biological problems early.

23 JUNE 2004 Enumclaw, WA, United States.

An attack in which Washington State cows were coated with a toxic substance went unreported to federal officials for 10 days, a performance that local and national officials said is unacceptable. The incident involved 10 dairy cows in Enumclaw, 35 miles southeast of Seattle. The animals were painted on June 5 with a sticky red substance that caused welts, oozing sores, and internal bleeding. Three of the cows died. Food and Drug Administration testing later identified the substance as chromium, used in dyes and as a wood preservative. No milk from the cows entered the food supply, the FDA said. Officials say the incident represents a failure to take potential agriculture contamination seriously.

15 JUNE 2004 Bogota, Colombia.

Rebels gunned down at least 34 farm workers after tying them up at a ranch in one of Colombia's biggest cocaine-producing regions. The workers were sleeping in hammocks at the ranch near La Gabarra, 310 miles northeast of the capital, Bogota, when a group of armed men burst through the doors at dawn, tied them up with the hammocks' ropes, and then shot them with automatic weapons. The attack appeared to be the work of FARC, Colombia's largest guerrilla group, which has been battling to topple the government for 40 years.

26 MAY 2004 United States.

A disease caused by tall fescue, one of the most common cool-season pasture grasses in the U.S., is taking a costly toll on livestock, including both cattle and horses. According to the Department of Agronomy, University of Missouri, the disease is costing U.S. livestock producers more than $600 million each year.

24 MAY 2004 California, United States.
Eighteen cases of salmonella enteritidis caused the FDA to expand its recall of raw almonds distributed by a California company to a total of 13 million pounds.

2003 Colombia.
Revolutionary Armed Forces of Colombia [FARC] killed eight farmers after they refused to sell their crop of coca leaves to the terrorist group. Another paramilitary group killed three farmers just a few days later.

19 FEBRUARY 1998 Dhamar, Yemen.
Yemeni al-Hadda tribesmen kidnapped a Dutch agricultural expert in Dhamar. The kidnappers demanded development projects in their area and released the hostage the next day.

1998 United States.
A company recalled 14 million kilograms of frankfurters and luncheon meats potentially contaminated with Listeria. The parent company closed the plant and estimated the total loss to be $50–$70 million.

1997 United States.
An outbreak of E. coli 0157:H7 infection resulted in the recall of 11 million kilograms of ground beef.

1997 New Zealand.
The hemorrhagic virus spread among the wild rabbit population in New Zealand. Local farmers were suspected.

MARCH 1997 Taiwan.
Foot-and-mouth disease led to the slaughter of 8 million pigs and the halting of pork exports. The ultimate costs to the nation were at least $19 billion.

1996–1997 United States.
Outbreaks of cyclosporiasis were linked to consumption of Guatemalan raspberries.

1996 Texas, United States.
An episode of food poisoning using shigella was reported in a Texas hospital. A hospital lab worker was suspected.

1996 Japan.
Eight thousand children became ill, including some deaths, with Escherichia C04 0157:H7 infection from contaminated radish sprouts served in school lunches.

1995 Kansas, United States.
A man was poisoned using ricin. His estranged wife, a physician, was suspected.

1994 United States.
An outbreak of a S. enteritidis infestation from contaminated pasteurized liquid ice cream transported as a premix in tanker trucks caused illness in 224,000 people in 41 states.

1991 Shanghai, China.
An outbreak of hepatitis A associated with the consumption of clams affected 300,000.

1989 United States.
Anti-Pinochet extremists laced fruit bound for the United States with sodium cyanide. Only a handful of grapes were actually contaminated, but import suspensions subsequently imposed by the U.S., Canada, Denmark, Germany, and Hong Kong cost Chile in excess of $200 million in lost revenue. More than 100 growers and shippers went bankrupt.

1989 United States.
Staphylococcal toad poisoning was associated with eating mushrooms that had been canned in China.

1984 Oregon, United States.
More than 750 people were taken ill when members of the Oregon Rajneeshee religious cult poisoned salad bars with Salmonella typhimurium at various restaurants. The attack appeared to be a trial run for a more extensive attack intended to disrupt local elections.

1984 Canada.
Two Canadians attempted to kill a racehorse with pathogens as part of an insurance scam.

1984 Australia.
A prison inmate threatened to introduce foot-and-mouth disease into wild pigs, which would then infect livestock.

1981 Spain.
Eight hundred people died and 20,000 were injured [many permanently] by a chemical agent present in cooking oil.

1970s Israel.
A plot by Palestinian terrorists to inject mercury into Jaffa oranges reduced Israel's exports of citrus fruit to Europe by 40 percent.

1970 Canada.
Canadian college students were food-poisoned, supposedly by an estranged roommate.

1964 Japan.
Salmonella and dysentery agents were used to poison food in Japan. A physician is suspected.

1952 Kenya.
African bush milk was used to kill livestock. Mau Mau [an insurgent organization] was suspected.

1939 Japan.
Food was poisoned using salmonella. A Japanese physician was suspected.

1916 New York, United States.
Various biological agents were used to poison food in New York. A dentist was suspected.

1913 Germany.
Cholera and typhus were used to poison food in Germany. A former chemist employee was suspected.

1912 France.
Salmonella and toxic mushrooms were used to poison food in France. A French druggist was suspected.

APPENDIX B: AGRICULTURE-AND-FOOD-SECTOR INITIATIVES

State and federal agencies have taken a number of steps to improve security. At least twenty states have passed or are considering legislation related to agricultural terrorism, according to data compiled by the Council of State Governments. Many have hired more farm and food inspectors, have developed guidelines or requirements for improving physical security at agricultural facilities, and are building more effective disease surveillance networks. Over the past five years agencies have taken steps to boost their inspection and analysis capabilities. The U.S.D.A. has hired 20 new import surveillance liaison inspectors, who now inspect imported meat and poultry products at various locations across the country. The agency is increasing the identification and diagnostic capacities at federal and state laboratories to respond quickly to an outbreak. In addition, as a result of the Public Health Security and Bioterrorism Preparedness and Response Act of 2002, the Food and Drug Administration has tightened food-safety regulations in several ways: requiring food-processing facilities to register with the agency, mandating that companies provide advance notice of imported food shipments, and maintaining better records to make it easier to trace tainted food to its source.

Additional agriculture-and-food-sector-protection initiatives include efforts to:

- **Evaluate overall security and identify and address vulnerabilities.** The Departments of Homeland Security, Agriculture, and Health and Human Services [HHS] are working in collaboration with state and local governments and industry to undertake a broad risk assessment of the agriculture and food sector in order to evaluate overall security and identify and address existing vulnerabilities.

- **Enhance detection and testing capabilities across the agricultural and food networks.** The DHS, the USDA, and the HHS, in collaboration with state and local governments and industry, are working to increase detection and testing capacity. Exploring mechanisms to improve detection capabilities ranging from technology development to increasing the number of veterinary, epidemiology, and technical specialists at the state level will facilitate earlier detection and response. Enhancing trace-back systems and increasing detection capabilities at borders and ports of origin will also significantly increase protection. Identifying, creating, and certifying additional laboratory capacity across the country would likewise increase the speed of analysis and response. Another goal is to increase the number of qualified personnel [veterinarians and lab technicians] and laboratories with the ability to diagnose and treat animal-disease outbreaks and crop contamination.

- **Access transportation-related security risks.** The DHS, the USDA, the IIHS, and the DOT are working with representatives from the agriculture and food industries to access security risks in food and commodity transport and develop appropriate solutions. The scope of the issues requires a thorough risk assessment integrating transportation-security measures into ongoing and newly initiated countermeasures undertaken by the food industry. Additional considerations include standardizing the methods by which the agriculture and food industries report truck hijackings and cargo thefts and then disseminating these reports within the food industry.

- **Identify potential infrastructure-protection incentives; identify and address existing disincentives.** The DHS, working with the USDA and the HHS, is exploring options for developing incentives or reducing disincentives to encourage the prompt reporting of problems.

- **Develop emergency-response strategies.** The DHS, the USDA, and the HSS, working with sector counterparts, are developing a strategy to coordinate risk communications and other emergency-response activities.

- **Enhance the detection of food tampering**. The DHS, the USDA, and the HSS are working together to develop new methods to detect deliberate food tampering.

- **Information-sharing strategies.** Information derived from the assessment of sector food-safety processes and procedures can provide a foundation for developing a food-sector critical-infrastructure protection system. For example, two major efforts to establish procedures for accidental outbreaks of animal disease have been reported in the "Animal Health Safeguarding Review: Results and Recommendations" by the National Association of State Departments of Agriculture Research Foundation (October 2001) and "The U.S. National Animal Health Emergency Management System" (2001 Annual Report). While plans for these studies were drafted with accidental introductions of disease contamination in mind, their findings and recommendations also apply to intentional acts. Another example is the implementation of recommendations from the 1999 Animal and Plan Health Inspection report "Safeguarding American Plan Resources." Further study and collaborative policy development are required to determine whether and how the food-safety system could be extended to deal with food-security issues.

Chapter 16

THE BANKING AND FINANCE SECTOR

This chapter outlines the:

- Banking and finance sector's value to America's economic and national security
- Banking-and-finance-sector attractiveness to terrorist and criminal elements
- Banking-and-finance-sector vulnerabilities
- Tailoring the S^3E Security Methodology to the banking and finance sector
- Banking-and-finance-sector challenges facing the security-assessment team
- Applying the security-assessment methodology to the banking and finance sector
- Preparing the banking-and-finance security-assessment report
- Historical overview of selected banking & finance incidents
- U.S. government banking & finance initiatives

INDISPENSABILITY TO AMERICA'S ECONOMIC AND NATIONAL SECURITY

According to the United States Treasury Department, there is as much as U.S. $450 billion in circulation worldwide, two-thirds outside of the United States. The commercial sectors in many foreign countries are customers of, suppliers for, and partners with American businesses. Our financial and securities markets are increasingly intertwined with foreign markets. A threat to the integrity of these markets is also a threat to business and financial institutions here at home.

An Attractive Target

In many nations American businesses and financial institutions are being targeted for securities fraud, extortion, racketeering, economic espionage, intellectual-property theft, corrupt business practices, and computer crime. In particular, **money laundering provides the fuel for terrorists, drug dealers, arms traffickers, and other criminals to operate and expand their activities.** The process of disguising or concealing illicit funds to make them appear legitimate is a serious crime, with an estimated $500 billion to $1 trillion laundered worldwide annually, according to the United Nations Office of Drug Control and Prevention.

Specifically, **money laundering is the process used to transform monetary proceeds derived from criminal activities into funds and assets that appear to have come from legitimate sources.** The process generally takes place in three stages: placement, layering, and integration. In the placement stage, cash is converted into monetary instruments such as money orders or traveler's checks or deposited into financial institution accounts. In the layering state, these funds are transferred or moved into other accounts or other financial institutions to further obscure their illicit origin. In the integration stage, the funds are used to purchase assets in the legitimate economy or to fund further activities. All financial sectors and certain commercial businesses can be targeted during one or more of these stages. Many of these entities are required to report transactions with certain characteristics to law enforcement if they appear to be potentially suspicious. The transactions would generally fall within either the placement or layering stage if they proved to be involved in money laundering.

International criminals move sums of money through the international financial system that are so huge they dwarf the combined economies of many nations. They are often organized in multicrime businesses and have capitalized on growth in international communications and transportation to expand their criminal operations. A historical overview of selected terrorist attacks, criminal incidents, industry mishaps, and government actions within the banking and finance sector is displayed at Appendix A.

Vulnerabilities

Terrorist financing is generally characterized by motives different than those typically used in money laundering, and the funds involved often originate from legitimate sources. However, **the techniques used by terrorists for obscuring the origin of funds and their ultimate destination are often similar to those used to launder money.** Therefore, the Treasury Department, law-enforcement agencies, and federal financial regulators often employ similar approaches and techniques in trying to detect and prevent both money laundering and terrorist financing. The range of potential adversaries includes disgruntled employees, disaffected individuals or groups, organized crime, domestic and international terrorists, and hostile nations.

In addition, the financial aspects of criminal and terrorist activities have become more complex due to rapid advances in technology and the globalization of the financial-services industry. Money laundering can have devastating effects on financial institutions and undermine the stability of democratic nations. Modern financial systems permit criminals to instantly transfer millions of dollars though currency-exchange houses, stock-brokerage houses, gold dealers, casinos, automobile dealerships, insurance companies, and trading companies. Private banking facilities, offshore banking, free-trade zones, wire systems, shell corporations, and trade financing all have the ability to mask illegal activities. The criminal's choice of money-laundering vehicles is limited only by his or her creativity. Ultimately, this laundered money flows into global financial systems, where it can undermine or subvert national economies, currencies, and the financial systems that are the cornerstone of legitimate international commerce. Money laundering is thus not only a law-enforcement issue but a serious national international security threat as well. Fighting such increasingly elusive, well-financed, and technologically adept criminals and terrorists not only reduces financial crime—it also deprives them of the means to commit other more serious crimes.

TAILORING THE S^3E SECURITY ASSESSMENT PROCESS FOR THE BANKING & FINANCE SECTOR

The exhibit on the next page highlights in grey shaded areas those specific elements that are tailored for special application to the banking and finance sector.

Strategic Planning
Tasks 1, 2, & 3
Operational Environment
Critical Assessment
Threat Assessment

Program Effectiveness
Task 4
Infrastructure Interdependency
Evaluation [P_{Ef}]

Operational Environment Task 1
Enterprise Goals
Corporate Image, Core Values, Areas of Responsibility, Performance Objectives Business Incentives, Investment
Commitments
Federal Statutes, State Laws, Municipal Codes, Confidentiality Laws, Civil Common Law, Willful Torts, Contractual Obligations, Agreements, Governing Authorities, Voluntary Commitments
Characterization
Mission, Location/Layout//Land Use, Neighboring Facilities, Terrain, Climate Resources, Operations, Processes
Operations – Logistics
Visitors, Customers, Vendors; Mail of Material Delivery, Receiving, Shipping; Staging, Storing, Distribution/Delivery; Time-Sensitive Processes; Communications, Brokers & Dealers Institutional Ownership, Original Identity, Exchange & Clearing Organizations, Market Participants, Transaction Identity and Facilitators

Enterprise
Roles & Responsibilities Operations, Processes, Protocols Functional Relationships, Security Awareness, Reporting Security Incident, Alert Notification System

Security Capabilities
Management Team, Organization & Composition, Staffing, Skills, Experience, Qualifications & Training, Operational Capabilities Dependency Programs, Administrative Inquiries, Bank Examination Process, Clearing & Settlements, Counterfeit Instruments, Cross-border Tracking, High Risk Accounts, Multi-Agency Investigative Units, Offshore Financial Centers Records & Reporting, Recovery & Resumption Reporting Suspicious Activity Shared Information Systems

Critical Assessment
Task 2
Resources, Operations, Functions Administrative, Aircraft/Fleet Facilities Board/Conference Rooms Cable/Communications Rooms Electrical/Mechanical Rooms Physical Infrastructure Telecommunications Centers IT Network Centers Training Facilities Warehouses, Loading Docks SCADA & Security Systems Infrastructure Sharing Critical Interdependencies Primary, Backup, Redundant Systems Accounts, Correspondence Accounts Beneficial Owner of Funds Bulk Transactions, Currency Exchange Data Collection Systems & Processes Due Diligence, Dissemination Systems Electronic Communications Network Operational Command Centers Foreign Accounts & Transactions Payable-Through Accounts Payment/Clearance/Settlement System Transaction Structure Composition Microwave/Telecommunications Sites Air, Water, Temperature Systems

Barriers & Delay Measures
Terrain, Approaches, Fences, Barriers, Walls, Gates Tamper Systems, Freeze Assets
Detection Measures
Interior/Exterior Sensors, Special Purpose Sensors, Tamper Systems Counterfeit Instruments Suspected, Tainted Accounts Inventory, Tracking Systems Recall Notification Systems
Access Control Measures
People/Vehicle/Delivery Controls Screening, Verification
Assessment Measures
Perimeter Security, Lighting Posts and Patrols, CCTV Suspicious Behavior Monitoring/Surveillance/Inspection Automation Review Policy
Communications Media
Radio, Telephone, PA, Hardwire, Wireless, Fiber Optic Information Network
Security Control Center[s]
Annunciation & Display, Event Storage, Reports, Alarm Integration Information Networks
Security Infrastructure
Primary/Secondary/Backup Redundancy

Threat Assessment
Task 3
International, National, Regional, Local Range & Levels of Threats Design Basis Threat Profile, Probability Threat Categories, Capabilities, Methods & Techniques

Response & Recovery
Tactics, Techniques, First/Second & Integrated Responders Recovery & Settlement, Closeout Actions

Quality Assurance/Quality Control Process

Exhibit 16.1 The S^3E Security Assessment Methodology for the Banking and Finance Sector

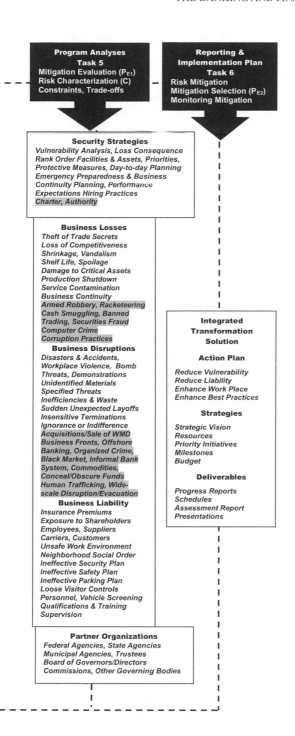

Program Analyses
Task 5
Mitigation Evaluation (P_{E1})
Risk Characterization (C)
Constraints, Trade-offs

Reporting &
Implementation Plan
Task 6
Risk Mitigation
Mitigation Selection (P_{E2})
Monitoring Mitigation

Security Strategies
Vulnerability Analysis, Loss Consequence
Rank Order Facilities & Assets, Priorities,
Protective Measures, Day-to-day Planning
Emergency Preparedness & Business
Continuity Planning, Performance
Expectations Hiring Practices
Charter, Authority

Business Losses
Theft of Trade Secrets
Loss of Competitiveness
Shrinkage, Vandalism
Shelf Life, Spoilage
Damage to Critical Assets
Production Shutdown
Service Contamination
Business Continuity
Armed Robbery, Racketeering
Cash Smuggling, Banned
Trading, Securities Fraud
Computer Crime
Corruption Practices

Business Disruptions
Disasters & Accidents,
Workplace Violence, Bomb
Threats, Demonstrations
Unidentified Materials
Specified Threats
Inefficiencies & Waste
Sudden Unexpected Layoffs
Insensitive Terminations
Ignorance or Indifference
Acquisitions/Sale of WMD
Business Fronts, Offshore
Banking, Organized Crime,
Black Market, Informal Bank
System, Commodities,
Conceal/Obscure Funds
Human Trafficking, Wide-
scale Disruption/Evacuation

Business Liability
Insurance Premiums
Exposure to Shareholders
Employees, Suppliers
Carriers, Customers
Unsafe Work Environment
Neighborhood Social Order
Ineffective Security Plan
Ineffective Safety Plan
Ineffective Parking Plan
Loose Visitor Controls
Personnel, Vehicle Screening
Qualifications & Training
Supervision

Integrated
Transformation
Solution

Action Plan

Reduce Vulnerability
Reduce Liability
Enhance Work Place
Enhance Best Practices

Strategies

Strategic Vision
Resources
Priority Initiatives
Milestones
Budget

Deliverables

Progress Reports
Schedules
Assessment Report
Presentations

Partner Organizations
Federal Agencies, State Agencies
Municipal Agencies, Trustees
Board of Governors/Directors
Commissions, Other Governing Bodies

Exhibit 16.1 *(continued)*

The banking and financial infrastructure consists of a variety of physical structures such as buildings and financial utilities as well as human capital. **Most of the industry's activities and operations take place in large commercial office buildings.** Physical structures to be protected include those that house retail or wholesale banking operations, financial markets, regulatory institutions, and physical repositories for documents and financial assets. Today's financial utilities such as payment, clearing, and settlement systems are primarily electronic, although the physical transfer of assets does still occur. The financial infrastructure also includes such electronics as computers, storage devices, and telecommunication networks. In addition to these key physical components, many financial-services employees have highly specialized skills and are therefore considered essential elements of the industry's critical infrastructure.

The financial industry depends on continued public confidence and involvement to maintain normal operations. Financial institutions maintain only a small faction of depositor assets in cash on hand. If depositors and customers were to seek to withdraw their assets simultaneously, severe liquidity pressures would be placed on the financial system. With this in mind, federal safeguards are in place to prevent liquidity shortfalls. In times of crisis or disaster, maintaining public confidence demands that financial institutions, markets, and payment systems remain operational or that their operations can be quickly restored.

With regard to retail financial services, physical assets are well distributed geographically throughout the industry. The sector's retail niche is characterized by a high degree of substitutability, allowing one type of payment mechanism or asset to be easily replaced with another during a short-term crisis. For example, in retail markets consumers can make payments through cash, checks, or credit cards.

Challenges Facing the Security-Assessment Team

The security-assessment team must recognize that international criminals engage in a wide range of dangerous activities, including acquisition and sale of weapons of mass destruction, transfer of sensitive American technology to rogue foreign states, trade in banned or dangerous substances, and traffic in drugs, women, and children.

These crimes pose a grave threat to the security, stability, values, and other interests of the global community. As jurisdictions take countermeasures, the security-assessment team must keep abreast of the criminals and terrorists who generate these proceeds and the money launderers who disguise those proceeds as they develop new and more sophisticated methods for moving money around the globe. As money launderers are driven out of traditional banking systems, they exploit alternatives such as offshore financial centers, the Hawala [Asian informal banking] system, the Colombian black-market peso-exchange system, or the nonbank financial sector to move their criminal proceeds. With lightning speed, money launderers can probe the financial system for vulnerabilities and adapt their methods to exploit these soft spots.

As with other critical sectors, the banking and financial sector depends on several other sectors such as electric power, transportation, and telecommunications for continuity of operations. For example, the equity-securities markets remained closed for four business days following 9/11 not because any markets or market systems were inoperable but because the telecommunications lines in lower Manhattan that connect key market participants were heavily damaged and could not be restored immediately. As a mitigation measure, financial institutions have made great strides to build redundancy and backup into their systems and operations.

Another challenge is that overlapping federal intelligence authorities involved in publicizing threat information cause confusion and duplication of effort for both industry and government. The Department of the Treasury organized the Financial and Banking Information Infrastructure Committee [FBIIC] as a standing committee of the President's Critical Infrastructure Protection Board [PCIPB]. The FBIIC is comprised of representatives from 13 federal and state financial-regulatory agencies [1]. The FBIIC is currently working with the National Infrastructure Protection Center, the Financial Services Information Sharing Analysis Center [ISAC] and the DHS to improve the information-dissemination and -sharing processes. The FBICC includes representatives of the federal and state financial regulatory agencies including the Commodity Futures Trading Commission, the Conference of State Bank Supervisors, the Federal Deposit Insurance Corporation, the Federal Housing Finance Board, the Federal Reserve Bank of New York, the Federal Reserve Board, the National Association of Insurance Commissioners, the National Credit Union Administration, the Office of the Comptroller of the Currency, the Office of Federal Housing Enterprise Oversight, the Offices of Homeland and Cyberspace Security, the Office of Thrift Supervision, and the Securities and Exchange Commission.

APPLYING THE S^3E SECURITY ASSESSMENT METHODOLOGY TO THE BANKING AND FINANCE SECTOR

The security-assessment team should use its collective expertise to:

- Identify the important mission and functions of the banking or financial facility [**Task 1—Operational Environment**] and incorporate the development of all business operations and continuity expectations. Examination focuses on services and the customer base to determine the business contributions to the community, its culture, and business objectives. This includes, but is not necessarily limited to identification of business goals and objectives, customer base, and commitments and characterization of facilities and boundaries, as shown in Exhibits 16.2 through 16.5.

Areas of Responsibility	Performance Objectives
Business Incentives & Investment	Promote Interagency Cooperation
Corporate Image & Core Values	Prosecute Money Laundering
Economic Stability	Organizations
Mission & Services	Public Safety & Public Confidence

Exhibit 16.2 Identify Banking and Finance Enterprise Business Goals and Objectives

Foreign Governments	Peer Banking and Financial Institutions
General Public [Domestic and	[Domestic and International]
International]	Public and Private Corporations
Non-governmental Organizations	[Domestic and International]
[Domestic and Abroad]	Size of Population Served
Nonprofit Organizations	U.S. Government, State, & Local Agencies
[Domestic and International]	

Exhibit 16.3 Identify Banking and Finance Enterprise Customer Base

15 U.S.C. 1644 - Fraudulent Use of Credit Cards	18 U.S.C. 641 - Embezzlement of Public Money
29 CFR 1910.36 - Means of Egress	18 U.S.C. 2511 - Technical Surveillance
29 CFR 1910.151 - Medical Services and First Aid	21 U.S.C. 801 - Drug Abuse and Control
	26 U.S.C. 7201 - Tax Matters
29 CFR 1910.155 - Fire Protection	29 U.S.C. 2601 - Family and Medical Leave Act

(continued on next page)

Exhibit 16.4 Identify Banking and Finance Enterprise Commitments

31 U.S.C. 5322 - Financial Transaction Reports

41 U.S.C. 51 - U.S. Contractor Kickbacks

41 U.S.C. 701 - Drug Free Workplace Act

46 U.S.C. 1903 - Maritime Drug Enforcement

50 U.S.C. 1801 - Foreign Intelligence Surveillance Act

Americans with Disabilities Act [42 U.S.C. 12101]

Architectural Barrier Act 1968 [As Amended]

Bank Protection Act 1968 [As Amended]

Bank Secrecy Act [As Amended]

Bioterrorism Preparedness and Response Act, 2002 [As Amended]

Carriers, Suppliers, Vendors, Insurance

Civil Rights Act 1964 [U.S.C. 200e] [As Amended]

Commerce Commission

Commodity Futures Trading Commission [CFTC]

Comprehensive Drug Abuse Prevention Act 1970 [As Amended]

Comprehensive Environmental Response, Compensation and Liability Act 1980 [As Amended]

Controlled Substance Act [As Amended]

Drug Free Workplace Act 1988 [As Amended]

Emergency Planning & Right-to-Know Act 1986 [As Amended]

Employee Polygraph Protection Act 1988 [29 U.S.C. 2001] [As Amended]

Exchange and Clearing Organizations

Fair Credit Reporting Act [15 U.S.C. 1681] [As Amended]

Family Education Rights and Privacy Act 1974 [As Amended]

Family and Medical Leave Act 1993 [As Amended]

Federal Anti-Tampering Act

Federal Polygraph Protection Act [As Amended]

Federal Privacy Act 1974 [As Amended]

Federal Reserve Board [FRB]

Foreign Corrupt Practices Act

Freedom of Information Act [U.S.C. 552] [As Amended]

National Labor Relations Act [29 U.S.C. 151] [As Amended]

Interagency Agreements

Internal Revenue Service

International Monetary Force [IMF]

International Program on Chemical Safety

International Emergency Economic Powers Act [IEEPA] [As Amended]

Labor Management Relations Act

Memorandum of Agreement

Memorandum of Understanding

Money Laundering Control Act [MILCA] [As Amended]

National Credit Union Administration [NCUA]

Omnibus Crime Control and Safe Streets Act, 1968 [As Amended]

Policy Coordination Committee [PCC]

Privacy Act [5 U.S.C. 522e] [As Amended]

Office of Thrift Supervision [OTS]

Rehabilitation Act, 1973 [As Amended]

Security Exchange Commission [SEC]

State and Local Industry Administrators and Authorities

State and Local Emergency Planning Officials and Committees

Uniform Trade Secrets Act

United National Security Council Resolutions 1373 and 1390

U.S. Department of Commerce

U.S. Department of Homeland Security [DHS]

U.S. Department of Justice [DOJ]

U.S. Patriot, 2002 [As Amended]

U.S. Postal System

U.S. Environmental Protection Agency [EPA]

U.S. Occupational Safety and Health Administration [OSHA]

Exhibit 16.4 *(continued)*

Commercially Leased Facilities	Neighboring Facilities
Dedicated site owned and operated by the enterprise.	Operated and maintained through reciprocal agreements with internal and external entities.
Dispersed Locations	
Facility Layout & Location	Terrain & Climate
Facility Land Use	Types of Assets at specific locations
Infrastructure Sharing	

Exhibit 16.5 Characterize Configuration of Banking and Finance Enterprise Facilities and Boundaries

Accounts, Correspondence Accounts	Infrastructure Sharing
Administrative, Aircraft/Fleet Facilities	Institutional Ownership & Transactions
Air, Water, Temperature Systems	IT Network Center
Anti-money Laundering Efforts	Mail & Material Delivery
Beneficial Owner of Funds	Microwave/Telecommunications Sites
Board/Conference Rooms	Mission, Service Capabilities
Brokers & Dealers, Transaction Facilitators	Offshore Banking & Market Participants
Bulk Transactions, Currency Exchange	Operational Command Centers
Cable/Communications Rooms	Original and Transaction Identity
Concentration of Key Staff	Payable-through Accounts
Counterfeit Detection Instruments	Payment, Clearance, and Settlement Systems
Critical Interdependencies	Physical Infrastructure
Data Collection and Process Systems	Population & Population Throughputs
Depositor Institute	Primary, Backup, Redundancy Systems
Distribution & Transmission Systems	Receiving & Shipping
Diverse Backup Facilities	Resources, Operations, Functions
Due Diligence and Dissemination Systems	SCADA & Security C3 Centers
Electrical/Mechanical Rooms	Shared Information Systems
Electronic Communications Networks	Staging, Storing, Distribution & Delivery
Exchange & Clearing Houses	Telecommunications Centers
Extradition & Prosecution	Teller Windows and Customer Service Areas
Foreign Accounts & Transactions	Time-Sensitive Processes
Foreign Correspondence Accounts	Training Facilities
Freezing & Confiscation of Assets	Transaction Structure Compositions
Geographic Diversity	Vaults, Safes, Depository Boxes
Hiring Practices	Visitors, Customers, Vendors
Hours of Operations	Warehouses, Loading Docks

Exhibit 16.6 Critical Assessment of Banking and Finance Enterprise Assets, Operations, Processes, and Logistics

- Describe the utility configuration, operations, and other elements **[Task 2—Critical Assessment]** by examining conditions, circumstances, and situations relative to safeguarding public health and safety and in reducing the potential for disruption of banking and financial services. Focus is on an integrated evaluation process to determine what assets need protection in order to minimize the impact of threat events. The security-assessment team identifies and prioritizes facilities, processes, and assets critical to banking and finance service objectives that might be subject to malevolent acts or disasters that could result in a potential series of undesired consequences. The process takes into account the impacts that could substantially disrupt the ability of the utility to provide safe services and strives to reduce security risks associated with the consequences of significant events. These include but are not necessarily limited to assessment of assets, operations, and processes and prioritization of assets and operations, as shown in Exhibits 16.6 and 16.7.
- Review the existing design-basis threat profile **[Task 3—Threat Assessment]** and update it as applicable to the client. If a profile has not been established, then the security-assessment team needs to develop one. Both the security-assessment team and the client must have a clear understanding of undesired events and consequences that may impact banking and financial services to the community and business continuance-planning goals. Emphasis is on prioritizing threats through the process of identifying methods of attack, recent or past events, and types of disasters that might result in significant consequences; the analysis of trends; and the assessment of the likelihood of an attack occurring. Focus is on threat characteristics, capabilities, and target attractiveness. Social-order demographics to develop trend analysis in order to support framework decisions are also considered. Under this task the security-assessment team should also validate previous banking and finance assessment studies and update conclusions and recommendations into a broader framework of strategic planning. These include but are not necessarily limited to determination of types of malevolent acts, assessment of other types

Least critical facilities based on value system	Most critical facilities based on value system
Least critical assets based on value system	Most critical assets based on value system

Exhibit 16.7 Prioritize Critical Banking and Finance Enterprise Assets and Operations in Relative Importance to Business Operations

of disruptions, identification of perpetrators, assessment of loss consequence, and assessment of threat attractiveness and likelihood of occurrence, as shown in Exhibits 16.8 through 16.12.

- Examine and measure the effectiveness of the security system [**Task 4—Evaluate Program Effectiveness**]. The security assessment evaluates business practices, processes, methods, and existing protective measures to assess their level of effectiveness against vulnerability and risk and to identify the nature and criticality of business areas and assets of greatest concern to banking and finance operations. Focus is on vulnerability and vulnerable access points where penetration may be accomplished and how. Emphasis is placed on the physical characteristics of vulnerable points, accessibility to the location, and the effectiveness of protocol obstacles an adversary would likely have to overcome to reach a critical asset. These include but are not necessarily limited to evaluation of existing operations, existing security organization, relationships with partner organizations, and SCADA and security-system performance, definition of adversary plan and path analysis, and assessment of response and recovery, as shown in Exhibits 16.13 through 16.18.

Acquisition or sale of weapons of mass destruction	Damage/destroy service capability
	Damage/destruction of interdependency systems resulting in production shutdown
Armed attack	
Arson	Damage/destruction of pipelines and cabling
Assassination, kidnapping, assault	Delay in processing transactions
Barricade/hostage incident	Email attacks
Bomb threats and bomb incidents	Encryption breakdown
Bulk financial transactions	Escaping asset freezing and confiscation
Business fronts	Explosion
Chemical, Biological, Radiological [CBR] contamination of assets, supplies, equipment, and areas	Extortionate credit transactions
	False personation, false statements
Computer system failures and viruses	FAX security
	Hacker attacks
Computer theft	Hardware failure, data corruption
Computer-based extortion	Hawala system
Concealing or obscuring the clandestine diversion of funds intended for legitimate purposes to terrorist organizations	High intensity money laundering related financial crime areas [HIFCA]
Counterfeiting and forgery of coins, currency, bonds, software/hardware and other misrepresentations	High risk accounts and secrecy havens
	Inability to access system or database
	Inability to perform routine transactions
Cyber attack on SCADA or other systems	Information banking systems, charities, bulk cash, commodities *(continued on next page)*
Cyber stalking, software piracy	

Exhibit 16.8 Determine Types of Malevolent Acts That Could Reasonably Cause Banking and Finance Enterprise Undesirable Events

Intentional shutdown of electrical power	Torture
Loss of high value commodities	Tracking effects of outage across related resources and dependent systems
Loss of key personnel	
Misuse of banking and finance supply chain	Tracking effects of outage overtime
Misuse of wire and electronic communications	Treason, sedition, and subversive activities
Misuse/damage of control systems	Trespassing and vandalism
Phishing, hacking, computer crime	Unauthorized transfer of American technology to rogue foreign states
Power failure	
Rape and sodomy	Unidentified materials or wrong deliveries
Sabotage	Unlawful access to stored communications
Sexual abuse, sexual exploitation	Unlawful access to stored wire and electronic communications and transactional records access
Social engineering	
Software piracy	Unlawful wire and electronic communications interception and interception of oral communications
Storage of criminal information on the system	
Supplier, courier disruptions	Use of cellular/cordless telephones to disrupt network communications or in support of other crimes
Telecommunications failure	
Terrorism	
Theft of trade secrets, intellectual propriety, confidential information, patents, and copyright infringements	Voice mail system
	Wide-scale disruption and disaster
	Wide-scale evacuation

Exhibit 16.8 *(continued)*

Bribery, graft, conflicts of interest	Insensitive terminations
Civil disorders, civil rights violations	Institutional corruption, fraud, waste, abuse
Conspiracy	Loss of competitiveness
Eavesdropping	Loss of proprietary information
Economic espionage censorship	Mail fraud
Embezzlement, extortion and threats	Maintenance activity
Historical rate increases	Natural disasters
Homicide	Obscenity
Human error	Obstruction of justice
Ignorance and indifference	Online pornography and pedophilia
Industrial accidents and mishaps	Planned production shutdown or outage
Ineffective hiring practices	Racketeer influenced and corrupt organizations
Ineffective personnel and vehicle controls	Racketeering
Ineffective safety and security plans	Release and misuse of personal information
Ineffective supervision and training	Resource constraints and competing priorities
Ineffective use of resources and time	Robbery and burglary
Ineffective visitor controls	Shrinkage, shelf-life, spoilage
Inefficiencies and waste	Sudden unexpected layoffs

Exhibit 16.9 Assess Other Disruptions Impacting Banking and Finance Enterprise Operations

Activist groups, subversive groups or cults	Terrorist organizations posing as legitimate entities using the system for criminal purposes
Disenfranchised individuals, hackers	
Legitimate entities serving as conduits for terrorist financing	
	Terrorist organizations posing as legitimate entities
Organized criminal groups, offshore accounts	
	Vandals, lone-wolfs, insiders
State-sponsored or independent terrorist groups	

Exhibit 16.10 Identify Category of Banking and Finance
Enterprise Perpetrators

Business fronts, organizations, and individuals posing as legal enterprises	Local civil unrest
	Loss of key personnel
Compromise of public confidence	Loss of life [customers and employees]
Cost of recovery and resumption	Loss of primary and backup power systems
Dealing with unanticipated crisis	
Duration of loss cycle and number of users impacted	Loss or disruption of communications
	Loss or disruption of information
Illness [customers and employees]	Production shutdown
Inability to detect fraud, corruption, and illegal transactions	Slowdown and interruption of delivery services
Inability to freeze and confiscate assets	Temporary closure of financial institutions
Inability to store, transfer, or distribute commodities and funds	

Exhibit 16.11 Assess Initial Impact of Banking and
Finance Enterprise Loss Consequence

Consequence analysis	National and international social demographics
Economic, social, political impact	
Impact of economic base	

Exhibit 16.12 Assess Initial Likelihood of Banking and
Finance Enterprise Threat Attractiveness and Likelihood of
Malevolent Acts of Occurrence

Alert Notification System	Reporting Security Incidents
Emergency Preparedness Planning	Roles and Responsibilities
Functional Interfaces	Security Awareness
Operations, Processes, Protocols	Security Plans, Policies & Procedures

Exhibit 16.13 Evaluate Existing Banking and Finance
Enterprise Operations and Protocols [P_{E1}]

Administrative Inquiries	Multi-Agency Investigative Units
Automated Review Process	Offshore Financial Centers
Bank Examination Process	Operational Capabilities
Clearing & Settlements	Organization & Composition
Cooperative public-private efforts to prevent money laundering	Primary & Backup Power Sources
	Qualifications & Training
Counterfeit Instruments	Records & Reporting
Criminal Investigations	Recovery & Resumption
Cross-border Tracking	Reporting Suspicious Activity
Dependency Program	Shared Information Systems
Disrupt the domestic flow of illicit money	Staffing, Skills, Experience
Disrupt the global flow of illicit money	State and local government efforts to fight money laundering
Freezing and confiscation of assets	
Hacker Prevention	Structured Accounts & Transactions
Management Team	Supervision
Monitoring, Surveillance, Inspection	Track High Risk Accounts

Exhibit 16.14 Evaluate Existing Banking and Finance
Enterprise Security Organization [P_{E1}]

Agency for Toxic Substances and Disease Registry	International Criminal Police Organization [INTERPOL]
American Electronics Association [AEA]	
American Institute of Certified Public Accountants	International Electronics Supply Group [IESG]
American National Standards Institute	International Monetary Force [IMF]
American Psychological Association	International Parking Institute
American Society for Industrial Security	International Organization for Standardization [ISO]
American Society for Testing and Materials	
American Society for Testing Materials, Vaults	Interstate Commerce Commission [ICC]
Centers for Disease Control and Prevention [CDC]	Joint Terrorism Task Force [JTTF]
Asian-Pacific Economic Cooperation [APEC]	Law Enforcement Support Center [LESC]
Association of Southeast Asian Nations [ASEAN]	Library of Congress, Congressional Research Service
Banking and Finance Information Sharing and Analysis Centers [ISAC]	Medical Practitioners and Medical Support Personnel
Commerce Commission	Multi-lateral Expert Groups
Commodity Futures Trading Commission	Mutual Aid Associations
Computer Security Institute	National Advisory Commission on Civil Disorder
Critical Infrastructure Protection Board [PCIPB]	National Crime Information Center [NCIC]
Defense Central Investigation Index [DCII]	National Credit Union Administration [NCUA]
El Paso Intelligence Center [EPIC]	National Electrical Manufacturing Association [NEMA]
Executive Office for U.S. Attorneys [EOUSA]	
Exchange and Clearing Organizations	National Fire Protection Association [NFPA]
Factory Mutual Research Corporation	National Information Infrastructure [NII]
Federal Deposit Insurance Corporation [FDIC]	National Infrastructure Protection Center
Federal Home Loan Bank Board	National Institute of Law Enforcement and Criminal Justice
Federal Reserve Board [FRB]	
Federal Reserve System	National Institute of Standards and Technology [NIST]
Federal, state, and local agency responders	
Finance and Banking Information Infrastructure Committee [FBIIC]	National Insurance Crime Bureau [NICB] Online
	National Labor Relations Board
Financial Action Task Force [FATF]	National Parking Association
Illuminating Engineers Society of North America	National Safety Council
Institute for a Drug Free Workplace	National White Collar Crime Center [NWCCC]
Institute of Electrical and Electronic Engineers [IEEE]	Occupational Safety and Health Review Commission [OSHRC]
Internal Revenue Service	Office of National Drug Control Policy
International Association of Chiefs of Police	Office of the Comptroller of the Currency [OCC]
International Association of Professional Security Consultants	Office of Thrift Supervision [OTS]
	Policy Coordination Committee [PCC]
International Computer Security Association	Private Security Advisory Council

(continued on next page)

Exhibit 16.15 Evaluate Banking and Finance Enterprise Interface and Relationship with Partner Organizations [P_{E1}]

Regional Information Sharing System [RISS]:	U.S. Central Intelligence Agency [CIA]
Middle Atlantic-Great Lakes Organized Crime Law Enforcement Network [MAGLOCLEN]	U.S. Department of Commerce [DOC]
	U.S. Department of Homeland Security [DHS]
	U.S. Department of Justice [DOJ]
Mid-States Organized Crime Information Center [MOCIC]	U.S. Environmental Protection Agency [EPA]
New England State Police Information Network [NESPIN]	U.S. Federal Bureau of Investigation [FBI]
	U.S. Government Accounting Office [GAO]
Regional Organized Crime Information Center [ROCIC]	U.S. Postal System
	U.S. Secret Service
Rocky Mountain Information Network [RMIN]	U.S. Treasury Department
	U.S. Treasury Office of Enforcement Financial Crime [FinCen]
Western States Information Network [WSN]	
Safe Manufactures National Association	U.S. Treasury Terrorist Financing Operations Section [TFOS]
Security Exchange Commission [SEC]	
The Terrorism Research Center	Underwriters Laboratories [UL]
Treasury Enforcement Communications Systems [TECS]	

Exhibit 16.15 *(continued)*

Alert Advisory Notification Systems	Modem & Internet Access
Assessment Capability	Monitoring, Surveillance, Inspection
Badge Controls	Performance Standards & Vulnerabilities
Barrier/Delay Systems	Personnel & Vehicle Access Control Points
Chemical & Other Vendor Deliveries	Physical Protection System Features
Communications Capability	Post Orders & Security Procedures
Cyber Intrusions, Firewalls, Other Protection Features	Response Capability
Display & Annunciation Capability	SCADA & Security C3 Centers
Evacuation Response Plans	Security Awareness
Intrusion Detection Capability	Security Training/Exercises/Drills
Inventory, Tracking, Recall Systems	System Capabilities & Expansion Options
IT Network Computer Centers	System Configuration & Operating Data
Lock & Key Controls, Fencing & Lighting	Tamper Systems

Exhibit 16.16 Evaluating Existing Banking and Finance Enterprise SCADA and Security-System Performance Levels [P_{E1}]

Counterfeit instruments	Offshore transactions
Deception actions	Target selection and alternative targets
Delay-penetration obstacles	Payable through accounts
Front organizations	Time-delay, detection, assessment
Interception analysis	analysis
Likely techniques to be employed	

Exhibit 16.17 Define the Banking and Finance Enterprise Adversary Plan, Distractions, Sequence of Interruptions, and Path Analysis

Automation Review Policy	Recall Notification Systems
Closeout Actions	Recovery & Resumption
Freezing Assets	Reporting Suspicious Activity
Inventory, Tracking Systems	Suspected, Tainted Accounts

Exhibit 16.18 Assess Effectiveness of Banking and Finance Enterprise Response and Recovery [P_{E1}]

- Recognize the importance of bringing together the right balance of people, information, facilities, operations, processes, and systems **[Task 5—Program Analyses]** to deliver a cost-effective solution. The security-assessment team develops an integrated approach to measurably reduce security risk by reducing vulnerability and consequences through a combination of workable solutions that bring real, tangible enterprise value to the table. Emphasis is placed on assessing the current status and level of protection provided against desired protective measures to reasonably counter the threat. Operational, technical, and financial constraints are examined, as well as future expansion plans and constraints. These include but are not necessarily limited to analysis of security strategies and operations, refinement of previous consequence analysis and likelihood of threat occurrence, analysis of risk reduction and mitigation solutions, development of long- and short-term mitigation solutions, evaluation of solutions and residual vulnerability, and development of cost estimates, as shown in Exhibits 16.19 through 16.25.

Business Continuity Planning	Loss Consequence Approach
Charter and Authority	Performance Expectations
Day-to-Day Planning	Priorities & Protective Measures
Emergency Response Planning	Rank Order Facilities & Assets
Hiring Practices	Vulnerability Analysis

Exhibit 16.19   Analyze Effectiveness of Banking and Finance Enterprise Security Strategies and Operations [P_{E1}]

Business fronts, organizations, and individuals posing as legal enterprises	Local civil unrest
	Loss of key personnel
Compromise of public confidence	Loss of life [customers and employees]
Cost of recovery and resumption	Loss of primary and backup power systems
Dealing with unanticipated crisis	
Duration of loss cycle and number of users impacted	Loss or disruption of communications
	Loss or disruption of information
Illness [customers and employees]	Production shutdown
Inability to detect fraud, corruption, and illegal transactions	Slowdown and interruption of delivery services
Inability to freeze and confiscate assets	Temporary closure of financial institutions
Inability to store, transfer, or distribute commodities and funds	

Exhibit 16.20   Refine Previous Analysis of Banking and Finance Enterprise Undesirable Consequences That Can Affect Functions

Consequence analysis	National and international social demographics
Economic, social, political impact	
Impact of economic base	

Exhibit 16.21   Refine Previous Analysis of Banking and Finance Enterprise Likelihood of Threat Attractiveness and Malevolent Acts of Occurrence

| Creation and/or revision or modification of emergency operation plans including dependency support requirements. | Development, revision, or modification of interdependency requirements. |
| Creation and/or revision or modification of sound business practices and security polices, plans, protocols, and procedures. | SCADA/security system upgrades to improve detection and assessment capabilities. |

Exhibit 16.22 Analyze Selection of Specific Risk Reduction Actions Against Current Risks, and Develop Prioritized Plan for Banking and Finance Enterprise Mitigation Solutions [P_{E2}]

| Mirrors business culture image | Reasonable and prudent mitigation options |

Exhibit 16.23 Develop Short- and Long-Term Banking and Finance Enterprise Mitigation Solutions

| Effectiveness calculations | Residual vulnerability consequences |

Exhibit 16.24 Evaluate Effectiveness of Developed Banking and Finance Enterprise Mitigation Solutions and Residual Vulnerability [P_{E2}]

| Practical, cost-effective recommendations | Reasonable return on investments |

Exhibit 16.25 Develop Cost Estimate for Short- and Long-Term Banking and Finance Enterprise Mitigation Solutions

PREPARING THE BANKING AND FINANCE SECURITY-ASSESSMENT REPORT

Chapter 10 provides a recommended approach to reporting security-assessment results.

U.S. GOVERNMENT BANKING AND FINANCE INITIATIVES

The banking and financial-services industry is highly regulated and highly competitive. Industry professionals and government regulators regularly engage in identifying sector vulnerabilities and take appropriate protective measures, including sanctions for institutions that do not consistently meet standards. Government and industry efforts to enhance banking and finance security include a variety of local, national, and international efforts. Appendix B summarizes some significant actions taken or currently underway.

APPENDIX A: A HISTORICAL OVERVIEW OF SELECTED TERRORIST ATTACKS, CRIMINAL INCIDENTS, AND INDUSTRY MISHAPS WITHIN THE BANKING AND FINANCE SECTOR

The author has compiled the information in this appendix in a chronological sequence to share the historical perspective of incidents with the reader interested in conducting further research. The listing, along with the main text of Chapter 2, presents a capsule review of terrorist activity and other criminal acts perpetrated around the world and offers security practitioners specializing in particular regions of the world a quick reference to range and level of threat.

Source: *The information is a consolidated listing of events, activities, and news stories as reported by the U.S. Department of Homeland Security, U.S. Department of State, and various other government agencies. Contributions from newspaper articles and news media reports are also included. They are provided for a better understanding of the scope of terrorism and other criminal activity and further research for the interested reader.*

13 SEPTEMBER 2006 Moscow, Russia.
The First Deputy Chairman of the Russian Bank, Andrei Kozlov, was shot dead along with his driver in an apparent contract killing. Kozlov had headed high-profile campaigns to flush out financial crime in Russia, resulting in the closure of several banks.

20 APRIL 2006 Indonesia.
Four Indonesian nationals were arrested on charges of raising money for el J'ama Islamiya.

9 AUGUST 2005 Fortaleza, Brazil.
Thieves tunneled into a bank in NE Brazil and stole $68 million, the largest bank robbery in Brazilian history.

13 JUNE 2005 Kirkuk, Iraq.
Suicide bombers killed 20 people outside a bank in Kirkuk.

12 APRIL 2005 United Kingdom and the United States.
Three men held on terrorism charges in the United Kingdom were also indicted in the United States on charges that they planned to blow up financial buildings in Washington, DC, New York, and New Jersey. The four-count indictment filed in U.S. District Court in Manhattan alleges the men took part in months of methodical reconnaissance of financial targets between August 2000 and April 2001. Evidence included surveillance on a computer seized in Pakistan. Authorities said the group was also planning on targeting London's Heathrow Airport.

14 MARCH 2005 Spain.

Prosecutors in at least seven countries confirmed they are opening investigations into a money-laundering network cracked open by Spanish police that is believed to have links to criminal gangs operating across Europe and North America. Investigators from France, the Netherlands, the United Kingdom, and Russia were expected to arrive in Spain to begin collaborating with Spanish police, while prosecutors in the United States, Canada, and Germany were also opening investigations into the network, which allegedly funneled up to $800 million from drugs and arms trafficking, prostitution, theft, and fraud through the Del Valle Abogados law firm in Marbella, Spain. More than 40 people of five nationalities were arrested. At least nine known criminal gangs are believed to have used Del Valle Abogados to create front companies through which to launder money by investing it in real estate along the Costa del Sol and moving the gains to tax havens such as the British colony of Gibraltar.

9 MARCH 2005 Boca Raton, FL, United States.

Hackers gained access to sensitive personal information of about 32,000 U.S. citizens on databases owned by publisher Reed Elsevier. The Federal Bureau of Investigation and the Secret Service were investigating. The breach was found after a customer's billing complaint in the previous week led to the discovery that an identity and password had been misappropriated. The information accessed included names, addresses, and Social Security and driver's license numbers but not credit history, medical records, or financial information.. The unit that was compromised collects data from government agencies, building large databases and ways to extract information from them. Many of the customers are law-enforcement agencies and financial institutions.

7 MARCH 2005 Los Angeles, CA, United States.

A Nigerian national who stole the identities of thousands of people was sentenced to 5 1/2 years in federal prison. Adedayo Benson was also ordered to pay nearly $155,000 in restitution to ten financial companies. He and his sister were arrested in 2002 on charges of tapping into several public-records databases, gaining access to records of over 7,000 people and using their identities to buy at least $1 million in merchandise. Authorities said the siblings posed as real-estate agents and opened accounts with Choice Point, Advantage Financial, and Equifax. Benson's sister was sentenced to 4 1/2 years in prison.

4 MARCH 2005 Cincinnati, OH, United States.
Credit-card information from customers of more than 100 DSW Shoe Warehouse stores was discovered stolen from a DSW computer database over a period of three months. The Secret Service was investigating. A credit-card company noticed suspicious activity and alerted DSW.

18 FEBRUARY 2005 Colombia and the United States.
Officials from Colombia and the U.S. create finance database. In a continuing effort to combat money laundering and other financial crimes, U.S. Immigration and Customs Enforcement [ICE] agents are working with Colombia's customs officials to develop a joint database to exchange trade and financial information. Agents from ICE, which is part of the Department of Homeland Security, are helping Colombia's National Tax and Customs Directorate [DIAN] to develop a Trade Transparency Unit. Officials from both agencies are developing an unprecedented joint database to exchange trade and financial information. ICE agents also delivered 215 computers and other equipment to DIAN in that effort.

17 FEBRUARY 2005 United States.
Internet fraudsters, motivated by money and armed with sophisticated technology, pose an increased economic threat as they steal private data from companies and individuals, said the director of the U.S. Secret Service. Security analysts warned that Internet hackers, once motivated by the thrill of shutting down computer systems, are joining forces with organized-crime groups as they seek to profit from hacking into databases and stealing personal data through a variety of tactics. Increased cooperation and information sharing between U.S. agencies, foreign governments, technology companies, and the financial community has helped mitigate online fraud. Experts said companies and individuals are better protected now than ever before and are also more aware of online fraud risks. However, Internet fraudsters are increasingly targeting less-protected small businesses rather than large companies that can spend millions of dollars on security software to protect their computer systems.

11 FEBRUARY 2005 The Cook Islands, Indonesia, and the Philippines.
Three countries removed from list of those not helping combat money laundering. It was decided at the Financial Action Task Force [FATF] Plenary XVI that the Cook Islands, Indonesia, and the Philippines can be removed from the list of Non-Cooperative Countries and Territories [NCCTs]. Recent FATF visits to those countries confirmed that they are

effectively implementing anti-money laundering [AML] measures to remedy deficiencies that were identified by the FAT, including strict customer identification, suspicious transaction reporting, bank examinations, and legal capacities to investigate and prosecute money laundering. All three countries have developed financial intelligence units—specialized units that analyze financial data, coordinate national efforts, and facilitate international cooperation. The FATF will now monitor the implementation of these measures in the three countries to ensure that they sustain their recent commitments and progress. The current NCCT list includes Myanmar, Nauru, and Nigeria.

11 FEBRUARY 2005 Athens, Greece.

A series of homemade bombs damaged the entrances of five Greek banks and two local offices of the ruling Conservative party. The blasts, triggered by camping-gas canisters and gasoline, in central and eastern Athens targeted four branches of Eurobank and one of National Bank within an hour. There was no claim of responsibility. Fringe groups mainly identified as anarchists have in the past regularly staged similar attacks in the capital. Due to increased policing and installation of surveillance cameras before the Athens Olympics in August 2004, as well as the capture and breakup of the deadly November 17 guerrilla group in mid-2002, authorities had seen a drop in such attacks. However, in previous months the attacks appeared to be on the rise again, with several bomb attacks mainly against police convoys and guards. Police sources said the new attacks raised fears that a new guerrilla group may have been formed.

5 FEBRUARY 2005 London, United Kingdom.

The international community must step up the financial war on terrorism and choke off funding for extremist groups, Department of Treasury Undersecretary John Taylor told a gathering of finance ministers in London at a summit of the world's seven wealthiest nations [G-7]. The United States has stepped up action against foreign financial institutions it suspects of money laundering and, using powers under the 2001 Patriot Act, has cut them off from the U.S. financial system. Taylor called on other countries to adopt such measures. Since the 9/11 terror attacks, more than $140 million in financial assets belonging to suspected terrorist financiers has been frozen worldwide, according to the Department of Treasury.

1 FEBRUARY 2005 South Pacific.

South Pacific tries to tackle organized crime, drugs, terror. The region can become a weak link in the global fight against international crime and terror, a regional leader warned. Drug hauls in Fiji, Tonga,

Kiribati, the Federated States of Micronesia, Vanuatu, and the Marshall Islands have shown the extent of the peril facing the region's governments. Also cited were alleged gang-related slayings in Fiji and possible money laundering reported by the Reserve Bank of Fiji. Criminal gangs and terror networks can rapidly adapt to the opportunities that globalization has provided in a region once regarded as isolated. Those same operating conditions also could be used to support terrorism.

25 JANUARY 2005 Moscow, Russia.

A Moscow State University student was detained in connection with a bomb blast outside a Bank of Moscow branch. The student reportedly belonged to a radical left-wing group. The device was sent to the bank in a package with a message inside saying a blast would take place in 40 minutes. The bank's security officials carried the package out of the building and called the police. Shortly afterwards the package exploded. No one was injured, and the building was not damaged. Officials said that there was information about a series of similar terrorist attacks being planned.

25 JANUARY 2005 United States.

The Treasury Department designated an individual financially aiding the Iraqi insurgency and al Qaeda. Sulayman Khalid Darwish was designated under Executive Order 13224 for providing financial and material support to the al-Zarqawi network and al Qaeda. The U.S. submitted Darwish to the United Nations 1267 Committee, which will consider adding him to the consolidated list of terrorists tied to al Qaeda, Osama bin Laden, and the Taliban. According to information available to the U.S. Government, Darwish is a member of the Advisory (Shura) Council of the Zarqawi organization and served as one of Zarqawi's operatives. Darwish, also involved in fundraising and recruiting for the organization, is a close associate of Zarqawi and one of the most prominent members of the Zarqawi network in Syria.

23 JANUARY 2005 Wisconsin, United States.

Secret Service notices change in manufacturers, technology of counterfeiting. Teenagers arrested in recent months for passing fake currency in southeast Wisconsin represent the new face of counterfeiting in the United States, U.S. Secret Service officials say. The Secret Service said bills generated with home-computer equipment have been a rising problem within the last ten years. Large, old-fashioned counterfeiting operations haven't totally disappeared, however. In 2004, the Secret Service estimated that in the United States, about $43.5 million in counterfeit bills were passed. That compares with about $25.3 million in 1994. The Secret Service is seeing an increasing number of high-school students who are trying to copy currency.

11 January 2005 United Nations.

Countries have seized or frozen $147 million in assets belonging to 435 individuals and groups linked to al Qaeda or the Taliban, crimping terrorist cash flows, a U.S. Treasury official said. United Nations sanctions require all member states to impose a travel ban and arms embargo against a list of those linked to Osama bin Laden's terror network and to freeze their financial assets. The list currently includes 320 individuals and 115 groups. The Assistant Treasury Secretary said the sanctions have forced terrorist groups like al Qaeda to use more informal ways of moving and raising money, adding that the U.S. has identified the use of cash couriers by al Qaeda to move money across borders.

11 January 2005 Ulm, Germany.

German police arrested 11 after raids on mosques and flats suspected of being used to support and finance Islamic terror networks. Ulm is considered an ideological center.

27 December 2004 Moscow, Russia.

Russia's Prime Minister announced that he will participate with the Finance Ministry and Financial Action Task Force on money laundering. The FATF is an intergovernmental body whose purpose is the development and promotion of policies, at both national and international levels, to combat money laundering and terrorist financing. At the present time 31 countries and two organizations are members of FATF.

13 December 2004 Bermuda.

Bermuda's lawmakers gave final approval to a bill to prevent terrorists from hiding money in bank accounts or property in the offshore finance center. The Anti-Terrorism Act follows a law against money laundering and an agreement by the British colony to share financial information with U.S. tax authorities. The law will require businesses to report immediately to police if they believe money is being used for terrorist purposes.

6 December 2004 Santa Fe, Argentina.

A bomb exploded in an Argentine branch of Citibank, but no one was hurt. A bomb squad removed a second package suspected to contain a bomb before it exploded. No group claimed responsibility. The explosion took place at an ATM machine of the bank, causing minor damage.

30 NOVEMBER 2004 Middle East and North Africa.

Arab states agreed to work together to try to keep money out of the hands of terrorists through the creation of a 14-member Middle East-North Africa Financial Action Task Force. The task force is the first of its kind in the region, which has come under increased scrutiny following the terrorist attacks of September 11, 2001. The United Arab Emirates have been identified by U.S. investigators as a major money-transfer center for al Qaeda. The task force brings together Bahrain, Saudi Arabia, Syria, Lebanon, Qatar, Kuwait, Tunisia, Jordan, Algeria, Morocco, Egypt, Oman, the United Arab Emirates, and Yemen into a regional version of the Paris-based Financial Action Task Force that was set up in 1987 to monitor and fight money laundering and expanded its role in 2001 to combat the financing of terror.

29 NOVEMBER 2004 Bahrain.

A new task force was established in Bahrain to combat money laundering and terrorist financing across the region, called the Middle East and North Africa Financial Action Task Force [MENAFATF]. Sixteen countries signed up for membership, but that number was expected to increase. The MENAFATF—a part of an international web of regional task forces being established around the world—is an extension of the Financial Action Task Force [FATF], which was established by the G7 Summit in 1989 as an intergovernmental body to combat money laundering and terrorist financing. FATF has come up with 40 recommendations for its 31 member countries and territories and two regional organizations. The new regional task force will attempt to implement those recommendations, as well as implementing United Nations treaties and agreements in its member countries.

23 NOVEMBER 2004 Canada.

Canadian police cannot prosecute most of the money-laundering and terrorism financing cases referred by the agency that monitors such activities because the country's privacy laws require too much key information to be withheld. The Financial Transactions and Reports Analysis Center [FINTRAC] passed 160 new cases of suspected terrorist financing or money laundering onto the Royal Canadian Mounted Police and other agencies in the 12 months that ended in March 2004. That information led to no prosecutions and rarely resulted in investigations.

22 NOVEMBER 2004 United States.

Phishing attacks rose more than threefold from August to October, according to the Anti-Phishing Working Group [APWG], an industry association focused on identity theft and fraud. In October 2004, 6,597 new, unique phishing e-mail messages were reported to the association—up from 2,158 in August. The group considers messages unique when they involve a single e-mail blast with a unique subject line, sent out one at a time and targeting one company or organization. The baiting sites, which were placed within the messages, also increased to 1,142 in October—up from 727 in August, according to the group. Digital Impact, which manages e-mail marketing campaigns, says that phishing has a response rate of 3 percent. The APWG places the response rate at an even higher rate of 5 percent. Like spam, phishing messages cost next to nothing to produce, but the returns are far greater—limited only by the amount of money in a user's bank account. APWG estimates that in the United States, phishing was a $1.2 billion industry last year.

22 November 2004 Singapore.

An FBI antiterrorism expert told a conference held in Singapore that internal auditors can play a critical role in identifying and stopping terrorism financing, because all transaction records will have to pass through them for inspection. With their expertise and training, internal auditors can detect suspicious records in the same way that a police officer spots a suspicious driver on the streets. Some indicators of possible terrorism-financing activity include customers frequently changing their addresses, accounts that get multiple deposits from varied and unknown sources, or dormant accounts that suddenly receive a huge deposit followed by a series of small withdrawals until the sum is exhausted.

21 November 2004 United Kingdom.

UK consumers were warned to be on their guard against phishing e-mails purporting to be from the American bank BB&T. BB&T, which has 1,400 branches across the U.S., confirmed the e-mails received in the UK were the work of phishers. American banks are often the subject of phishing operations because they have the most customers. The rise in phishing is, conversely, linked to the improving security of online banking. Because the banks' Websites are so secure now, fraudsters are targeting customers for information to get in.

19 November 2004 United States.

A 1,200 percent increase in phishing attacks since January 2004 has forced companies not only to redouble their efforts fighting the attacks but to change the way they use the Internet to communicate with their customers, each other, and law-enforcement officials. Without those changes, experts said, phishing will contribute to an erosion in consumer confidence at a time when online businesses cannot afford the loss. For their part, banks are pooling their resources. In September 2004, members of the Financial Services Technology Consortium—a group of banks, financial-services firms, universities, and government agencies—began compiling a database of phishing sites that they plan to make accessible to banks and federal authorities. The goal is to allow investigators to share intelligence and quickly determine whom to contact at various ISPs and law-enforcement agencies.

17 NOVEMBER 2004 Buenos Aires, Argentina.

A series of blasts shook three banks in the Argentine capital, Buenos Aires, killing one man and injuring another. A branch bank of U.S.-based Citibank was targeted in two separate explosions, while a third hit a branch of Argentine bank Banco Galicia. The bombs were not big enough to cause major destruction and appeared to be homemade devices. It was not clear who was responsible.

31 OCTOBER 2004 Romania.

Romania's public prosecutor announced that militant Muslim sympathizer groups in Romania are providing cash to help finance terrorist activities in the West. It was reported that certain Arab businessmen belonging to the Boerica financial group are regularly transferring large sums of money abroad.

29 OCTOBER 2004 United Kingdom.

Britain's financial watchdog plans to tighten scrutiny of banks, trade financiers, and high-risk investment funds in what a senior official concedes is a near-impossible battle to intercept terrorist funds. The head of the crime division at the Financial Services Authority [FSA] said the increased supervision is necessary to counter the increasingly inventive ways in which militants are laundering dirty money.

13 OCTOBER 2004 United States.

The U.S. designated the Islamic African Relief Agency [IARA], along with five senior officials, pursuant to E.O. 13224. This action blocked all accounts, funds, and assets of IARA in the United States and criminalized the provision of money or other types of support to any of its offices. IARA is headquartered in Khartoum, Sudan, and maintains over 40 offices throughout the world, including the United States. IARA is also known as Islamic Relief Agency [ISRA], Islamic American Relief Agency, Al-Wakala al-Islamiya l'il-Ighatha, and Al-Wakala al-Islamiya al-Afrikia l'il-Ighatha. Information available to the U.S. indicated that international offices of IARA provided direct financial support for Osama bin Laden and engaged in a joint program with an institute controlled by bin Laden that was involved in providing assistance to Taliban fighters as well as providing financial support for other activities relating to terrorism. The U.S. has designated 393 individuals and entities as terrorists, financiers, or facilitators. In addition, the global community has frozen over $142 million in terrorist-related assets.

29 SEPTEMBER 2004 United States.

More than three years after the 9/11 attacks, the United States is still struggling to understand how al Qaeda gets its funding, how much it earns, and where it spends the money. Two members of the bipartisan September 11 commission, which investigated the 2001 attacks, told a Senate hearing that U.S. officials and the international community had managed to choke off some of the militant network's financing but said stopping the flow of funds had proved impossible. The complex web of raising and moving money, the mix of terrorist and legitimate funds, and problems of international cooperation were among the obstacles to stopping terrorist financing. The September 11 Commission said in its final report that al Qaeda and its leader, Osama bin Laden, were largely funded by diverting money from Islamic charities, generating some $30 million per year. They said the September 11 attacks probably cost between $400,000 and $500,000.

29 SEPTEMBER 2004 Kathmandu, Nepal.

A bomb ripped through a bank in the Nepalese capital Kathmandu, and a suspected Maoist rebel was shot dead as a guerrilla strike shut down shops and transport in much of the Himalayan kingdom. The blast shattered windows of the Nepal-Bangladesh Bank and of several houses at Lalitpur on the capital's outskirts but caused no casualties.

28 SEPTEMBER 2004 Istanbul and Adana, Turkey.

Two small bombs exploded in front of branches of the British HSBC Bank in Istanbul, and a third blast hit a Turkish-American Association in the capital, Ankara. While no one was injured in the attacks, the percussion bombs shattered windows of nearby buildings and cars. No one claimed immediate responsibility for the blasts. However, Islamist militants, Kurdish separatists, and leftist guerrillas have all carried out bomb attacks in Turkish cities.

23 SEPTEMBER 2004 United States and European Union.

The U.S. and the European Union seek to coordinate action against a growing problem of terrorist sympathizers using charities and cash couriers as covers to fund violence. The U.S. reports terrorists are forced to use such methods by a successful international clampdown on funding through conventional banking. The U.S. is working with 32 other nations and regional groups through a Paris-based international watchdog, the Financial Action Task Force, to develop rules and standards on dealing with charities.

20 SEPTEMBER 2004 Lagos, Nigeria.

The Economic and Financial Crimes Commission [EFCC], in joint operations with the FBI, arrested 28 Internet fraudsters in Lagos, Nigeria. Approximately $3.5 million was recovered from the crooks in fraudulent cashier checks and goods bought over the Internet and shipped to Nigeria by credit-card scammers. The year-long cooperation between the EFCC and the FBI was targeted at stemming rising cases of cyber crime by Nigerians.

15 SEPTEMBER 2004 United States.

U.S. aid bolsters terrorist-finance fight. Terrorists move money through channels as diverse as major banks, charities, and alternative remittance systems. United Nations member states are obligated to apply sanctions against designated terrorists and their financial supporters, including freezing assets, banning travel, and enforcing arms embargos. However, gaps in enforcing sanctions exist, and the United States and its international partners are working to address how to deal with informal financial systems and nongovernmental organizations through which terrorists collect and move their funds. The United States is providing substantial assistance to other governments to help them attain the technical ability and skills to clamp down on terrorist-financing activity.

9 SEPTEMBER 2004 United States.

The Department of Treasury announced the designation of the U.S. branch of the Saudi Arabia-based Al Haramain Islamic Foundation [AHF] as linked to terror. In addition, the AHF branch located in the Union of the Comoros was also designated. The assets of the U.S. AHF branch, headquartered in Oregon, were blocked pending investigation in February 2004. The investigation showed direct links between the U.S. branch and Osama bin Laden. Additionally, the affidavit alleged the U.S. branch of AHF criminally violated tax laws and engaged in other money-laundering offenses. Information showed that individuals associated with the branch tried to conceal the movement of funds intended for Chechnya by omitting them from tax returns and mischaracterizing their use, which they claimed was for the purchase of a prayer house in Springfield, MO. Other information available to the U.S. showed that funds that were donated to AHF with the intention of supporting Chechen refugees were diverted to support mujahideen, as well as Chechen leaders affiliated with the al Qaeda network.

25 AUGUST 2004 Jersey City, NJ, United States.
Police evacuated a building containing financial-company offices in Jersey City, NJ, after a bomb threat was called into one of its tenants. The Department of Homeland Security said Al Qaeda was targeting financial buildings in New York, Washington, DC, and Newark, NJ, for attack and raised security near financial centers.

20 AUGUST 2004 United States.
Authorities indicted three suspected members of the Palestinian militant group Hamas on charges that they participated in a lengthy racketeering conspiracy to provide money for terrorist acts in Israel. The three activists allegedly used bank accounts in the U.S. to launder millions of dollars to support Hamas, which the U.S. government has designated as a terrorist organization. The three were charged with racketeering conspiracy for allegedly joining with 20 others since at least 1988 to conduct business for Hamas, which included conspiracies to commit murder, kidnapping, passport fraud, and other crimes.

12 AUGUST 2004 United States.
Militants find alternative ways to fund terrorism. Members of Osama bin Laden's al Qaeda and other militants are turning increasingly to crime—from dealing drugs to selling knockoff shampoos and pirated CDs—to pay for attacks amid a crackdown on the movement of terrorist funds through world banks, security officials said. As terrorist cells become more self-reliant, they are calling into question the notion that they need an international financial support network to stage attacks. Department of Treasury officials acknowledge the shift and say it is a symptom of their success, as their efforts have made it harder and costlier for terrorist groups like al Qaeda to move and raise money. Al Qaeda and its affiliates have been linked to the heroin trade in Afghanistan, credit-card fraud in Europe, and gem smuggling out of Africa.

1 AUGUST 2004 United States.
The terrorist threat level was raised to orange based on information that the U.S. is subject to attack by suicide bombings on financial buildings in New York, New Jersey, and Washington, DC. Targets include the New York Stock Exchange, NY World Financial Center, Citigroup, Prudential, the International Monetary Fund, and the World Bank.

27 JULY 2004 United States.

Five former leaders of the Holy Land Foundation, once the biggest Islamic charity in the United States, were arrested on charges that they funneled $12.4 million to Palestinian terrorists. Within months of the 9/11 attacks, the Bush administration froze several million dollars in Holy Land's assets. FBI officials said they suspected that founding members of Holy Land in the early 1990s devised a scheme to funnel money to Hamas in support of jihad in Palestine. The indictment charged that Holy Land used hospitals, Islamic committees, and other organizations in the West Bank and Gaza that were controlled by Hamas to funnel money to terrorist causes. Payments were then distributed to family members of individuals who were "martyred" or jailed in terrorist attacks, the government charged.

24 JULY 2004 Woonsocket, RI, United States.

A bomb squad blew up a blinking device that was intended to look like a bomb. The device was located behind the City Bank in Woonsocket, RI.

20 JULY 2004 Russia.

The UK's National Hi-Tech Crime Unit [NHTC] arrested members of a Russian gang believed to have extorted hundreds of thousands of pounds from online-betting sites in Britain. A joint operation between the NHTCU and its counterparts in the Russian Federation resulted in the arrest of three men in Russia. Online bookies in the UK have been subject to attacks and demands for money since October 2003. The arrested men are part of a gang that is suspected of launching attacks on the servers and websites of online sports books. The attacks bombarded the servers with thousands of messages, effectively shutting down the company and costing millions of pounds in lost business. The gang then sent e-mail demands asking for money in order for the attacks to be stopped for one year. As part of the investigation, ten members of the gang were arrested in Riga, Latvia, in November 2003. Officers were able to identify the financial trail, which led to the gangsters arrested in Russia.

20 JULY 2004 Charlotte, NC, United States.

A Pakistani citizen was taken into custody after videotaping the 60-story Bank of America headquarters and another skyscraper in Charlotte, NC. The officer who arrested Kamran Akhtar said he tried to walk away when officers approached and gave conflicting statements about what he was doing and where he was going. Videotapes in Akhtar's possession also showed buildings in Atlanta, Houston, Dallas, New

Orleans, and Austin, as well as transit systems in some of those cities and a dam in Texas. The federal prosecutor listed Akhtar as a resident of the New York City borough of Queens and said he also went under the name Kamran Shaikh.

10 JULY 2004 Canada.

Canada's spy agency [CSIS] warned that Eastern European crime groups could jeopardize the north's emerging diamond industry by slipping foreign stones tainted through bloody conflicts onto the market. Police and intelligence agencies have long been concerned about the trade in what are known as conflict diamonds—the black-market sale of gems mined primarily in western Africa for cash to fuel bloody civil wars. Individuals allegedly connected to Eastern European crime groups were implicated in a diamonds-for-arms trade with African rebel groups in the early 1990s. CSIS also points out that there have been allegations that Osama bin Laden's al Qaeda terrorist network converted $20 million into conflict diamonds from Sierra Leone.

9 JULY 2004 South Korea.

The number of suspected money-laundering reports in South Korea quadrupled to around 2,000 in the first six months of 2004, from 517 in the same period the previous year. The number of cases reached 1,744 in 2003, a 7 percent increase from 2002, the Korea Financial Intelligence Unit [KFIU] reported. At present, local financial-service companies are required to report all transactions over a set sum if they believe unlawful activities may be involved. The KFIU referred 851 cases to the public prosecutor's office, the police, the National Tax Service, and the Financial Supervisory Service. It said a third of those investigated were found to have been illegal.

8 JULY 2004 United States.

The financial industry will be the target of 68 percent of phishing attacks in 2004, according to a new study by Radicati Group. At the present time, most phishing attacks target the financial industry in the form of e-mails to bank customers requesting their account number and PIN for "verification" purposes. Radicati Group predicts that the number of phishing attacks per month will more than double by 2008, reaching an average of 110 per month, up from 51 per month in 2004. In response to the heightened threat of phishing attacks, an antiphishing market has emerged. Companies take two approaches to fighting phishing—solutions that aim to stop customers from giving phishers their personal information and protective measure in place for fraud resulting from a phishing attack. The overall industry is expected to bring in revenues of $202 million in 2004 and $880 million by 2008.

8 July 2004 United Kingdom.

Over a million United Kingdom consumers have been victims of security breaches while shopping via the Internet. More than one in 20 consumers experienced attempted or actual theft of financial or personal details while shopping on the Internet. The research showed that 24 percent of those affected defected to an alternative online brand, while 23 percent decided not to buy anything from that company again; 73 percent said security is more important than price, quality, or convenience when shopping online, and 70 percent would boycott a website even if they only had word-of-mouth evidence that the brand had been involved in a security scare. When asked if they would continue to use a website if their financial data was stolen, 79 percent claimed they would stop shopping online. Statistics from IT security organizations show there are many recorded instances where a new website has been successfully attacked within 15 minutes of being launched.

7 July 2004 Russia.

Russia's National Duma [the Council of Representatives] approved a law providing the legal framework for fighting money laundering and the funding of terrorist groups. According to the law, 0 percent loans worth over 600,000 rubles [$20,000] by noncredit institutions and real-estate contracts worth at least three million rubles [$100,000] are subject to strict control. Credit institutions are prevented from opening bank accounts for clients or their representatives without their presence at the bank. In addition, credit institutions are not allowed to set up or maintain relations with banks that do not have representative offices in the country in which they are operating.

7 July 2004 United States.

A battle's breaking out over phishing prevention. As phishing and spoofing scams continue to rise on the Web and identity theft becomes a primary concern for financial-services institutions, some credit-card issuers have begun experimenting with new card technology to combat the problem. Identity theft is among the most common security threats that banks and other businesses face today. More than half of all identity-theft incidents occur in the banking industry, with about 33 percent from credit-card fraud alone.

30 JUNE 2004 European Union.

The European Commission proposed widening the net to catch money launderers and terrorists with new transaction-reporting duties for attorneys, insurance brokers, and other businesses beyond the banking system. The European Union's [EU] executive arm in Brussels, Belgium, is seeking to require that businesses verify customer identities, watch for suspicious transaction patterns, and report any cash purchase of more than 15,000 euros [$18,000]. The rules would cover businesses such as service providers, trusts, jewelers, auction houses, and casinos in the 25 EU countries. The proposal would be the EU's third set of rules on money laundering. The first was a 1991 directive aimed at drug dealers. The laws were broadened with a second directive after the 2001 terrorist attacks, which implied rather than specified terrorism as a target.

25 JUNE 2004 Paris, France.

Police seized 2 1/2 million Euros in counterfeit notes in raids on the biggest forgery scheme since the launch of the EU single currency.

18 JUNE 2004 United States.

Identity theft is more prevalent in NFL than other pro sports. The National Football League said cases involving identity theft and impersonation fraud have increased dramatically in recent years and are now the most common way that NFL players are victimized. Impersonation fraud and identity theft aren't unique to the NFL, but National Basketball Association [NBA] and Major League Baseball [MLB] security officials confirmed that they don't handle the volume of cases the NFL does. Impersonators usually target lesser-known players instead of more recognizable superstars. They are well versed on details of the player's career and often carry forged documents. Identity-theft victims spend an average of 600 hours recovering from the crime, according to statistics provided by the Identity Theft Resource Center.

16 JUNE 2004 United States.

More than a dozen corporate giants in the retail, telecommunications, financial-services, banking, and technology industries joined forces to combat phishing, spoofing, and other methods of online identify fraud. The Trusted Electronic Communications Forum [TECF] will focus on eliminating phishing's threat to e-mail and e-commerce. The companies are concerned about virtual threats that have impeded the progress of Internet communications and damaged the trust between enterprises and their customers. The TECF is a cross-industry, cross-geographic consortium dedicated to the standardization of technologies, techniques, and best practices in the fight against phishing, spoofing, identity theft cyber crime.

15 June 2004 United States.

New research published by Gartner indicates that illegal access to checking accounts, often gained via technology-borne schemes such as "phishing," has become the fastest-growing form of consumer theft in the United States. Roughly 1.98 million people experienced some form of security breach into their checking accounts during the previous 12 months. The research company said that crimes such as phishing accounted for a staggering $2.4 billion in fraud, or an average of $1,200 per victim, during the previous 12 months. Those most often targeted were people who had just begun to utilize online accounts to do business. Of the 4 million consumers who encountered fraud last year when opening a new online account, approximately half said they also received a phishing e-mail. Gartner said that checking account attacks ranked second only to physical credit card thefts in its study, which polled 5,000 people and was based on a 12-month period ending in April 2004. The research examined five types of consumer fraud: new account fraud, check forgery, unauthorized access to checking accounts, illegal credit-card purchases, and fraudulent cash advances on credit cards.

10 June 2004 Santiago, Chile.

A bomb exploded outside a bank in the Chilean capital. The group, calling itself "Julio Guerra, Southern Operation," took responsibility for the blast in a phone call to Santiago, Chile's, Bio Bio radio station. The predawn explosion caused considerable structural damage to a Banco del Estado office in the La Cisterna municipality, some six miles from downtown Santiago. The caller said the attack had been staged in retaliation for delays in passing a bill on behalf of people imprisoned for crimes sanctioned in antiterrorist legislation. Large chunks of concrete broke off the second story of the Banco del Estado building, and three ATMs were destroyed.

9 June 2004 Taiwan.

The numbers and personal codes of more than 100,000 Internet banking and auction-site clients were feared to have been stolen by hackers from across the Taiwan Strait. Criminal Investigation Bureau officials in Taiwan said they had arrested a Taiwanese man named Chen Chung-shun and seized a huge amount of confidential data, including 45 million e-mail addresses, almost 200,000 bank and auction-site account numbers with their corresponding personal secret codes, and information on three figurehead bank accounts. Investigators believe Chen had been collaborating with Chinese hackers since February 2004 to steal Internet bank codes by planting "shell" or "revised" versions of "Trojan horse" programs into the personal computers of customers using Internet banking services. Chen had reportedly gathered 45 million Taiwanese e-mail addresses and in mid-February, started sending advertising e-mails containing shell or revised Trojan horses to those e-mail addresses. By mid-March, he had sent out over 18 million e-mails. Officials said the ring withdrew the money from the International Commercial Bank of China ATM machines in China or transferred it to hundreds of figurehead accounts that had been established in the names of 10 Taiwanese people.

1 JUNE 2004 Philippines.
The Philippine government's wish to have the country removed from the international blacklist of money laundering-friendly countries was not granted by the influential Financial Action Task Force [FATF] due to delays in complying with the group's requirements. The FATF executive committee was still reviewing the law's implementing rules and regulations, which the government submitted earlier in the year. FATF considers the Philippines to be part of a group of noncooperative countries and territories where money laundering may either be rampant or relatively easy to perpetrate. Before a law was passed in 2001, money laundering was not considered a crime in the Philippines. The government's failure to have the country removed from the notorious blacklist means more advanced economies like the U.S. and U.K. will have to continue scrutinizing international financial transactions going to and coming from the Philippines.

29 MAY 2004 Russia.
The Central Bank of Russia recalled the general banking license from the commercial, privately owned Novocherkassk City Bank for numerous violations of the law on Counteraction of Money Laundering and Sponsorship of Terrorism. The CB also blamed the bank for failing to satisfy cash claims from creditors and to carry out obligatory payments. Additionally, the bank's management is accused of falsifying returns data and of failing to comply with a number of laws that regulate banking activities.

27 MAY 2004 Las Vegas, NV, United States.
The FBI continued to search for 11 people named in two federal racketeering indictments alleging two Romanian crime rings used stolen credit cards and false identification to bilk hundreds of thousands of dol-

lars from Las Vegas, NV, casinos and ATMs.

23 MAY 2004 Jiutepec, Mexico.

Bombs exploded outside three banks, heavily damaging them but causing no injuries. The explosions occurred around midnight in an industrial area of Jiutepec, about 35 miles from Mexico City. Authorities found a note near the bombing sites signed by a group calling itself Comando Jaramillista Morelense 23 de Mayo. The note lashes out at President Vicente Fox and "neoliberal counter-reforms," while calling for the departure of the governor of Morelos state, where Jiutepec is located.

28 JANUARY 2004 Washington, DC, United States.

The FBI launched an investigation into the source of the fastest-spreading email virus in history—"My Doom." Microsoft and SCO both offered a $250,000 reward for information leading to the arrest of the virus' creator.

NOVEMBER 2003 Istanbul, Turkey.

Four suicide bombers exploded trucks outside the HSBC base in Istanbul, the British consulate, and two synagogues, killing more than 60 people in attacks blamed on a group linked to al Qaeda.

11 NOVEMBER 2003 Athens, Greece.

Authorities neutralized an explosive device detected outside Athens' Citibank branch. An unidentified person phoned the Athens newspaper and announced that a bomb was going to explode at the bank. The Organization Khristos Kassimis was suspected.

13 AUGUST 2003 Bandipora, Kashmir.

A bomb attached to a bicycle exploded outside the State Bank of India, injuring 31 persons.

18 JULY 2003 McAllen, TX, United States.

An electric meter on a bank billboard was smashed. The Frogs were responsible.

13 JULY 2003 Athens, Greece.

Three Molotov cocktails were thrown at a branch office of the Eurobank, causing minor damage.

12 JUNE 2003 Thessaloniki, Greece.

Unidentified culprits entered the front lobby of the U.S.-owned Citibank and doused the ATM in a flammable liquid, placed a gas canister in it, and set it on fire. The explosion destroyed the ATM and caused extensive damage to the lobby and office equipment.

8 APRIL 2003 Izmir, Turkey.
Concussion hand grenades placed before the Bornova Court, Citibank, and British Consulate exploded, causing material damage. The MLK-P was suspected.

27 MARCH 2003 Santiago, Chile.
Antiwar protesters exploded a small bomb at a branch of the U.S.-based Bank of Boston. The bomb smashed windows, destroyed an ATM, and caused minor damage to two adjacent stores. Police found a pamphlet at the site that said "death to the empire," which they took as a reference to the U.S.

22 MARCH 2003 Koropi, Greece.
A makeshift incendiary device exploded in an ATM outside a Citibank branch. The explosion and subsequent fire caused severe damage to the ATM.

20 MARCH 2003 Knolargos, Greece.
Terrorists placed four gas canisters at the entrance to the Citibank and then set them on fire, causing minor damage.

11 FEBRUARY 2003 Tribuvan Airport, Nepal.
A U.S. citizen attempting to leave the country was found in possession of $494,000 and large sums of other foreign currencies.

10 OCTOBER 2001 Washington, DC, United States.
Officials discovered a $100,000 transaction belonging to Mohamed Atta, ring-leader of the September 11 hijackers.

1994-2001 United States.
During this period 18,500 defendants were charged in U.S. district courts with money laundering as the sole charge. Over the same period, 10,610 were charged with money laundering as the most serious offence; 9,169 were convicted.

NOVEMBER 2000 Athens, Greece.
The Athens offices of Citigroup and Barclays Bank and the studio of a Greek-American sculptor were bombed. RN was responsible.

JULY 1999 Athens, Greece.
A U.S. insurance company and a local bank were bombed. RN was suspected.

1998 Chicago, IL, United States.
Intelligence agencies reported that the Global Relief Foundation in Chicago, IL, knowingly or unknowingly raised more than $5 million per year to be given to Osama bin Laden. Also reported is the Al-Haramain Islamic Foundation [presence in Oregon and Missouri], claimed to have raised over $30 million a year, with part of these funds being directed to Osama bin Laden.

DECEMBER 1998 Athens, Greece.
European interests were attacked, including Barclays Bank. RN was responsible.

17 NOVEMBER 1998 Athens, Greece.
A bomb exploded outside a Citibank branch in Athens, causing major damage. An unidentified telephone caller to a local newspaper claimed the attack was to protest arrests made during a student march.

18 MAY 1998 Los Angeles, CA, United States.
A three-year undercover money-laundering scheme resulted in the arrest of 167 individuals and seizures of over $103 million, four tons of marijuana, and two tons of cocaine. Involved were 26 Mexican bank officials and other bankers from two Venezuelan banks.

OCT 1997 Urbana, IL, United States.
The President of Bio Products Division of the Archer Daniels Midlands Company [ADM] pleaded guilty to 37 counts of money laundering, totaling $9 million.

JANUARY 1996 Colombo, Sri Lanka.
A truck packed with explosives rammed into the central bank in Colombo, Sri Lanka, igniting towering fires in the business and tourist district. At least 53 people were reported killed and some 1,400 wounded. The truck-bomb attack was followed up by a ground frontal assault by a terrorist group, which engaged in a firefight with bank guards using automatic weapons and rocket-propelled grenade launchers.

1995–96 Mexico.
Authorities arrested Raul Salinas de Gortari, brother of former Mexican President Carlos Salinas de Gortari, for the assassination of Jose Francisco Ruiz Massiel, Secretary General of Mexico's Institutional Revolutionary Party. They uncovered over $300 million in seventy different bank accounts in seven countries, including several U.S. banks, all allegedly belonging to Raul. It was suspected that the money came from Raul's links to drug cartels, extortion, and bribery.

1995 United States.
U.S. Post Office estimates suggest that $50 billion annually is being laundered through international mail.

1991–1994 Pennsylvania, United States.
Two men were convicted of defrauding foreign governments of $10 million.

1990 United States.
Money laundering in the U.S. was estimated at $500 billion a year out of $2.4 trillion in bank transactions.

1985 Boston, MA, United States.
The Bank of Boston was fined $500,000 for failing to file currency-transaction reports on transactions totaling $1.2 billion.

APPENDIX B: UNITED STATES GOVERNMENT AND INDUSTRY BANKING AND FINANCE INITIATIVES

The attacks in New York City on September 11, 2001, showed that the financial-services industry is highly resilient—the safeguards and backup systems in place performed well. Since 1998 the sector has been working with the Department of the Treasury to organize itself and address the risks of emerging threats, particularly cyber intrusions. It was also the first sector to establish an ISAC to share security-related information among industry members.

Major institutions in this sector continue to perform ongoing assessments of their security programs. Since 2001, the industry and its associations have initiated lessons-learned reviews to identify corrective actions for the improvement of security and response and recovery programs, as well as to provide a forum for sharing best practices through trade associations and other interdisciplinary groups. The sector as a whole, with the support of the Department of the Treasury, has also initiated a sector-wide risk review. In addition, individual institutions have stepped up their own investments due to a better understanding of the threat.

Additional banking and finance-sector protection initiatives include efforts to:

- **Identify and address the risks of sector dependencies on electronic networks and telecommunications services.** The financial sector's reliance on information systems and networks has resulted in a number of concerns for the industry. The Department of the Treasury, in concert with the DHS, will convene a working group consisting of representatives from the telecommunications and financial-services sectors, as well as other federal agencies, to study and address the risks that arise from the sector's dependencies on electronic networks and telecommunications services.
- **Enhance the exchange of security-related information.** The DHS will work with the Department of Treasury, the FBIIC, and the FS-ISAC to improve federal-government communications with sector members and streamline the mechanisms through which they exchange threat information on a daily basis as well as during security incidents.
- **Freezing of terrorist assets:**
  A cornerstone of the fight against terrorism has been the continuing concerted interagency effort to target and interdict the financial

structure of terrorist operations around the world. Weeks after September 11, 2001, President Bush signed Executive Order 13224, giving the United States government a strong tool for eliminating the financial supporters and networks of terrorism. In part, the order allowed the Secretary of the Treasury to freeze the assets of 27 individuals and organizations known to be affiliated with the al Qaeda network. The order gives the Secretary of the Treasury broad powers to impose sanctions on banks around the world that provide criminal entities access to the international financial system. It also provides for designation of additional entities as terrorist organizations. It expands government authority to permit the designation of individuals and organizations that provide support or financial or other services to or associate with designated terrorists. By late October 2002, the freeze list had expanded to include designated terrorist groups, supporters, and financiers of terror. The order allows the United States government, as well as coalition partners acting in concert, to block tens of millions of dollars intended to bankroll the murderous activities of al Qaeda and other terrorist groups.

In addition, on September 28, 2001, the United Nations Security Council adopted Resolution 1373, which requires all states to "limit the ability of terrorists and terrorist organizations to operate internationally" by freezing their assets and denying them safe haven. The Security Council also set up a Counter Terrorism Committee to oversee implementation of Resolution 1373. United Nations Security Council Resolution 1390 of January 16, 2002, obligated member states to freeze funds of "individuals, groups, undertakings, and entities" associated with the Taliban and al Qaeda.

As of May 2003, about $143 million in terrorist funds had been frozen worldwide as a result of these initiatives, according to United States and United Nations financial data. As of December 2003, the United States government had blocked more than $35.5 million in assets of the Taliban, al Qaeda, and other terrorist entities and supporters. More than 160 other nations participated in blocking more than $112 million in assets. As of December 2003, the total number of individuals and entities designated under this order was 334. The overall effect of frozen funds is largely unknown because most of the flow of terrorist funds reportedly takes place outside of formal banking channels. Much of al Qaeda's money is reportedly held not in banks but in untraceable assets such as gold and diamonds.

- **Money Laundering Suppression Action [PL 103-325]:**
  The Money Laundering and Financial Crimes Strategy Act of 1998 required the President—acting through the Secretary of the Treasury and in consultation with the Attorney General and other relevant federal, state, and local law-enforcement and regulatory officials—to develop and submit to the Congress an annual national money-laundering strategy by February 1 of each year from 1999 through 2003. The goal of the Strategy Act was to increase coordination and cooperation among the various law-enforcement and regulatory agencies and to effectively distribute resources to combat money laundering and related financial crimes. The 1998 Strategy Act required that each national money-laundering strategy [NMLS] define comprehensive, research-based goals, objectives, and priorities for reducing money laundering and related financial crimes in the United States. The annual NMLS generally has included multiple priorities to combat money laundering to guide federal agencies' activities. In the aftermath of the 9/11 attacks, the Strategy Act of 1998 was adjusted in 2002 to reflect a new federal priority—combating terrorist financing. In addition, the Combating Money Laundering and Terrorist Financing Act of 2003, among other purposes, extends the requirement for the annual NMLS required by 1998 federal legislation to 2006.
- Another provision of the Strategy Act authorized the Secretary of the Treasury to designate high-intensity money-laundering and related financial crime areas [HIFCA] in which federal, state, and local law enforcement would work cooperatively to develop a focused and comprehensive approach to targeting money-laundering activity. The act also urges states to enact uniform laws to license and regulate certain financial services. By doing so, states have protected consumers by providing increased stability and transparency to an industry prone to abuse, while at the same time offering state and local law enforcement the necessary tools to dismantle informal and unlicensed money-transaction networks.
- As envisioned by the Strategy Act, HIFCAs were to represent a major NMLS initiative and were expected to have a flagship role in the United States government's effort to disrupt and dismantle large-scale money-laundering operations. They were intended to improve the coordination and quality of federal money-laundering investigations by concentrating the investigative expertise of federal, state, and local agencies in unified task forces, thereby leveraging resources and creating investigative synergies.

- The former United States Customs Service [now part of the United States Immigration and Customs Enforcement [ICE]—a component of the Department of Homeland Security—and the FBI both have a long history of investigating money laundering and other financial crimes. In response to the terrorist attacks of 9/11, the Treasury and Justice Departments jointly established a multi-agency review groups and regional task forces dedicated to combating terrorist financing. The review group investigates suspicious financial transactions in order to uncover models to help identify future illegal financing. The Treasury established Operation Green Quest, led by the Customs Service, to augment existing counterterrorist efforts by targeting current terrorist funding sources and identifying possible future sources. It works to freeze the accounts and seize the assets of individuals and organizations that finance terrorist groups. In addition, Operation Green Quest was designed to attack the shipment of bulk currency and the financial systems that may be used by terrorists to raise and move funds, such as fraudulent charities. In January 2003, the Customs Service expanded Operation Green Quest by doubling the personnel commitment to approximately 300 agents and analysts nationwide working solely on terrorist financing matters. In March 2003, Operation Green Quest was transferred to ICE within the Department of Homeland Security.

- On September 13, 2001, the FBI formed a multi-agency task force—now known as the Terrorist Financing Operations Section [TFOS]—to combat terrorist financing. The mission of TFOS has evolved into a broad role to identify, investigate, prosecute, disrupt, and dismantle all terrorist-related financial and fundraising activities. The FBI also took action to expand the antiterrorist financing focus of its Joint Terrorism Task Forces [JTTF]—teams of local and state law-enforcement officials, FBI agents, and other federal agents and personnel whose mission is to investigate and prevent acts of terrorism. According to the FBI, the first JTTF came into being in 1980, and the total number of task forces has nearly doubled since September 11, 2001. Today, there is a JTFF in each of the FBI's 56 main field offices, and additional task forces are located in designated smaller FBI offices. In 2002, the FBI created a national JTFF to collect terrorism information and intelligence and funnel it to the field JTTFs and partner agencies.

- As with any joint federal operation, the ICE and the JTTFs experienced structural and leadership problems in delineating antiterrorist financing roles and responsibilities, including resolving jurisdictional issues. As such, general efforts to resolve crucial departmental issues were largely unsuccessful until May 2003, when the Attorney General and the Secretary of Homeland Security signed a Memorandum of Agreement that contained a number of provisions designed to resolve jurisdictional issues and enhance interagency coordination of terrorist financing investigations. According to the agreement, the FBI is to lead terrorist financing investigations and operations, using the intergovernmental and interagency national JTTF at FBI headquarters and the JTTFs in the field. The agreement also specified that through TFOS the FBI is to provide overall operational command to the national JTTF and the field JTTFs. Further, to increase information sharing and coordination of terrorist financing investigations, the Agreement required the FBI and ICE to detail appropriate personnel to each other's agency and develop specific collaborative procedures to determine whether applicable ICE investigations or financial-crimes leads may be related to terrorism or terrorist financing. Also, the agreement required the FBI and ICE to produce a joint written report on the status of the implementation of the agreement four months from its effective date.

- According to the Government Accounting Office [GAO], from a strategic perspective the annual NMLS has had mixed results in guiding the efforts of law enforcement in the fight against money laundering and, more recently, terrorist financing. Although expected to have a flagship role in the United States government's efforts to disrupt and dismantle large-scale money-laundering operations, HIFCA task forces generally are not yet structured and operating as intended. The Treasury and Justice Departments are in the process of reviewing the HIFCA task forces, which ultimately could result in program improvements. Also, most of the NMLS initiatives designed to enhance interagency coordination of money-laundering investigations have not yet achieved their expectations. Federal law-enforcement agencies recognize that they must continue to develop and use interagency coordination mechanisms to leverage existing resources to investigate money laundering and terrorist financing. Since the Memorandum of Agreement signed by the Attorney General and the

Secretary of Homeland Security, progress has been made in waging a coordinated campaign against sources of terrorist financing, and continued progress depends largely on the agencies' ability to establish and maintain effective interagency relationships and meet various other operational and organizational challenges.

Chapter 17

THE TELECOMMUNICATIONS SECTOR

This chapter outlines the:

- Criticality of the telecommunications sector to national security
- Telecommunications-sector attractiveness to terrorist and criminal elements
- Telecommunications-sector vulnerabilities
- Tailoring the S^3E Security Methodology to the telecommunications sector
- Telecommunications-sector challenges facing the security assessment team
- Applying the security assessment methodology to the telecommunications sector
- Preparing the telecommunications security-assessment report
- Historical overview of selected telecommunications incidents
- U.S. government telecommunications initiatives

A LINK TO ALL OTHER SECTORS IS VITAL TO OUR NATIONAL SECURITY

We live in a world that is increasingly more dependent on information and the technology that allows us to communicate and do business globally at the speed of light. Today, more information is delivered faster and in greater volume than ever before. Information has always been time-dependent but is more so today then ever before because of advances in communication technologies. The composition of the telecommunications sector evolves continuously due to technology advances, business and competitive pressures, and changes in the regulatory environment. Despite its dynamic nature, the sector has consistently provided robust and reliable communications and processes to meet the needs of businesses and governments.

As the government and critical infrastructure industries rely heavily on the public telecommunications infrastructure for vital communications services, the sector's protection initiatives are particularly important. The telecommunications sector provides voice and data service to public and private users through a complex and diverse public-network infrastructure encompassing the public switched telecommunications network [PSTN], the Internet, and private-enterprise networks. The PSTN provides switched circuits for telephone, data, and leased point-to-point services. It consists of physical facilities [including over 20,000 switches], access tandems, and other equipment. These components are connected by nearly two billion miles of fiber and copper cable. The physical PSTN remains the backbone of the infrastructure, with cellular, microwave, and satellite technologies providing extended gateways to the wire-line network for mobile users. Supporting the underlying PSTN are operations, administration, maintenance, and provisioning systems that provide the vital functions such as billing, accounting, configuration, and security management.

Advances in data-network technology and the increasing demand for data services have spawned the rapid proliferation of the Internet infrastructure. The Internet consists of a global network of packet-switched networks that use a common suite of protocols. Internet service providers [ISPs] provide end users with access to the Internet. Larger ISPs use network operation centers [NOCs] to manage their high-capacity networks by linking them through Internet peering points or network access points. Smaller ISPs usually lease their long-haul transmission capacity from the larger ISPs and provide regional and local Internet access to end users via the PSTN. Internet-access providers interconnect with the PSTN through points of presence, typically a switch or a router, located at carrier central offices. International PSTN and Internet traffic travels via underwater cables that reach the United States at various cable landing points. In addition to the PSTN and the Internet, enterprise networks are an important component of the telecommunications infrastructure—they support the voice and data needs and operations of large enterprises. These networks comprise a combination of leased lines or services from the PSTN or Internet providers.

The Telecommunications Act of 1996 opened local PSTN service to competition. It required incumbent carriers to allow their competitors to have open access to their networks. As a result, carriers began to concentrate their assets in collection facilities and other buildings known as telcom hotels, collocation sites, or peering points instead of laying down new cable. ISPs also gravitated to these facilities to reduce the costs of exchanging traffic with other ISPs. Open competition, therefore, has caused the operation of the PSTN and the Internet [including switching, transport, signaling, routing, control, security, and management] to become increasingly interconnected, software-driven, and remotely managed, while the industry's physical assets are increasingly concentrated in shared facilities.

An Attractive Target

The importance of information protection continues to grow as our society becomes more dependent on high-technology systems. Increased threats to the systems and the information they store, process, display, and transmit due to expanded connectivity on a global basis means that maintaining adequate security requires constant new development and implementation, especially in light of the overload of new technologies within the telecommunications sector. More importantly, the smooth and orderly functioning of our society as it is today is utterly dependent on our ability to communicate quickly through our telephone lines, radio frequencies, Internet connections, and wireless networks. Any attacker knows that severely disabling these systems can cause social disorder, panic, loss of vital services, the inability of government to function effectively, and massive economic repercussions if businesses are unable to operate. Two recent examples of such breakdowns in the telecommunications sector with destructive consequences are the World Trade Center attacks in September 2001 and Hurricane Katrina in August 2005.

Telecommunications Vulnerabilities

The tentacle-like structural nature of the telecommunications sector makes it inherently vulnerable on all fronts. Damage at a single point can have major sector-wide repercussions. During the 2001 terrorist attacks, significant damage impaired telecommunications facilities, lines, and equipment. The loss of telecommunications service, as well as damage to the power and transportation infrastructure, delayed the reopening of the financial markets. Much of the disruption to voice and data communications services throughout lower Manhattan [including the financial district] occurred when a building in the World Trade Center complex collapsed into an adjacent Verizon communications center, which served as a major local communications hub within the public network. Approximately 34,000 businesses and residences in the surrounding area lost services. This also resulted in disruptions to customers in other service areas, as various carriers had equipment collocated at the site that linked their networks to Verizon. In addition, considerable amounts of telecommunications traffic that originated and terminated in other areas also passed through this location. AT&T's local network service in lower Manhattan was also significantly disrupted following the attacks. Although not directly targeted, the telecommunications sector became a victim of significant collateral damage. This was not only an inconvenience for the average person but hindered speedy response and recovery operations. However, overall the telecommunications sector demonstrated remarkable resiliency as the damage to its assets was offset by **diverse, redundant, and multifaceted communications capabilities—a key aspect of security for this sector.**

On the Internet front, the threat is not only to physical assets but also stems from an intruder's ability to hack into the system from the outside and be capable of causing disruption or destruction from "within." The most widely reported Internet security problems of the past few years are "denial of service" attacks against large commercial sites, in which intruders place Trojan-horse programs on computers that have persistent high-speed Internet access and relatively little security. At a given signal, the attacker's Trojan-horse programs conduct a coordinated attack against other sites, sending messages at a rate too high for the sites to handle. With the explosive growth in broadband services, high-speed Internet access for home users makes it likely that future denial-of-service attacks may use Trojan-horse programs planted on home computers as well.

Home broadband architectures face a variety of threats that, while present on dial-up connections, are easier to exploit using the faster, "always on" qualities of broadband connections. The relatively short duration of most dial-up connections makes it more difficult for attackers to compromise home users connected to the Internet. The advent of "always on" broadband connections provides attackers the speed and communications bandwidth necessary to compromise home computers and networks. **Almost all users face the risk that intruders can read, change, or delete files on their personal computers.** Another concern is the potential for an intruder to hijack the average user's computer, establishing a "backdoor" that can be activated anytime the machine is online, giving the intruder control over the user's machine. Ironically, as governmental and corporate organizations have hardened their networks and become more sophisticated at protecting their computing resources, they have driven malicious entities to pursue other targets of opportunity. Home users with broadband connections are these new targets of opportunity both for their own computing resources and as an alternative method for attacking and gaining access to government and corporate networks.

Due to growing interdependencies among the various critical infrastructures, a direct or indirect attack on any of them could result in cascading effects across the others. Such interdependencies increase the need to identify critical assets and secure them against both physical and cyber threats. Critical infrastructures rely upon a secure and robust telecommunications infrastructure. **Redundancy within the infrastructure is critical to ensure that single points of failure in one infrastructure will not adversely impact others.** It is vital that the security-assessment team works to characterize this state of diversity within the telecommunications architecture. It must also collaborate to understand the topography of the physical components of the architecture in order to establish a foundation for defining a strategy to ensure physical and logical diversity.

TAILORING THE S^3E SECURITY ASSESSMENT METHODOLOGY FOR THE TELECOMMUNICATIONS SECTOR

The exhibit on the next page highlights in grey shaded areas those specific elements that are tailored for application to the telecommunications sector.

Strategic Planning
Tasks 1, 2, & 3
Operational Environment
Critical Assessment
Threat Assessment

Program Effectiveness
Task 4
Infrastructure Interdependency
Evaluation [P_{E1}]

Operational Environment Task 1
Enterprise Goals
Corporate Image, Core Values, Areas of Responsibility, Performance Objectives Business Incentives, Investment
Commitments
Federal Statutes, State Laws, Municipal Codes, Confidentiality Laws, Civil Common Law, Willful Torts, Contractual Obligations, Agreements, Governing Authorities, Voluntary Commitments
Characterization
Mission, Location, Facility Layout Facility/Land Use, Neighboring Facilities, Terrain, Climate Resources, Operations, Processes
Operations – Logistics
Visitors, Customers, Vendors; Mail and Material Delivery, Receiving, Shipping; Staging, Storing, Distribution & Delivery; Time-Sensitive Processes; Communications, Institutional Ownership, Original Identity, Administration & Provisioning, Carrier Central Office Services, Broadband Packet-based IP Networks, Network Operations Center, Operations & Maintenance

Critical Assessment Task 2
Resources, Operations, Functions Administrative, Aircraft/Fleet Facilities Board/Conference Rooms Cable/Communications Rooms Electrical/Mechanical Rooms, Physical Infrastructure, Telecommunications Centers, IT Network Centers Training Facilities, Warehouses, Loading Docks, SCADA & Security Systems Infrastructure Sharing, Critical Interdependencies Primary, Backup, Redundant Systems Data Collection Systems & Processes Due Diligence, Dissemination Systems Primary/Backup Power Sources Distribution & Transmission Systems High Capability Networks, Base Stations Modem/Cable Connections, Telecommunications Architecture Switching, Transport, Signaling, Routing System Configuration Switch Circuits Land Haul Transmission Capability Electronic Communications Network Transaction Structure & Composition Microwave/Telecommunications Sites Air, Water, Temperature, HVAC Systems

Threat Assessment Task 3
International, National, Regional, Local Range & Levels of Threats Design Basis Threat Profile, Probability Threat Categories, Capabilities, Methods & Techniques

Enterprise
Roles & Responsibilities Operations, Processes, Protocols Functional Relationships, Security Awareness, Reporting Security Incident, Alert Notification System

Security Capabilities
Management Team, Organization & Composition, Staffing, Skills, Experience, Qualifications & Training, Operational Capabilities Dependency Programs Power Sources, Redundant/Backup Systems, Interoperability, Information Systems, Emergency Preparedness & Business Continuity Planning, Internet Service Providers, Next Generation Network, Public Switched Telecommunications Network

Barriers & Delay Measures
Terrain, Approaches, Fences, Barrier, Walls, Gates, Tamper Systems Cryptographic Key Management, Encryption, Diverse Telecommunications Architecture, Switching, Transport, Routing
Detection Measures
Interior/Exterior Sensors, Special Purpose Sensors, Tamper Systems System Protective Features Network Configuration
Access Control Measures
People/Vehicle/Deliver Controls Screening, Verification, Internet, Routers/Modems, Browsers, Mainframe, LAN Servers, WAN, Access, Network Access Points
Assessment Measures
Perimeter Security, Lighting Posts and Patrols, CCTV, Suspicious Behavior, Monitoring, Surveillance, Inspection
Communications Media
Radio, Telephone, PA, Hardwire, Wireless, Fiber Optic, Information Network
Security Control Center[s]
Annunciation & Display, Event Storage, Reports, Alarm Integration Information Networks
Security Infrastructure
Primary/Secondary/Backup Redundancy

Response & Recovery
Tactics, Techniques, First/Second & Integrated Responders, System Distribution, Diverse Technologies, Diverse Backup Facilities, System Switchover, Closeout Actions

Quality Assurance/Quality Control Process

Exhibit 17.1 The S^3E Security-Assessment Methodology for the Telecommunications Sector

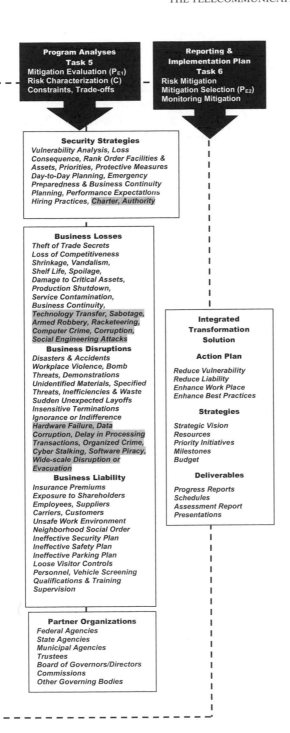

Program Analyses
Task 5
Mitigation Evaluation (P_{E1})
Risk Characterization (C)
Constraints, Trade-offs

Reporting &
Implementation Plan
Task 6
Risk Mitigation
Mitigation Selection (P_{E2})
Monitoring Mitigation

Security Strategies
Vulnerability Analysis, Loss
Consequence, Rank Order Facilities &
Assets, Priorities, Protective Measures
Day-to-Day Planning, Emergency
Preparedness & Business Continuity
Planning, Performance Expectations
Hiring Practices, Charter, Authority

Business Losses
Theft of Trade Secrets
Loss of Competitiveness
Shrinkage, Vandalism,
Shelf Life, Spoilage,
Damage to Critical Assets,
Production Shutdown,
Service Contamination,
Business Continuity,
Technology Transfer, Sabotage,
Armed Robbery, Racketeering,
Computer Crime, Corruption,
Social Engineering Attacks

Business Disruptions
Disasters & Accidents
Workplace Violence, Bomb
Threats, Demonstrations
Unidentified Materials, Specified
Threats, Inefficiencies & Waste
Sudden Unexpected Layoffs
Insensitive Terminations
Ignorance or Indifference
Hardware Failure, Data
Corruption, Delay in Processing
Transactions, Organized Crime,
Cyber Stalking, Software Piracy,
Wide-scale Disruption or
Evacuation

Business Liability
Insurance Premiums
Exposure to Shareholders
Employees, Suppliers
Carriers, Customers
Unsafe Work Environment
Neighborhood Social Order
Ineffective Security Plan
Ineffective Safety Plan
Ineffective Parking Plan
Loose Visitor Controls
Personnel, Vehicle Screening
Qualifications & Training
Supervision

Integrated
Transformation
Solution

Action Plan

Reduce Vulnerability
Reduce Liability
Enhance Work Place
Enhance Best Practices

Strategies

Strategic Vision
Resources
Priority Initiatives
Milestones
Budget

Deliverables

Progress Reports
Schedules
Assessment Report
Presentations

Partner Organizations
Federal Agencies
State Agencies
Municipal Agencies
Trustees
Board of Governors/Directors
Commissions
Other Governing Bodies

Exhibit 17.1 *(continued)*

Telecommunications Challenges Facing the Security-Assessment Team

Today, technology is continually changing rapidly. A significant challenge facing the security-assessment team is to: ensure the diversity of telecommunications services; to improve the reliability and efficiency of networks, telecommunications carriers, and the physical network facilities they use to route circuits; and to support critical government and industry operations to withstand the effects of wide-scale network disruptions.

A more common obstacle is that as organizations increasingly demand remote connectivity to corporate and government networks, the security of these remote end points becomes increasingly critical to the overall protection of the network. Contributing to this trend is the explosive growth in the popularity of broadband connections for home users. These developments complicate the process of securing organizational and home networks. Other hurdles involve assuring the priorities of telecommunications carriers: service reliability, cost balancing, security, and effective risk-management postures.

The government places high priority on the consistent application of security across the entire infrastructure. Although private- and public-sector stakeholders share similar objectives, they have different perspectives on what constitutes acceptable risk and how to achieve security and reliability. Therefore, an agreement on a sustainable security threshold and corresponding security requirements remains elusive.

APPLYING THE S^3E SECURITY ASSESSMENT METHODOLOGY TO THE TELECOMMUNICATIONS SECTOR

The security-assessment team should use its collective expertise to:

- Identify the important mission and functions of the telecommunications sector [**Task 1—Operational Environment**] and incorporate the development of all business operations, configuration, and continuity expectations. Examination focuses on services and the customer base to determine the business contributions to the community, its culture, and business objectives. This includes but is not necessarily limited to identification of business goals and objectives, customer base, and commitments and characterization of facilities and boundaries, as shown in Exhibits 17.2 through 17.5.

Areas of Responsibility	Performance Objectives
Business Incentives & Investment	Promote Industry Cooperation
Corporate Image & Core Values	Prosecute Major Criminal Infractions
Mission & Services	Public Safety & Public Confidence

Exhibit 17.2 Identify Telecommunications Enterprise
Business Goals and Objectives

Local, Regional, National, International	Peer Telecommunications Institutions, [Domestic and International]
General Public [Domestic and International]	Public and Private Corporations [Domestic and International]
Nongovernmental Organizations [Domestic and Abroad]	Size of Population Served
Nonprofit Organizations [Domestic and International]	U.S. Government, State, and Local Agencies

Exhibit 17.3 Identify the Telecommunications
Enterprise Customer Base

29 CFR 1910.36 - Means of Egress	50 U.S.C. 1801 - Foreign Intelligence Surveillance Act
29 CFR 1910.151 - Medical Services and First Aid	Americans with Disabilities Act [42 U.S.C. 12101]
29 CFR 1910.155 Fire Protection	Architectural Barrier Act 1968 [As Amended]
18 U.S.C. 1343 - Wire Fraud	
18 U.S.C. 2511 - Technical Surveillance	Bioterrorism Preparedness and Response Act, 2002 [As Amended]
21 U.S.C. 801 - Drug Abuse and Control	Carriers, Suppliers, Vendors, Insurance
26 U.S.C. 7201 - Tax Matters	Civil Rights Act 1964 [U.S.C. 200e] [As Amended]
29 U.S.C. 2601 - Family and Medical Leave Act	Commerce Commission
31 U.S.C. 5322 - Financial Transaction Reports	Comprehensive Drug Abuse Prevention Act 1970 [As Amended]
41 U.S.C. 51 - U.S. Contractor Kickbacks	Comprehensive Environmental Response, Compensation and Liability Act 1980 [As Amended]
41 U.S.C. 701 - Drug Free Workplace Act	
46 U.S.C.1903 - Maritime Drug Enforcement	

(continued on next page)

Exhibit 17.4 Identify Telecommunications Enterprise Commitments

Computer Security Act 1987 [As Amended]	Internal Revenue Service
	Labor Management Relations Act
Controlled Substance Act [As Amended]	Memorandum of Agreement
Drug Free Workplace Act 1988 [As Amended]	Memorandum of Understanding
Electronic Communications Privacy Act [As Amended]	Omnibus Crime Control and Safe Streets Act, 1968 [As Amended]
Emergency Planning & Right-to-Know Act 1986 [As Amended]	Policy Coordination Committee [PCC]
	Privacy Act [5 U.S.C. 522e] [As Amended]
Employee Polygraph Protection Act 1988 [29 U.S.C. 2001] [As Amended]	Rehabilitation Act, 1973 [As Amended]
	State and Local Industry Administrators and Authorities
Fair Credit Reporting Act [15 U.S.C. 1981] [As Amended]	State and Local Emergency Planning Officials and Committees
Family Education Rights and Privacy Act 1974 [As Amended]	Telecommunication Act 1996 [As Amended]
Family and Medical Leave Act 1993 [As Amended]	
	Uniform Trade Secrets Act
Federal Communications Commission [FCC]	U.S. Department of Commerce
	U.S. Department of Homeland Security [DHS]
Federal Privacy Act 1974 [As Amended]	
Foreign Corrupt Practices Act	U.S. Department of Justice [DOJ]
Freedom of Information Act [U.S.C. 552] [As Amended]	U.S. Patriot Act, 2002 [As Amended]
	U.S. Postal System
National Labor Relations Act [29 U.S.C. 151] [As Amended]	U.S. Environmental Protection Agency [EPA]
Information Technology Management Reform Act 1996 [As Amended]	U.S. Occupational Safety and Health Administration [OSHA]
Interagency Agreements	

Exhibit 17.4 *(continued)*

- Describe the configuration, operations, and other elements [**Task 2— Critical Assessment**] by examining conditions, circumstances, and situations relative to the potential for disruption of services. Focus is on an integrated evaluation process to determine which assets need protection in order to minimize the impact of threat events. The security-assessment team identifies and prioritizes facilities, processes, and assets critical to telecommunication service objectives that might

Commercially Leased Facilities	Neighboring Facilities
Dedicated site owned and operated by enterprise.	Operated and maintained through reciprocal agreements with internal and external entities.
Dispersed Locations	
Facility Layout & Location	Terrain & Climate
Facility Land Use	Types of assets at specific locations.
Infrastructure Sharing	

Exhibit 17.5 Characterize Configuration of Telecommunications Enterprise Facilities and Boundaries

be subject to malevolent acts or disasters that could result in a potential series of undesired consequences. The process takes into account the impacts that could substantially disrupt the ability of the utility to provide safe telecommunication services and strives to reduce security risks associated with the consequences of significant events. These include but are not necessarily limited to assessment of assets, operations, and processes and prioritization of assets and operations, as shown in Exhibits 17.6 and 17.7.

- Review the existing design-basis threat-profile **[Task 3—Threat Assessment]** and update it as applicable to the client. If a profile has not been established, then the security-assessment team needs to develop one. Both the security-assessment team and the client must have a clear understanding of undesired events and consequences that may impact telecommunication services to the community and business continuance-planning goals. Emphasis is on prioritizing threats through the process of identifying methods of attack, past or recent events, and types of disasters that might result in significant consequences; the analysis of trends; and the assessment of the likelihood of an attack occurring. Focus is on threat characteristics, capabilities, and target attractiveness. Social-order demographics to develop trend analysis in order to support framework decisions are also considered. Under this task the security-assessment team should also validate previous telecommunication assessment studies and update conclusions and recommendations into a broader framework of strategic planning. These include but are not necessarily limited to determination of types of malevolent acts, assessment of other disruptions, identification of perpetrator categories, and assessment of loss consequences and threat attractiveness and likelihood, as shown in Exhibits 17.8 through 17.12.

Administrative, Aircraft/Fleet Facilities	Long-haul transmission capability
Analysis data, historical data base	Mainframe, LAN Servers, WAN access
Base Stations, cable modem connections	Microwave/Telecommunications Sites
Broadband connections, cable landing points	Mission, Service Capabilities
Business impact analysis	Network access points, points of presence
Cable/Communications Rooms & Centers	Offsite storage or backup media, non-
Cold, Warm, Hot, Mirrored, Mobile, Fixed Sites	electronic records, and system documentation
Concentration of key staff, supervision, staffing, qualifications, training	Online Security Assessment
	Operational Command Centers
Convergence of technologies	Physical Infrastructure
Crisis Management Centers	Physical protection system features
Cryptographic key management, encryption	Population & Population Throughputs
Distribution and Transmission Systems	Pre-employment screening
Diverse backup facilities	Primary and redundant operating systems
Electronic Communications Network	Primary Power Sources
E-mail servers, e-mail, digital mobile phones and cameras	Proprietary Information
Emergency master system shutdown switch	Reporting Suspicious Activity
Emergency preparedness and business continuity planning	Satellite, radio-cell towers, mobile wireless
	SCADA & Security C3 Centers
Fire suppression systems, fire and smoke detectors, plastic tarps, water sensors	Security plans, policies, protocols
	Short-term/long-term backup power sources
Firewall networks, tamper systems, cyber intrusions	Switching, transport, signaling, routing, control
	System Configuration Switch Circuits
Heat-resistant and water proof containers for backup media and non-electronic records	System design and procurement
High capability networks	Telcom hotels, collection sites, peering points
Hiring Practices	telephone
Hours of Operation	Telecommunications architecture
HVAC units	Vendor deliveries
Internet, routers, modems, web browsers	Voice and data networks, virtual private
Interoperability capability, shared information systems	networks
	Warehousing, material inventories
IT Network Computer Centers	Wireless and networking technologies
Least privilege access points	Wireless local area networks

Exhibit 17.6 Critical Assessment of Telecommunications Enterprise Assets, Operations, Processes, and Logistics

Least critical facilities based on value system	Most critical facilities based on value system
Least critical assets based on value system	Most critical assets based on value system

Exhibit 17.7 Prioritize Critical Telecommunications Enterprise Assets and Operations in Relative Importance to Business Operations

Armed attack	Local exchange carrier
Arson	Loss of key personnel
Assassination, kidnapping, assault	Misuse/damage of control systems
Barricade/hostage incident	Misuse of wire and electronic communications
Bomb threats and bomb incidents	Phishing, hacking, computer crime
Calling card fraud	Power failure
Carrier slamming	Planned production shutdown or outage
Chemical, Biological, Radiological [CBR] contamination of assets, supplies, equipment, and areas	Rape and sodomy
	Remote maintenance access points
	Routing
Computer-based extortion	Sabotage
Computer system failures and viruses	Sexual abuse, sexual exploitation
Computer theft	Social engineering
Cyber attack on SCADA or other systems	Software piracy
Cyber stalking, software piracy	Storage of criminal information on the system
Damage/destruction of interdependency systems resulting in production shutdown	Unlawful access to stored wire and electronic communications and transactional records access
Damage/destroy service capability	
Damage/destruction of pipelines and cabling	Supplier, courier disruptions
Delay in processing transactions	Telecommunications failure
Denial of service attacks	Terrorism
Dialed number recorder	Torture
Digital systems	Treason, sedition, and subversive activities
Direct inward system access	Trespassing and vandalism
Email attacks	Unauthorized transfer of American technology to rogue foreign states
Encryption breakdown	
Explosion	Unlawful access to stored communications
FAX security	Unlawful wire and electronic communications interception and interception of oral communications
Hacker attacks	
Hardware failure, data corruption	Use of cellular/cordless telephones to disrupt network communications or in support of other crimes
Inability to access system or database	
Inability to perform routine transactions	
Industrial accidents and mishaps	Voice mail system
Intentional shutdown of electrical power	Wide-scale disruption and disaster
Inter exchange carrier	Wide-scale evacuation

Exhibit 17.8 Determine Types of Malevolent Acts That Could Reasonably Cause Telecommunications Enterprise Undesirable Events

Administration	Mail fraud
Auditing software	Maintenance activity
Bribery, graft, conflicts of interest	Natural disasters
Civil disorders, civil rights violations	Obscenity
Conspiracy	Obstruction of justice
Eavesdropping	Online pornography and pedophilia
Economic espionage censorship	Racketeering
Embezzlement, extortion and threats	Racketeer influenced and corrupt organizations
Historical rate Increases	
Homicide	Release and misuse of personal information
Human error	
Inefficiencies and waste	Resource constraints and competing priorities
Ineffective hiring practices	Robbery and burglary
Ineffective personnel and vehicle controls	Shrinkage, shelf-life, spoilage
Ineffective safety and security plans	Sudden unexpected layoffs
Ineffective supervision and training	Telemarketing fraud
Ineffective use of resources and time	Trunk restrictions
Ineffective visitor controls	Tracking effects of outage overtime
Ignorance and indifference	Tracking effects of outage across related resources and dependent systems
Insensitive terminations	
Institutional corruption, fraud, waste, abuse	Theft of trade secrets, intellectual property, confidential information, patents, and copyright infringements
Loss of competitiveness	
Loss of proprietary information	Unidentified materials or wrong deliveries

Exhibit 17.9 Assess Other Disruptions Impacting
Telecommunications Enterprise Operations

Activist groups, subversive groups or cults	purposes
Disenfranchised individuals, hackers	Terrorist organizations posing as legitimate entities using the system for criminal purposes
Legitimate entities serving as conduits for terrorist financing	
Organized criminal groups	Vandals, lone wolves, insiders
State-sponsored or independent terrorist groups using the system for criminal	

Exhibit 17.10 Identify Category of Telecommunications
Enterprise Perpetrators

- Examine and measure the effectiveness of the security system [**Task 4—Evaluate Program Effectiveness**]. The security assessment evaluates business practices, processes, methods, and existing protective measures to assess their level of effectiveness against vulnerability and risk and to identify the nature and criticality of business areas and assets of greatest concern to telecommunication operations. Focus is on vulnerability and vulnerable access points where penetration may be accomplished and how. Emphasis is placed on the physical characteristics of vulnerable points, accessibility to the location, and the effectiveness of protocol obstacles an adversary would likely have to overcome to reach a critical telecommunication asset. These include but are not necessarily limited to evaluation of security operations and protocols, organization, relationships, and SCADA and security performance; definition of adversary plan; and assessment of response and recovery, as shown in Exhibits 17.13 through 17.18.

Business fronts, organizations, and individuals posing as legal enterprises	Local civil unrest
	Loss of key personnel
Compromise of public confidence	Loss of life [customers and employees]
Cost of recovery and resumption	Loss of primary and backup power systems
Dealing with unanticipated crisis	
Duration of loss cycle and number of users impacted	Loss or disruption of communications
	Loss or disruption of information
Illness [customers and employees]	Production shutdown
Inability to detect fraud, corruption, and illegal transactions	Slowdown and interruption of delivery services
Inability to store, transfer, distribute commodities and funds	Temporary closure of financial institutions
	Wide-scale disaster

Exhibit 17.11   Assess Initial Impact of Telecommunications Enterprise Loss Consequence

Consequence analysis	National & international social demographics
Economic, social, political impact	
Impact of economic base	

Exhibit 17.12   Assess Initial Likelihood of Telecommunications Enterprise Threat Attractiveness and Likelihood of Malevolent Acts of Occurrence

Alert Notification Systems	Reporting Security Incidents
Emergency Preparedness Planning	Roles & Responsibilities
Functional Interfaces	Security Awareness
Operations, Processes, Protocols	Security Plans, Policies & Procedures

Exhibit 17.13 Evaluate Existing Telecommunications Enterprise Security Operations and Protocols [P_{E1}]

Business Continuity Planning	Organization & Compositions
Emergency Preparedness	Power Sources
Group Dependency Programs	Protective Systems
Information Systems	Public Switched Telecommunications Networks
Information Technology Management	
Internet Service Providers	Qualifications and Training
Interoperability	Redundant & Backup Systems
Management Team	Staffing, Skills, Experience
National Information Infrastructure	Supervision
Next Generation Network	System Configuration
Operational Capabilities	

Exhibit 17.14 Evaluate Existing Telecommunication Enterprise Security Organization [P_{E1}]

Agency for Toxic Substances and Disease Registry	Critical Infrastructure Protection Board [PCIPB]
	Defense Central Investigation Index [DCII]
American Electronics Association [AEA]	El Paso Intelligence Center [EPIC]
American Institute of Certified Public Accountants	Factory Mutual Research Corporation
	Federal, state, and local agency responders
American National Standards Institute	Government Network Security Information Exchanges
American Psychological Association	
American Society for Industrial Security	Illuminating Engineers Society of North America
American Society for Testing and Materials	
American Society for Testing Materials, Vaults	Institute for a Drug Free Workplace
Centers for Disease Control and Prevention [CDC]	Institute of Electrical and Electronic Engineers [IEEE]
Commerce Commission	International Association of Chiefs of Police
Computer Security Institute	International Association of Professional Security Consultants
Counter Terrorist Center [CTC]	

(continued on next page)

Exhibit 17.15 Evaluate Enterprise Interface and Relationship with Partner Organizations [P_{E1}]

International Computer Security Association	Occupational Safety and Health Review Commission [OSHRC]
International Criminal Police Organization [INTERPOL]	Office of National Drug Control Policy
International Electronics Supply Group [IESG]	Policy Coordination Committee [PCC]
International Parking Institute	Private Security Advisory Council
International Organization for Standardization [ISO]	Public Switched Telecommunications Network [PSTN]
Internet Service Providers [IPS]	Regional Information Sharing System [RISS]:
Interstate Commerce Commission [ICC]	Middle Atlantic-Great Lakes Organized Crime Law Enforcement Network [MAGLOCLEN]
Joint Terrorism Task Force [JTTF]	
Law Enforcement Support Center [LESC]	Mid-States Organized Crime Information Center [MOCIC]
Library of Congress, Congressional Research Service	New England State Police Information Network [NESPIN]
Medical Practitioners and Medical Support Personnel	Regional Organized Crime Information Center [ROCIC]
Multi-lateral Expert Groups	Rocky Mountain Information Network [RMIN]
Mutual Aid Associations	
Mutual Legal Assistance Treaties	Western States Information Network [WSN]
National Advisory Commission on Civil Disorder	Regional Task Forces
National Crime Information Center [NCIC]	Safe Manufactures National Association
National Electrical Manufacturing Association [NEMA]	Telecommunications Information Sharing and Analysis Centers [ISAC]
National Fire Protection Association [NFPA]	The Terrorism Research Center
National Information Infrastructure [NII]	U.S. Central Intelligence Agency [CIA]
National Infrastructure Protection Center	U.S. Department of Commerce [DOC]
National Institute of Law Enforcement and Criminal Justice	U.S. Department of Defense [DOD]
	U.S. Department of Homeland Security [DHS]
National Institute of Standards and Technology [NIST]	U.S. Department of Justice [DOJ]
	U.S. Department of State [DOS]
National Insurance Crime Bureau [NICB] Online	U.S. Environmental Protection Agency [EPA]
National Labor Relations Board	U.S. Federal Bureau of Investigation [FBI]
National Parking Association	U.S. Federal Communications Commission [FCC]
National Reliability and Interoperability Council [NRIC]	U.S. Government Accounting Office [GAO]
National Safety Council	U.S. Postal System
National Telecommunications Advisory Committee	U.S. Secret Service
	Underwriters Laboratories [UL]
National White Collar Crime Center [NWCCC]	World Health Organization [WHO]

Exhibit 17.15 *(continued)*

Assessment Capability	Lock & Key Controls, Fencing & Lighting
Badge Controls	Long-haul Transmission Capability
Barrier/Delay Systems	Network Access Points, Points of Presence
Base Stations, Cable Modem Connections	
Broadband Connections, Cable Landing Points	Operational Command Centers
	Performance Standards & Vulnerabilities
Cable/Communications Rooms & Centers	Personnel & Vehicle Access Control Points
Communications Capability	SCADA & Security C3 Centers
Convergence of Technologies	Switching, Transporting, Signaling, Routing, Control
"Cyber Intrusions, Firewalls, Other Protection Features"	
	System Capabilities & Expansion Options
Display & Annunciation Capability	Tamper Systems
Distribution & Transmission Systems	Telecommunications Architecture
Electronic Communications Network	Virtual Private Networks
High-capability Networks	Voice & Data Networks
Intrusion Detection Capability	Wireless Local Area Networks
IT Network Computer Centers	

Exhibit 17.16 Evaluate Existing Telecommunications Enterprise SCADA and Security System Performance Levels [P_{E1}]

Deception actions	Target selection and alternative targets
Delay-penetration obstacles	Time-delay, detection, assessment analysis
Interception analysis	
Likely techniques to be employed	

Exhibit 17.17 Define the Telecommunications Enterprise Adversary Plan, Distractions, Sequence of Interruptions and Path Analysis

Convergence of technologies	Network assess points, points of presence
Cryptographic key management, encryption	Physical protection system features
Diverse backup facilities and switchover capabilities	Switching, transport, signaling, routing, control
	System configuration switch circuits
Firewall networks, tamper systems, cyber intrusions	Voice and data networks, virtual private networks
Monitoring, surveillance, inspection	Wireless and networking technologies

Exhibit 17.18 Assess the Effectiveness of Existing Telecommunications Enterprise Response and Recovery [P_{E1}]

- Recognize the importance of bringing together the right balance of people, information, facilities, operations, processes, and systems [**Task 5—Program Analyses**] to deliver a cost-effective solution. The security-assessment team develops an integrated approach to measurably reduce security risk by reducing vulnerability and consequences through a combination of workable solutions that bring real, tangible enterprise value to the table. Emphasis is placed on assessing the current status and level of protection provided against desired protective measures to reasonably counter the threat. Operational, technical, and financial constraints are examined as well as future expansion plans and constraints. These include but are not necessarily limited to analysis of security strategies and operations, refinement of previous consequences analysis and threat attractiveness and likelihood analysis, development of mitigation plan and short- and long-term solutions, evaluation of developed plan, and cost estimation, as shown in Exhibits 17.19 through 17.25.

Business Continuity Planning	Hiring Practices
Charter and Authority	Performance Expectations
Day-to-Day Planning	Priorities, Protective Measures
Emergency Preparedness Planning	Rank Order Facilities & Assets
Loss Consequence Approach	Vulnerability Analysis

Exhibit 17.19 Analyze Effectiveness of Telecommunications Enterprise Security Strategies and Operations [P_{E1}]

Business fronts, organizations, and individuals posing as legal enterprises	Loss of key personnel
	Loss of life [customers and employees]
Compromise of public confidence	Loss of primary and backup power systems
Cost of recovery and resumption	
Dealing with unanticipated crisis	Loss or disruption of communications
Duration of loss cycle and number of users impacted	Loss or disruption of information
	Production shutdown
Illness [customers and employees]	Slowdown and interruption of delivery services
Inability to detect fraud, corruption, and illegal transactions	
	Temporary closure of financial institutions
Inability to store, transfer, distribute commodities and funds	Wide-scale disaster
Local civil unrest	

Exhibit 17.20 Refine Previous Analysis of Telecommunications Enterprise Undesirable Consequences That Can Affect Functions

Consequence analysis	National & international social
Economic, social, political impact	demographics
Impact of economic base	

Exhibit 17.21 Refine Previous Analysis of Telecommunications Enterprise Likelihood of Threat Attractiveness and Malevolent Acts of Occurrence

Creation and/or revision or modification of sound business practices and security policies, plans, protocols, and procedures.	Development, revision, or modification of interdependency requirements.
Creation and/or revision or modification of emergency operation plans including dependency support requirements.	SCADA/security system upgrades to improve detection and assessment capabilities.

Exhibit 17.22 Analyze Selection of Specific Risk Reduction Actions Against Current Risks, and Develop Prioritized Plan for Telecommunications Enterprise Mitigation Solutions [P_{E2}]

Mirrors business culture image	Reasonable and prudent mitigation options

Exhibit 17.23 Develop Short- and Long-Term Telecommunications Enterprise Mitigation Solutions

Effectiveness calculations	Residual vulnerability consequences

Exhibit 17.24 Evaluate Effectiveness of Developed Telecommunications Enterprise Mitigation Solutions and Residual Vulnerability [P_{E2}]

Practical, cost-effective recommendations	Reasonable return on investments

Exhibit 17.25 Developing Cost Estimate for Short- and Long-Term Telecommunications Enterprise Mitigation Solutions

PREPARING THE TELECOMMUNICATIONS SECURITY-ASSESSMENT REPORT

Chapter 10 provides a recommended approach to reporting security-assessment results.

TELECOMMUNICATIONS INITIATIVES

Efforts to enhance telecommunications security include a variety of local, national, and international efforts. Appendix B summarizes various significant actions taken or currently underway by the U.S. Department of Homeland Security and industry.

APPENDIX A: A HISTORICAL OVERVIEW OF SELECTED TERRORIST ATTACKS, CRIMINAL INCIDENTS, AND INDUSTRY MISHAPS WITHIN THE TELECOMMUNICATIONS SECTOR

The author has compiled the information in this appendix in a chronological sequence to share the historical perspective of incidents with the reader interested in conducting further research. The listing, along with the main text of Chapter 2, presents a capsule review of terrorist activity and other criminal acts perpetrated around the world and offers security practitioners specializing in particular regions of the world a quick reference to range and level of threat.

Source: *The information is a consolidated listing of events, activities, and news stories as reported by the U.S. Department of Homeland Security, U.S. Department of State, and various other government agencies. Contributions from newspaper articles and news-media reports are also included. They are provided for a better understanding of the scope of terrorism and other criminal activity and further research for the interested reader.*

7 MARCH 2005 Delhi, India.

India's software and services outsourcing industry is a likely target for a terrorist group operating in the country, local police warned. Indian outsourcing and software companies said they are prepared to cope with the threat. Documents seized from three members of the Lashkar-e-Toiba (LeT) terrorist group killed in an encounter with the police on Saturday, March 5, revealed that they planned to carry out suicide attacks on software companies in Bangalore. LeT is demanding independence for the Indian states of Jammu and Kashmir. The terrorists planned to hit these companies in an effort to hinder the economic development of the country. IBM, Intel, Texas Instruments, Accenture, Wipro, and Infosys Technologies are among those with operations in Bangalore. Most of the technology companies in the city have already set up disaster-recovery plans and special disaster-recovery sites that could be used in the event of a terrorist attack.

15 SEPTEMBER 2004 Los Angeles, CA, United States.

The FBI seized $87 million worth of illegal software. A two-year investigation by U.S. law-enforcement authorities resulted in one of the largest seizures of fake software ever in the U.S. and charges against 11 individuals. The defendants from California, Washington, and Texas were indicted with conspiring to distribute counterfeit computer software and documentation with a retail value of more than $30 million. When arresting the defendants and searching their homes, offices, and storage facilities, FBI agents uncovered an additional stockpile of more than $56 million in fake Microsoft, Symantec and Adobe Systems products. Microsoft worked closely with the authorities on the case.

9 AUGUST 2004 United States.

The widespread availability of sensitive information on corporate Websites appears to have been largely overlooked by IT and security managers, who responded last week to the Department of Homeland Security [DHS] warning of a heightened terrorist threat against the financial-services sector. Freely available on the Web, for example, are 3-D models of the exterior and limited portions of the interior of the Citigroup headquarters building in Manhattan—one of the sites specifically named in the latest terror advisory issued by the DHS. Likewise, details of the Citigroup building's history of structural-design weaknesses, including its susceptibility to toppling over in high winds, the construction of its central support column, and the fire rating of the materials used in the building, are readily available on the Web. Similarly, the website of the Chicago Board of Trade includes photographs of the facility's underground parking garages, floor plans of office suites, and contact names and phone numbers for the telecommunications service providers that serve the building. The DHS said that, while companies have the right to post whatever information they want, it encourages all companies to add website reviews to their list of preventive security measures.

7 AUGUST 2004 United States.

Al Qaeda suicide bombers and ambush units in Iraq routinely depend on the Web for training and tactical support, relying on the Internet's anonymity and flexibility to operate with near impunity in cyberspace. In Qatar, Egypt, and Europe, cells affiliated with al Qaeda that have recently carried out or seriously planned bombings have relied heavily on the Internet. Among other things, al Qaeda and its offshoots are building a massive and dynamic online library of training materials— some supported by experts who answer questions on message boards or in chat rooms—covering such varied subjects as how to mix ricin poison, how to make a bomb from 19 commercial chemicals, how to pose as a fisherman and sneak through Syria into Iraq, how to shoot at a U.S. soldier, and how to navigate by the stars while running through a night-shrouded desert. Jihadists seek to overcome in cyberspace specific obstacles they face from armies and police forces in the physical world. In planning attacks, radical operatives are often at risk when they congregate at a mosque or cross a border with false documents. They are safer working on the Web.

28 OCTOBER 2003 Lal Chowk, Kashmir.

A bomb exploded at the customer-billing counter in a telegraph office building, injuring 36 persons.

APPENDIX B: UNITED STATES GOVERNMENT AND INDUSTRY TELECOMMUNICATIONS INITIATIVES

The telecommunications infrastructure is undergoing a significant transformation that involves the convergence of traditional circuit-switched networks with broadband packet-based IP networks, including the Internet. Eventually, the packet networks will subsume the circuit-switched networks, leading to the establishment of a public, broadband, diverse, and scaleable packet-based network known as the next-generation network [NGN]. Additionally, the evolution of the telecommunications infrastructure has included steady growth in mobile wireless services and applications. Wireless telecommunications providers transmit messages using an infrastructure of base stations and radio-cell towers located throughout the wireless provider's service area. Wireless services consist of digital mobile phones and emerging data services including Internet communications, wireless local-area networks, and advanced telephone services.

Convergence, the growth of the NGN, and the emergence of new wireless capabilities continue to introduce new physical components to the telecommunications infrastructure. Government and industry have consistently worked together to develop strategies that ensure that the evolving infrastructure remains reliable, robust, and secure. Public-private partnerships and organizations currently addressing telecommunications security include the President's National Security Telecommunications Advisory Committee and Critical Infrastructure Protection Board [PCIPB], the Government Network Security Information Exchanges, the Telecommunications ISAC, and the Network Reliability and Interoperability Council of the FCC. Recommendations by these bodies and collaboration among industry and government shape the security and reliability of the evolving infrastructure.

Given the reality of the physical and cyber threats to the telecommunications sector, government and industry must continue to work together to understand vulnerabilities, develop countermeasures, establish policies and procedures, and raise awareness necessary to mitigate risks. The telecommunications sector has a long, successful history of collaboration with government to address concerns over the reliability and security of its infrastructure. The sector has recently undertaken a variety of new initiatives to further ensure both reliability and quick recovery and reconstruction. Within this environment of increasing emphasis on protection issues, public-private partnership can be further leveraged to address a number of key telecommunications initiatives, including efforts to:

- **Define an appropriate threshold for security.** The DHS has agreed to work with industry to define an appropriate security threshold for the sector and develop a set of requirements derived from that definition. Moreover, the DHS has agreed to work with industry to close the gap between respective security expectations and requirements. Reaching agreement on a methodology for ensuring physical diversity is a key element of this effort.

- **Expand infrastructure diverse-routing capability.** The DHS has agreed to enhance the government's capabilities to define and map the overall telecommunications architecture. This effort could identify critical intersections among the various infrastructures and lead to strategies that better address security and reliability.

- **Understand the risks associated with vulnerabilities of the telecommunications infrastructure.** The DHS has agreed to work with the private sector to conduct studies to understand physical vulnerabilities within the telecommunications infrastructure and their associated risks. Studies will focus on facilities where many different types of equipment and multiple carriers are concentrated.

- **Coordinate with key allies and trading partners.** The DHS has agreed to work with other nations to consider innovative communications paths that provide priority communications processes to link our governments, global industries, and networks in such a manner that vital communications are assured.

GENERAL GLOSSARY

This appendix contains some terms that do not actually appear in this book. They have been included to present a comprehensive list that pertains to the preparation of a security assessment.

A

Access Control Any combination of barriers, gates, electronic security equipment, and/or guards that can deny entry to unauthorized personnel or vehicles.

Access-Control Point [ACP] A station at an entrance to a building or a portion of a building where identification is checked and people and hand-carried items are searched. At the scene of a disaster or event, a checkpoint established to control entry and exit.

Access Controls Procedures and controls that limit or detect access to minimum essential infrastructure resource elements [e.g., people, technology, applications, data, and/or facilities], thereby protecting these resources against loss of integrity, confidentiality, accountability, and/or availability.

Access-Control System [ACS] Also referred to as an electronic entry-control system; an electronic system that controls entry and egress from a building or area.

Access Group software configuration of an access-control system that groups together access points or authorized users for easier arrangement and maintenance.

Access Road Any roadway such as a maintenance, delivery, service, emergency, or other special limited-use road that is necessary for the operation of a building or structure.

Accountability The explicit assignment of responsibilities for oversight of areas of control to executives, managers, staff, owners, providers, and users of minimum essential infrastructure resource elements.

Acetyl Cholinesterase An enzyme that hydrolyzes the neutron-transmitter acetylcholine. The action of this enzyme is inhibited by nerve agents.

Acoustic Eavesdropping The use of listening devices to monitor voice communications or other audibly transmitted information with the objective of compromising information.

Active Vehicle Barrier An impediment placed at an access-control point that may be manually or automatically deployed in response to detection of a threat.

Activist Group A group intent on making a statement or pressing a cause. Examples include the Earth Liberation Front, which claimed responsibility for burning down a $12 million ski lodge near Vail, CO, or the Oregon cult that poisoned a salad bar and a water system with Salmonella.

Acute Radiation Syndrome Consists of three levels of effects: hernatopoletic [blood cells, most sensitive]; gastrointestinal [GI cells, very sensitive]; and central nervous system [brain/muscle cells, insensitive]. The initial signs and symptoms are nausea, vomiting, fatigue, and loss of appetite. Below about 200 rems, these symptoms may be the only indication of radiation exposure.

Aerosol Fine liquid or solid particles suspended in a gas [e.g., fog or smoke].

Aggressor Any person seeking to compromise a function or structure.

Airborne Contamination Chemical or biological agents introduced into and fouling the source of supply of breathing or conditioning air.

Airlock A building-entry configuration with which airflow from the outside can be prevented from entering a toxic-free area. An airlock uses two doors, only one of which can be opened at a time, and a blower system to maintain positive air pressures and purge contaminated air from the airlock before the second door is opened.

Alarm Assessment Verification and evaluation of an alarm alert through the use of closed-circuit television or human observation. Systems used for alarm assessment are designed to respond rapidly, automatically, and predictably to the receipt of alarms at the security center.

Alarm Printers Alarm printers provide a hard copy of all alarm events and system activity as well as limited backup in case the visual display fails.

Alarm Priority A hierarchy of alarms by order of importance. This is often used in larger systems to give priority to alarms with greater importance.

All-Hazards Preparedness Preparedness for domestic terrorist attacks, major disasters, and other emergencies.

Alpha Particle A particle with a very short range in air and a very low ability to penetrate other materials but also with a strong ability to ionize materials. Alpha particles are unable to penetrate even the thin layer of dead cells of human skin and consequently are not an external radiation hazard. Alpha-emitting nuclides inside the body as a result of inhalation or ingestion are a considerable internal radiation hazard.

Annunciation A visual, audible, or other indication by a security system of a condition.

Antibiotic A substance that inhibits the growth of or kills microorganisms.

Antisera The liquid part of blood containing antibodies that react against disease-causing agents such as those used in biological warfare.

Antiterrorism [AT] Defensive measures used to reduce the vulnerability of individuals, forces, and property to terrorist acts.

Area Commander A military commander with authority in a specific geographical area or military installation.

Area Lighting Lighting that illuminates a large exterior area.

Areas of Potential Compromise Categories where losses can occur that will impact either a department's or an agency's minimum essential infrastructure and its ability to conduct core functions and activities.

Assessment The annunciation of remotely dispersed intrusion-detection and surveillance devices that creates a need for the security organization to be aware of the validity, severity, and nature of the event that triggers an alarm. This may require the strategic placement of area patrols and Close Circuit Television [CCTV] cameras particularly installed to be activated in conjunction with a sensor in alarm status. This capability significantly enhances the safety and response effectiveness of first responders and reduces the need for expensive use of patrols and fixed posts. Security lighting and advanced infrared red [IR] technology also enhances safety, security surveillance, and video-assessment capabilities. Beyond the technical capability to visually annunciate an alarm or provide the video surveillance of areas, a security monitoring station and supervisory element to direct the response are needed. Also referred to as the evaluation and interpretation of measurements and other information to provide a basis for decision-making.

Assessment-System Element A detection measure used to assist guards in visual verification of intrusion detection-system alarms and access-control-system functions and to assist in visual detection by guards. Assessment-system elements include Closed Circuit Television [CCTV] and protective lighting.

Asset A resource requiring protection. It includes contracts, facilities, property, records, unobligated or unexpended balances of appropriations, and other funds or resources [other than personnel]. An asset can be tangible [e.g., buildings, facilities, equipment, activities, operations, and information] or intangible [e.g., a company's information and reputation].

Asset Protection Security program designed to protect personnel, facilities, and equipment, in all locations and situations, accomplished through planned and integrated application of combating terrorism, physical security, operations security, and personal protective services, and supported by intelligence, counterintelligence, and other security programs.

Asset Value The degree of debilitating impact that would be caused by the incapacity or destruction of an asset.

Attack A hostile action resulting in the destruction, injury, or death to the civilian population or damage or destruction to public and private property.

Atropine A compound used as an antidote for nerve agents.

Audible Alarm Device An alarm device that produces an audible announcement [e.g., bell, horn, siren, etc.] of an alarm condition.

B

Backup A copy of files and programs made to facilitate recovery if necessary.

Bacteria Single-celled organisms that multiply by cell division and that can cause disease in humans, plants, or animals.

Balanced Magnetic Switch [BMS] A switch typically used to detect the opening of a door. These sensors can also be used on windows, hatches, gates, or other structural devices that can be opened to gain entry. Typically, the BMS has a three-position reed switch and an additional magnet [called the bias magnet] located adjacent to the switch. When the door is closed, the reed switch is held in the balanced or center position by interacting magnetic fields. If the door is opened or an external magnet is brought near the sensor in an attempt to defeat it, the switch becomes unbalanced and generates an alarm.

Ballistics Attack An attack in which small arms [e.g., pistols, submachine guns, shotguns, and rifles] are fired from a distance and rely on the flight of the projectile to damage the target.

Barbed Tape or Concertina A coiled tape or coil of wires with barbs or blades, deployed as an obstacle to human trespass or entry into an area.

Barbed Wire A double strand of wire with four-point barbs equally spaced along its length, deployed as an obstacle to human trespass or entry into an area.

Bar Code A black bar printed on white paper or tape that can be easily read with an optical scanner. This type of coding is not widely used for entry-control applications because it can be easily duplicated. It is possible to conceal the code by applying an opaque mask over it. In this approach, an infrared [IR] scanner is used to interpret the printed code. For low-level security areas, the use of bar codes can provide a cost-effective solution for entry control. Coded strips and opaque masks can be attached to existing ID badges, alleviating the need for complete badge replacement.

Beta Particles High-energy electrons emitted from the nucleus of an atom during radioactive decay. They normally can be stopped by the skin or a very thin sheet of metal.

Biochemicals The chemicals that make up or are produced by living things.

Biological Agent A living organism or the material derived from one that causes disease in or harm to humans, animals, or plants or deterioration of material. Biological agents may be used as liquid droplets, aerosols, or dry powders.

Biological Warfare The intentional use of biological agents as weapons to kill or injure humans, animals, or plants or to damage equipment.

Biological-Warfare Agent A living organism or the material derived from one that causes disease in or harm to humans, animals, or plants or deterioration of material. Biological agents may be used as liquid droplets, aerosols, or dry powders.

Biometric Device A technique used to control entry that is based on the measurement of one or more physical or personal characteristics of an individual. Because most entry-control devices based on this technique rely on measurements of biological characteristics, they have become commonly known as biometric devices. Characteristics such as fingerprints, hand geometry, voiceprints, handwriting, and retinal blood-vessel patterns have been used for controlling entry. Typically, in enrolling individuals several reference measurements are made of the selected characteristic and then stored in the device's memory or on a card. When that person attempts entry, a scan of the characteristic is compared with the reference-data template. If a match is found, entry is granted. Rather than verifying an artifact, such as a code or credential, biometric devices verify a person's physical characteristics, thus providing a form of identity verification. Because of this, biometric devices are sometimes referred to as personnel identity-verification devices.

Biometric Reader A device that gathers and analyzes biometric features.

Biometrics The use of physical characteristics of the human body as a unique identification method.

Bioregulator A biochemical that regulates bodily functions. Bioregulators that are produced by the body are termed "endogenous." Some bioregulators can be chemically synthesized.

Bioterrorism/Chemical/Radiation Contamination The introduction of toxic chemicals including release of or dumping of sewage, animal carcasses, or other hazardous materials into reservoirs, wells, and other water-supply systems. The introduction or release of biological agents into the environment or near critical assets. The introduction of or release of radioactive materials including a "dirty bomb" at designated locations.

Blast Curtain A heavy curtain made of blast-resistant materials that can protect the occupants of a room from flying debris.

Blast-Resistant Glazing Window glazing that is resistant to blast effects because of the interrelated function of the frame and glazing-material properties, frequently dependent upon tempered glass, polycarbonate, or laminated glazing.

Blast-Vulnerability Envelope The geographical area in which an explosive device will cause damage to assets.

Blister Agent A substance that causes blistering of the skin. Exposure is through liquid or vapor contact with any exposed tissue [eyes, skin, and lungs]. Examples are distilled mustard [HD], nitrogen mustard [HN], lewisite [L], mustard/lewisite [HL], and phenodichloroarsine [PD].

Blood Agent A substance that injures a person by interfering with cell respiration [the exchange of oxygen and carbon dioxide between blood and tissues]. Examples are arsine [SA], cyanogen chloride [CK], hydrogen chloride [HCI], and hydrogen cyanide [AC].

Blowback The intentional opening of one valve before the closure of another, deliberately causing pressure in the system to divert the forced flow in a direction other than designed.

Bollard A vehicle barrier consisting of a cylinder, usually made of steel and sometimes filled with concrete, placed on end in the ground and spaced about 3 feet apart to prevent vehicles from passing but allowing entrance of pedestrians and bicycles.

Boundary-Penetration Sensor Sensors designed to detect penetration or attempted penetration through perimeter barriers. These barriers include walls, ceilings, duct openings, doors, and windows. *See also* **Structural-Vibration Sensor, Glass-Breakage Sensor, Passive Ultrasonic Sensor, Balanced Magnetic Switch,** and **Grid-Wire Sensor**.

Building Hardening Enhanced construction that reduces vulnerability to external blast and ballistic attacks.

Building Separation The distance between closest points on the exterior walls of adjacent buildings or structures.

Buried-Line Sensor   A system consisting of detection probes or cable buried in the ground, typically between two fences that form an isolation zone or clear zone. These devices are wired to an electronic processing unit. The processing unit generates an alarm if an intruder passes through the detection field.

Business-Continuity Program [BCP]   An ongoing process supported by senior management and funded to ensure that the necessary steps are taken to identify the impact of potential losses, maintain viable recovery strategies and recovery plans, and ensure continuity services through personnel training, plan testing, and maintenance.

Business-Impact Analysis [BIA]   An analysis of system or program requirements, processes, and interdependencies used to characterize contingency requirements and priorities in the event of a significant disruption.

C

Cable Barrier   Cable or wire rope anchored to and suspended off the ground or attached to chain-link fence to act as a barrier to moving vehicles.

Capacitance Proximity Sensor   A sensor measuring the electrical capacitance between the ground and an array of sense wires. Any variations in capacitance, such as that caused by an intruder approaching or touching one of the sense wires, initiates an alarm. These sensors usually consist of two or three wires attached to outriggers along the top of a fence, wall, or roof edge.

Capacitance Sensor   A sensor that detects an intruder approaching or touching a metal object by sensing a change in capacitance between the object and the ground. A capacitor consists of two metallic plates separated by a dielectric medium. A change in the dielectric medium or electrical charge results in a change in capacitance. In practice, the metal object to be protected forms one plate of the capacitor, and the ground plane that surrounds the object forms the second plate. The sensor processor measures the capacitance between the metal object and the ground plane. An approaching intruder alerts the dielectric value, thus changing the capacitance. If the net capacitance change satisfies the alarm criteria, an alarm is generated.

Card Reader A device that gathers or reads information when a card is presented as an identification method.

Casualty [Toxic] Agent An agent that produces incapacitation, serious injury, or death and can be used to incapacitate or kill victims, including the blister, blood, choking, and nerve agents.

Causative Agent An organism or toxin that is responsible for causing a specific disease or harmful effect.

CCTV Pan-Tilt-Zoom [PTZ] Camera A Closed Circuit Television [CCTV] camera that can move side to side, up and down, and in or out.

CCTV Pan-Tilt-Zoom [PTZ] Control The method of controlling the PTZ functions of a camera.

CCTV Pan-Tilt-Zoom [TPZ] Controller The operator interface for performing PTZ control.

CCTV Switcher A piece of equipment capable of presenting multiple video images to various monitors, recorders, etc.

Central-Nervous-System Depressant A compound that has the predominant effect of depressing or blocking the activity of the central nervous system. The primary mental effects include disruption of the ability to think, sedation, and lack of motivation.

Central-Nervous-System Stimulant A compound that has the predominant effect of flooding the brain with too much information. The primary mental effect is loss of concentration, causing indecisiveness and the inability to act in a sustained, purposeful manner.

Cesium-137 [Cs-137] A strong gamma-ray source that can contaminate property, entailing extensive cleanup. It is commonly used in industrial measurement gauges and for irradiation of material. Its half-life is 30.2 years.

Chemical Agent A substance intended to kill, seriously injure, or incapacitate people through its physiological effects. Excluded from consideration are riot-control agents and smoke and flame materials. The agent may appear as a vapor, aerosol, or liquid; it can be either a casualty/toxic agent or an incapacitating agent.

Chimney Effect Air movement in a building between floors caused by differential air temperature [differences in density] between the air inside and outside the building. It occurs in vertical shafts, such as elevators, stairwells, and conduit/wiring/piping chases. Hotter air inside the building will rise and be replaced by colder outside air through the lower portions of the building. Conversely, reversing the temperature will reverse the flow [down the chimney]. Also known as stack effect.

Choking/Lung/Pulmonary Agent A substance that causes physical injury to the lungs. Exposure is through inhalation. In extreme cases, membranes swell and lungs become filled with liquid. Death results from lack of oxygen; hence, the victim is "choked." Examples are chlorine [CL], diphosgene [DP], cyanide [KCN], nitrogen oxide [NO], perfluororisobutylene [PHIB], phosgene [CG], red phosphorous [RP], sulfur trioxide chlorosulfonic acid [FS], Teflon, amphibious [PHIB] titanium tetrachloride [FM], and zinc oxide [HC].

Clear Zone An area that is clear of visual obstructions and landscape materials that could conceal a threat or perpetrator.

Closed-Circuit Television [CCTV] An electronic system of cameras, control equipment, recorders, and related apparatus used for surveillance or alarm assessment.

Cobalt-60 [Co-60] A strong gamma-ray source, extensively used as a radio therapeutic for cancer, food and material irradiation, gamma radiography, and industrial measurement gauges. Its half-life is 5.27 years.

Coded Device A device operating on the principle that a person has been issued a code to enter into an entry-control device. This code will match the code stored in the device and permit entry. Depending on the application, a single code can be used by all persons authorized to enter the controlled area, or each authorized person can be assigned a unique code. Group codes are useful when the group is small and controls are primarily for keeping out the general public. Individual codes are usually required for control of entry to more critical areas. Coded devices verify the entered code's authenticity, and any person entering a correct code is authorized to enter the controlled area. Electronically coded devices include electronic and computer-controlled keypads.

Collateral Damage Injury or damage to assets that are not the primary target of an attack.

Cold Site A backup facility that has the necessary electrical and physical components of a computer facility but does not have the computer equipment in place. The site is ready to receive the necessary replacement computer equipment in the event that the user has to move from the main computing location to an alternate site.

Combating Terrorism The full range of federal programs and activities applied against terrorism, domestically and abroad, regardless of the source or motive.

Community A political entity that has the authority to adopt and enforce laws and ordinances for the area under its jurisdiction. In most cases, the community is an incorporated town, city, township, village, or area of a country; however, each state defines its own political subdivisions and forms of government.

Components and Cladding Elements of the building envelope that do not qualify as part of the main wind-force-resisting system.

Computer A device that accepts digital data and manipulates the information based on a program or sequence of instructions for how data is to be processed.

Computer-Controlled Keypad Device A device similar to an electronic keypad device, except that it is equipped with a microprocessor in the keypad or in a separate enclosure at a different location. The microprocessor monitors the sequence in which the keys are depressed and may provide additional functions such as personal ID and digit scrambling. When the correct code is entered and all conditions are satisfied, an electric signal unlocks the door.

Confidentiality The protection of sensitive information against unauthorized disclosure and sensitive facilities from physical, technical, or electronic penetration or exploitation.

Consequence Criticality Rating The impact of a loss resulting from a terrorist attack, criminal activity, or natural disaster, typically calculated as the net cost of such an undesirable event. Impact can range from fatal—resulting in a total recapitalization, abandonment, or long-term discontinuance of services—to relatively unimportant.

Consequence Management [CM] Measures to protect public health and safety, restore essential government services, and provide emergency relief to governments, businesses, and individuals affected by the consequences of terrorism. State and local governments exercise the primary authority to respond to the consequences of terrorism. *See also* **Crisis Management**.

Constraint A political or cultural consideration imposed by the enterprise on the security-assessment team for achieving acceptable solutions. Constraints and limitations such as environmental and operational considerations that affect the performance capability of a security system or impact on security enhancements to processes, protocols, and human behaviors are technical constraints.

Contagious Capable of being transmitted from one person to another.

Contamination The undesirable deposition of a chemical, biological, or radiological material on the surface of structures, areas, objects, or people.

Continuity of Services and Operations Controls to ensure that, when unexpected events occur, departmental/agency minimum essential infrastructure services and operations, including computer operations, continue without interruption or are promptly resumed and that critical and sensitive data are protected through adequate contingency and business-recovery plans and exercises.

Control Center A centrally located room or facility staffed by personnel charged with the oversight of specific situations and/or equipment.

Controlled Area That portion of a restricted area usually near or surrounding a limited or exclusion area. Entry to the controlled area is restricted to personnel with a need for access. Movement of authorized personnel within this area is not necessarily controlled because mere entry to the area does not provide access to the security interest. The controlled area is provided for administrative control, for safety, or as a buffer zone for in-depth security for the limited or exclusion area.

Controlled Lighting Illumination of specific areas or sections.

Controlled Perimeter A physical boundary at which vehicle and personnel access is restricted at the perimeter of a site. Access control at a controlled perimeter should demonstrate the capability to search individuals and vehicles.

Control Valve A valve used to regulate water flow or pressure at different points of the system by creating head loss or pressure differential between upstream and downstream sections.

Conventional Construction Building construction that is not specifically designed to resist weapons, explosives, or chemical, biological, and radiological effects. Conventional construction is designed only to resist common loadings and environmental effects such as wind, seismic, and snow loads.

Coordinate To advance systematically an exchange of information among principals who have or may have a need to know certain information in order to carry out their roles in a response.

Counterintelligence Information gathered and activities conducted to protect against: espionage, other intelligence activities, sabotage, or assassinations conducted for or on behalf of foreign powers, organizations, or persons; or international terrorist activities, excluding personnel, physical, document, and communications-security programs.

Counterterrorism [CT] Offensive measures taken to prevent, deter, and respond to terrorism.

Covert Entry Attempts to enter a facility by using false credentials or stealth.

Crash Bar A mechanical egress device located on the interior side of a door that unlocks the door when pressure is applied in the direction of egress.

Credential Device A device that identifies a person having legitimate authority to enter a controlled area. A coded credential [e.g., plastic card or key] contains a prerecorded, machine-readable code. An electric signal unlocks the door if the prerecorded code matches the code stored in the system when the card is read. Like coded devices, credential devices only authenticate the credential; it assumes a user with an acceptable credential is authorized to enter.

Crime Prevention Through Environmental Design [CPTED] A crime-prevention strategy based on evidence that the design and form of the built environment can influence human behavior. CPTED usually involves the use of three principles: natural surveillance [by placing physical features, activities, and people to maximize visibility]; natural access control [through the judicious placement of entrances, exits, fencing, landscaping, and lighting]; and territorial reinforcement [using buildings, fences, pavement, signs, and landscaping to express ownership].

Crisis Management [CM] The measures taken to identify, acquire, and plan the use of resources needed to anticipate, prevent, and/or resolve a threat or act of terrorism. *See also* **Consequence Management**.

Critical Asset An asset essential to the minimum operations of the organization and to ensure the health and safety of the general public.

Critical Infrastructure Systems and assets, whether physical or virtual, so vital to the United States that the incapacitation or destruction of such systems and assets would have a debilitating impact on national security, national economic security, national public health or safety, or any combination of these [Section 1016[e] of the Patriot Act of 2001 [42 U.S.C. 519c[e].

Critical Infrastructure Information Information not customarily in the public domain and related to the security of critical infrastructure or protected systems [6 U.S.C. 101]:

Actual, potential, or threatened interference with, attack on, compromise of, or incapacitation of critical infrastructure or protected systems by either physical or computer-based attack or other similar conduct [including the misuse of or unauthorized access to all types of communications and data transmission systems] that violates Federal, State, or local law, harms interstate commerce of the United States, or threatens public health or safety;

The ability of any critical infrastructure or protected system to resist such interference, compromise, or incapacitation, including any planned or pass assessment, projections, or estimate of the vulnerability of critical infrastructure or a protected system, including security testing, risk evaluation thereto, risk management planning, or risk audit; or

Any planned or past operational problem or solution regarding critical infrastructure or protect systems, including repair, recovery, reconstruction, insurance, or continuity, to the extent it is related to such interference, compromise, or incapacitation.

Critical Infrastructure Sector A government agency or private enterprise identified under the Homeland Security Act of 2002.

Culture A population of microorganisms grown in a medium.

Curie [Ci] A unit of radioactive decay rate defined as 3.7×10^{10} disintegrations per second.

Cutaneous Pertaining to the skin.

D

Damage Assessment The process used to appraise or determine the number of injuries and deaths, damage to public and private property, and the status of key facilities and services [e.g., hospitals and other healthcare facilities, fire and police stations, communications networks, water and sanitation systems, utilities, and transportation networks] resulting from a man-made or natural disaster.

Data-Gathering Panel A local processing unit that retrieves, processes, stores, and/or acts on information in the field.

Data-Transmission Equipment A path for transmitting data between two or more components [e.g., a sensor and alarm-reporting system, a card reader and controller, a CCTV camera and monitor, or a transmitter and receiver].

Decay The process by which an unstable element is changed to another isotope or another element by the spontaneous emission of radiation from its nucleus. The process can be measured by using radiation detectors such as Geiger counters.

Decontamination The process of reducing or removing chemical, biological, or radiological material by absorbing, destroying, neutralizing, making harmless, or removing contaminants from the surface of a structure, area, object, or person.

Defense Layer A building design or exterior perimeter barrier intended to delay attempted forced entry. It also includes protection-in-depth.

Defensive Measure A protective measure that delays or prevents attack on an asset or that shields the asset from weapons, explosives, and chemical, biological and radiological [CBR] effects. Defensive measures include site work and building design.

Delay A technique employed to slow down the adversary or prevent him or her from reaching a vulnerable point or asset until a response force capable of defeating the adversary arrives. Delay measures function differently depending on the mode of attack and tactics used. The delay-penetration time needed by the adversary to reach an objective is dictated by the configuration of the area, the types and quantity of physical barriers to be crossed, the distance of the adversarial path, and the type of tools, weapons, and equipment available for use. The capability of a timed response, if one is regulated, should not exceed the delay-penetration time needed by the adversary to complete his or assignment once detected.

Delay Rating A measure of the effectiveness of penetration protection of a defense layer.

Design-Basis Threat [DBT] The threat [e.g., tactics and associated weapons, tools, or explosives] against which assets within a building must be protected and upon which the security engineering design of the building, operations, processes, procedures, and staffing are based.

Design Constraint Anything that restricts the design options for a protective system or that creates additional problems for which the design must compensate.

Design Opportunity Anything that enhances protection, reduces requirements for protective measures, or solves a design problem.

Design Team A group of individuals from various disciplines responsible for the protection-system design.

Detection Layer A ring of intrusion detection sensors located on or adjacent to a defensive layer or between two defensive layers.

Detection Measure A protective measure that detects intruders, weapons, or explosives; assists in assessing the validity of detection-control access to protected areas; and communicates the appropriate information to the response force. Detection measures include detection systems, assessment systems, and access-control system.

Detection-System Element A measure that detects the presence of intruders, weapons, or explosives. Detection-system elements include intrusion detection systems, weapons and explosives detectors, and guards.

Deterrence An intangible measurement and byproduct of the overall security-system design. It includes administrative and procedural processes as well as protective measures and technology. When the adversary perceives an unacceptable risk associated with the damage or destruction of a critical asset, the effectiveness of deterrence impacts the adversary's mode of attack, tactics, and capability. The exception to the general rule is a suicide assault plan.

Disaster An occurrence of a natural catastrophe, technological accident, or human-caused event that results in severe property damage, death, and/or multiple injuries.

Disaster Field Office [DFO] The office established in or near the designated area of a presidentially declared major disaster to support federal and state response-and-recovery operations.

Disaster-Recovery Center [DRC] A place established in the area of a presidentially declared major disaster, as soon as practicable, to provide victims the opportunity to apply in person for assistance and/or obtain information relating to that assistance.

Domestic Terrorism The unlawful use or threatened use of force or violence by a group or individual based and operating entirely within the United States or Puerto Rico without foreign direction, committed against persons or property to intimidate or coerce a government, the civilian population, or any segment thereof in furtherance of political or social objectives.

Door Position Switch A switch that changes state based on whether or not a door is closed. Typically, a switch mounted in a frame that is actuated by a magnet in a door. Also referred to as a door contact switch.

Door Strike, Electronic An electromechanical lock that releases a door plunger to unlock the door. Typically, an electronic door strike is mounted in place of or near a normal door strike plate.

Dose Rate [Radiation] A general term indicating the quantity [total or accumulated] of ionizing radiation or energy absorbed by a person or animal per unit of time.

Dosimeter A portable instrument for measuring and registering the total accumulated dose of ionizing radiation.

Dual-Technology Sensor A sensor that combines two technologies in order to minimize the generation of alarms caused by sources other than intruders. Ideally, this is achieved by combining two sensors that individually have a high probability of detection $[P_d]$ and do not respond to common sources of false alarms. Available dual-technology sensors combine an active ultrasonic or microwave sensor with a passive infrared [PIR] sensor. The alarms from each sensor are logically combined in an "and" configuration [i.e., nearly simultaneous alarms from both active and passive sensors are needed to produce a valid alarm].

Duress Alarm Device Also known as a panic button, a device designated specifically to initiate a alarm when one's life is threatened or to provide alert notification that a serious event is happening that threatens the lives of others or business operations.

E

Economic Cost-Benefit Analysis The comparison of options and alternatives related to the decision to commit assets or funds. The process attempts to measure or analyze the value of all the benefits that accrue from a particular choice.

Economic Espionage The unlawful or clandestine targeting or acquisition of sensitive financial, trade, or economic policy information; proprietary economic information; or critical technologies. This definition excludes the collection of public-domain data and legally available information, which constitutes a significant majority of economic collection. Aggressive intelligence collection that is entirely in the public domain and is legal may harm U.S. industry but is not espionage. It may, however, help foreign intelligence services identify and fill information gaps that could be a precursor to economic espionage. For a conviction under the Economic Espionage Act [EEA] of 1996 [Title 18 U.S.C., Chapter 90], a person must convert a trade secret to an economic benefit in interstate or foreign commerce.

Effective Stand-Off Distance A distance at which the required level of protection can be shown to be achieved through analysis or can be achieved through building hardening or other mitigating construction or retrofitting.

Electromagnetic Pulse [EMP] A sharp pulse of energy radiated instantaneously by a nuclear detonation that may affect or damage electronic components and equipment. EMP can also be generated in lesser intensity by non-nuclear means in specific frequency ranges to perform the same disruptive function.

Electronic Emanation An electromagnetic emission from computers, communications, electronics, wiring, and related equipment.

Electronic-Emanation Eavesdropping Use of electronic-emanation surveillance equipment from outside a facility or its restricted area to monitor electronic emanations from computers, communications, and related equipment.

Electronic-Entry Control System [EECS] An electronic device that automatically verifies authorization for a person to enter or exit a controlled area.

Electronic Keypad Device The common telephone keypad [12 keys] is an example of an electronic keypad. This type of keypad consists of simple push-button switches that, when depressed, are decoded by digital logic circuits. When the correct sequence of buttons is pushed, an electric signal unlocks the door for a few seconds.

Electronic Security System [ESS] An integrated system that encompasses interior and exterior sensors, closed-circuit television systems for assessment of alarm conditions, electronic entry-control systems, data-transmission media, and alarm-reporting systems for monitoring, control, and display of various alarm and system information.

Emergency A natural or human-caused situation that results in or may result in substantial injury or harm to the population or substantial damage to or loss of property.

Any occasion or instance for which, in the determination of the President, Federal assistance is needed to supplement State and local efforts and capabilities to save lives and to protect property and public health and safety, or to lessen or avert the threat of a catastrophe in any part of the United States [42 U.S.C. 5122].

Emergency Alert System [EAS] A communications system of broadcast stations and interconnecting facilities authorized by the Federal Communications Commission [FCC]. The system provides the President and other national, state, and local officials the means to broadcast emergency information to the public before, during, and after disasters.

Emergency Environmental-Health Service A service required to correct or improve damaging environmental health effects on humans, including inspection for food contamination, inspection for water contamination, and vector control; providing for sewage and solid-waste inspection and disposal; cleanup and disposal of hazardous materials; and sanitation inspection for emergency shelter facilities.

Emergency Medical Service [EMS] A service including personnel, facilities, and equipment required to ensure proper medical care for the sick and injured from the time of injury to the time of final disposition, including medical disposition within a hospital, temporary medical facility, or special care facility; release from the site; or declaration of death. Further, EMS specifically includes those services immediately required to ensure proper medical care and specialized treatment for patients in a hospital and coordination of related hospital services.

Emergency Mortuary Service A service required to assure adequate death investigation, identification, and disposition of bodies; removal, temporary storage, and transportation of bodies to temporary morgue facilities; notification of next of kin; and coordination of mortuary services and burial of unclaimed bodies.

Emergency Operation Center [EOC] The protected site from which state and local civil government officials coordinate, monitor, and direct emergency-response activities during an emergency.

Emergency Operations Plan [EOP] A document that describes how people and property will be protected in disaster and disaster-threat situations; details who is responsible for carrying out specific actions; identifies the personnel, equipment, facilities, supplies, and other resources available for use in the disaster; and outlines how all actions will be coordinated.

Emergency Planning Zone [EPZ] An area around a facility for which planning is needed to ensure prompt and effective actions are taken to protect the health and safety of the public if an accident or disaster occurs. In the Radiological Emergency Preparedness Program, the two EPZs are:

- **Plume Exposure Pathway** A circular geographic zone with a 10-mile radius centered at Ground Zero for which plans are developed to protect the public against exposure to radiation emanating from a radioactive plume caused as a result of a nuclear accident.

- **Ingestion Pathway** A circular geographic zone with a 50 mile-radius centered at Ground Zero for which plans are developed to protect the public from the ingestion of water or food contamined as a result of nuclear accident.

In the Chemical Stockpile Emergency Preparedness Program [CSEPP], the EPZ is divided into three concentric circular zones:

- **Immediate Response Zone [IRZ]** A circular zone ranging from 6 to 9 miles from the potential chemical-event source, depending on the stockpile location. Emergency response plans developed for the IRZ must provide for the most rapid and effective protective actions possible, because the IRZ will have the highest concentration of agent and the least amount of warning time.

- **Protective Action Zone [PAZ]** An area that extends beyond the IRZ to approximately 10 to 30 miles from the stockpile location. The PAZ is that area where public protective actions may still be necessary in case of an accidental release of chemical agent, but where the available warning and response time is such that most people could evacuate. However, other responses [e.g., sheltering] may be appropriate for institutions and special populations that could not evacuate within the available time.

- **Precautionary Zone [PZ]** The outermost portion of the EPZ, extending from the PAZ outer boundary to a distance where the risk of adverse impacts to humans is negligible. Because of the increased warning and response time available for implementation of response actions in the PZ, detailed local emergency planning is not required, although consequence-management planning may be appropriate.

Emergency Public Information [EPI] Information that is disseminated primarily in anticipation of an emergency or at the actual time of an emergency and that, in addition to providing information, frequently directs actions, instructs, and transmits direct orders.

Emergency Response Providers Any Federal, State, and local emergency public safety, law enforcement emergency response, emergency medical [including hospital emergency facilities], and related personnel, agencies, and authorities [42 U.S.C. 5122]

Emergency-Response Team [ERT] An interagency team, consisting of the lead representative from each federal department or agency assigned primary responsibility for an emergency support function [ESF] and key members of the Federal Coordinating Officer's [FCO's] staff, formed to assist the FCO in carrying out his/her coordination responsibilities.

Emergency-Response Team Advance Element [ERT-A] For federal disaster response and recovery activities under the Stafford Act, the portion of the ERT that is first deployed to the field to respond to a disaster incident. The ERT-A is the nucleus of the full ERT.

Emergency-Response Team National [ERT-N] An ERT that has been established and rostered for deployment to catastrophic disasters where the resources of the Federal Emergency Management Agency [FEMA] have been or are expected to be overwhelmed. Three ERT-Ns have been established.

Emergency Support Function [ESF] In the Federal Response Plan [FRP], a functional area of response activity established to facilitate the delivery of federal assistance required during the immediate response phase of a disaster to save lives, protect property and public health, and to maintain public safety. ESFs represent those types of federal assistance that a state will most likely need because of the impact of a catastrophic or significant disaster on its own resources and response capabilities or because of the specialized or unique nature of the assistance required. ESF missions are designed to supplement state and local response efforts.

Emergency Support Team [EST] An interagency group operating from Federal Emergency Management Agency [FEMA] headquarters. The EST oversees the national-level response support effort under the Federal Response Plan [FRP] and coordinates activities with the Emergency Support Function [ESF] primary and support agencies in supporting federal requirements in the field.

Entity-Wide Security Planning and management that provides a framework and continuing cycle of activity for managing risk, developing security policies, assigning responsibilities, and monitoring the adequacy of the entity's physical and cyber-security controls.

Entry Control Point A continuously or intermittently manned station at which entry to sensitive or restricted areas is controlled.

Entry Control Station A station located at main perimeter entrances where security personnel are present. Entry control stations should be located as close as practical to the perimeter entrance to permit personnel inside the station to maintain constant surveillance over the entrance and its approaches. Also referred to as a security gate or guard station.

Equipment Closet A room where field control equipment such as data-gathering panels and power supplies are typically located.

Evacuation Organized, phased, and supervised dispersal of people from dangerous or potentially dangerous areas.

Evacuation, Mandatory or Directed This is a warning to persons within the designated area that an imminent threat to life and property exists and individuals must evacuate in accordance with the instructions of local officials.

Evacuation, Spontaneous Residents or citizens in the threatened areas observe an emergency event or receive unofficial word of an actual or perceived threat and, without receiving instructions to do so, elect to evacuate the area. This movement, means, and direction of travel are unorganized and unsupervised.

Evacuation, Voluntary This is a warning to persons within a designated area that a threat to life and property exists or is likely to exist in the immediate future. Individuals issued this type of warning or order are not required to evacuate; however, it would be to their advantage to do so.

Evacuee A person removed or moving from an area threatened or struck by a disaster.

Event Any act of terrorism, criminal activity, service disruption, emergency, or disaster that requires pre-emergency planning, response, or recovery action.

Exclusion Area A restricted area containing a security interest. Uncontrolled movement permits direct access to the item. *See also* **Controlled Area** and **Limited Area**.

Exclusion Zone An area around an asset that has controlled entry with highly restrictive access. *See also* **Control Area**.

Explosives-Disposal Container A small container into which small quantities of explosives may be placed to contain their blast pressures and fragments if the explosive detonates.

Exterior Intrusion-Detection Sensor A sensor customarily used to detect an intruder crossing the boundary of a protected area. It can also be used in clear zones between fences or around buildings for protecting materials and equipment stored outdoors within a protected boundary.

F

Facial Recognition A biometric technology that is based on features of the human face.

Federal Agency Any department, independent establishment, Government corporation, or other agency of the executive branch of the Federal Government, including the United States Postal Service, but shall not include the American National Red Cross [6 U.S.C. 101].

Federal Coordinating Officer [FCO] The person appointed by the FEMA director to coordinate federal assistance in a presidentially declared emergency or major disaster.

Federal On-Scene Commander The FBI official designated upon Joint Operations Center [JOC] activation to ensure appropriate coordination of the overall U.S. government response with federal, state, and local authorities until such time as the Attorney General transfers the lead federal agency [LFA] role to the Federal Emergency Management Agency [FEMA].

Federal Departments and Agencies Those executive departments enumerated in 5 U.S.C. 101 and the Department of Homeland Security; independent establishments as defined by 5 U.S.C. 104[1]; government corporations as defined by 5 U.S.C. 103[1]; and the United States Postal Service.

Federal Preparedness Assistance Federal department and agency grants, cooperative agreements, loans, loan guarantees, training, and/or technical assistance provided to state and local governments and the private sector to prevent, prepare for, respond to, and recover from terrorist attacks, major disasters, and other emergencies. Unless noted otherwise, the term "assistance" refers to federal assistance programs.

Federal Response Plan [FRP] The FRP establishes a process and structure for the systematic, coordinated, and effective delivery of federal assistance to address the consequences of any major disaster or emergency.

Fence Protection An intrusion-detection technology that detects a person crossing a fence by various methods such as climbing, crawling, cutting, etc.

Fence Sensor A sensor that detects attempts to penetrate a fence around a protected area. Penetration attempts to climb over, cut through, or otherwise disturb the fence generate mechanical vibrations and stresses in fence fabric and posts that are usually different than those caused by natural phenomena like wind and rain. The basic types of sensors used to detect these vibrations and stresses are strain-sensitive cable, taut wire, fiberoptics, and capacitance.

Fiberoptics A method of data transfer by passing bursts of light through a strand of glass or clear plastic.

Fiberoptic-Cable Sensor A sensor functionally equivalent to a strain-sensitive cable sensor. However, rather than electrical signals, modulated light is transmitted down the cable, and the resulting received signals are processed to determine whether an alarm should be initiated. Because the cable contains no metal and no electrical signal is present, fiberoptic sensors are generally less susceptible to electrical interference from lightning or other sources.

Field Assessment Team [FAT] A small team of pre-identified technical experts that conducts an assessment of response needs [not a preliminary damage assessment (PDA)] immediately following a disaster.

Field of View The visible area in a video picture.

Flash Flood A flood that follows a situation in which rainfall is so intense and severe and runs so rapidly that it precludes recording and relating it to stream stages and other information in time to forecast a flood condition.

Flood A general and temporary condition of partial or complete inundation of normally dry land areas from overflow of inland or tidal waters, unusual or rapid accumulation or runoff of surface waters, or mudslides/mudflows caused by accumulation of water.

Forced Entry Entry to a denied area achieved through force to create an opening in fences, walls, doors, etc., or to overpower guards.

Fragment Retention Film [FRF] A thin, optically clear film applied to glass to minimize the spread of glass fragments when the glass is shattered.

Frame Rate In digital video, a measurement of the rate of change in a series of pictures, often measured in frames per second [fps].

Frangible Construction Building components that are designed to fail to vent blast pressures from an enclosure in a controlled manner and direction.

Fingerprint Verification Device A device using one of two approaches. One is pattern recognition of the whorls, loops, and tilts of the referenced fingerprint, which is stored in a digitized representation of the image and compared with the fingerprint of the prospective entrant. The second approach is minutiae comparison, which means that the endings and branching points of rides and valleys of the referenced fingerprint are compared with the fingerprint of the prospective entrant.

Function Any authority power, rights, privileges, immunities, programs, projects, activities, duties and responsibilities [6 U.S.C. 101].

Fungi Any of a group of plants mainly characterized by the absence of chlorophyll, the green-colored compound found in other plants. Fungi range from microscopic single-celled plants [such as molds and mildews] to large plants [such as mushrooms].

G

G-series Nerve Agent A chemical agent of moderate to high toxicity developed in the 1930s. Examples are tabun [GA], sarin [GB], soman

[GD], phosphonofluoridic acid, ethyl-, 1-methylethyl ester [GE], and cyclohexyl sarin [GF].

Gamma Ray A high-energy photon emitted from the nucleus of atoms, similar to an x-ray. It can penetrate deeply into body tissue and many other materials. Cobalt-60 and cesium-137 are both strong gamma emitters. Shielding against gamma radiation requires thick layers of dense materials, such as lead. Gamma rays are potentially lethal to humans.

General Support System An interconnected information resource under the same direct management control that shares common functionality. It usually includes hardware, software, information, data, applications, communications, facilities, and people and provides support for a variety of uses and/or applications. Individual applications support different mission-related functions. Users may be from the same or different organizations.

Glare Security Lighting Illumination projected from a secure perimeter into the surrounding area, making it possible to see potential intruders at a considerable distance while making it difficult to observe activities within the secure perimeter.

Glass-Breaking Sensor A sensor that detects the breaking of glass. The noise from breaking glass consists of frequencies in both the audible and ultrasonic range. Glass-breakage sensors use microphone transducers to detect the glass breakage. The sensors are designed to respond to specific frequencies only, thus minimizing such false alarms as may be caused by banging on the glass.

Glazing A material installed in a sash, ventilator, or pane [e.g., glass, plastic, etc., including material such as thin granite installed in a curtain wall].

Governor The chief executive of any State [6 U.S.C. 101].

Governor's Authorized Representative [GAR] The person empowered by a governor to execute, on behalf of a state, all necessary documents for disaster assistance.

Grid-Wire Sensor A sensor consisting of a continuous electrical wire arranged in a grid pattern. The wire maintains an electrical current. An alarm is generated when the wire is broken. The sensor detects forced entry through walls, floors, ceilings, doors, windows, and other barriers. The grid's maximum size is determined by the spacing between the wires, the wire's resistance, and the electrical characteristics of the source providing the current. The grid wire can be installed directly on the barrier, in a grill or screen that is mounted on the barrier, or over an opening that requires protection.

H

Half-life The amount of time needed for half the atoms of a radioactive material to decay.

Hand Geometry A variety of physical measurements of the hand, such as finger length, finger curvature, hand width, webbing between fingers, and light transmissivity through the skin, used in devices to verify identity. Both two- and three-dimensional units are available.

Hazard A source of potential danger or adverse condition.

Hazard Mitigation Any action taken to reduce or eliminate the long-term risk to human life and property from hazards. The term is sometimes used in a stricter sense to mean cost-effective measures to reduce the potential for damage to a facility or facilities from a disaster event.

Hazardous Material [HazMat] Any substance or material that, when involved in an accident and released in sufficient quantities, poses a risk to people's health, safety, and/or property. These substances and materials include explosives, radioactive materials, flammable liquids or solids, combustible liquids or solids, poisons, oxidizers, toxins, and corrosive materials.

High-Hazard Area A geographic location that, for planning purposes, has been determined through historical experience and vulnerability analysis to be likely to experience the effects of a specific hazard [e.g., hurricane, earthquake, hazardous-materials accident, etc.], resulting in vast property damage and loss of life.

Highly Enriched Uranium [HEU] Uranium that is enriched to above 20 percent Uranium-235 [U-235]. Weapons-grade HEU is enriched to above 90 percent U-235.

High-Risk Target Any material resource or facility that, because of mission sensitivity, ease of access, isolation, and symbolic value, may be an especially attractive or accessible terrorist target.

Homeland Security A concerted national effort to prevent terrorist attacks within the United States, reduce America's vulnerability to terrorism, and minimize the damage and recovery from attacks that do occur.

Host An animal or plant that harbors or nourishes another organism.

Hot Site A fully operational offsite data-processing facility equipped with both hardware and system software to be used in the event of a disaster.

Hurricane A tropical cyclone, formed in the atmosphere over warm ocean areas, in which wind speeds reach 74 miles per hour or more and blow in a large spiral around a relatively calm center or "eye." Circulation is counterclockwise in the Northern Hemisphere and clockwise in the Southern Hemisphere.

I

Impact Analysis A management-level analysis that identifies the impacts of losing the entity's resources. The analysis measures the effect of resource loss and escalating losses over time in order to provide the entity with reliable data upon which to base decisions on hazard mitigation and continuity planning.

Incapacitating Agent An agent producing temporary physiological and/or mental effects via action on the central nervous system. Effects may persist for hours or days, but victims usually do not require medical treatment; however, such treatment speeds recovery. *See also* **Vomiting Agent, Tear Agent, Central-Nervous-System Depressant, Central-Nervous-System Stimulant,** and **Industrial Agent**.

Incident Command System [ICS] A standardized organizational structure used to command, control, and coordinate the use of resources and personnel that have responded to the scene of an emergency. The concepts and principles of ICS include common terminology, modular organization, integrated communication, unified command structure, consolidated action plan, manageable span of control, designated incident facilities, and comprehensive resource management.

Industrial Agent A chemical developed or manufactured for use in industrial operations or research by industry, government, or academia. These chemicals are not primarily manufactured for the specific purpose of producing human casualties or rendering equipment, facilities, or areas dangerous for use by humans. Hydrogen cyanide, cyanogen chloride, phosgene, chloropicrin, and many herbicides and pesticides are industrial chemicals that also can be chemical agents.

Infectious Agent A biological agent capable of causing disease in a susceptible host.

Infectivity [1] The ability of an organism to spread. [2] The number of organisms required to cause an infection to secondary hosts. [3] The capability of an organism to spread out from the site of infection and cause disease in the host organism. Infectivity also can be viewed as the number of organisms required to cause an infection.

Information Sharing and Analysis Organization Any formal or informal entity or collaboration created or employed by public or private sector organizations, for the purposes of [6 U.S.C. 101]:

Gathering and analyzing critical infrastructure information in order to better understand security problems and interdependencies related to critical infrastructure and protected systems, so as to ensure the availability, integrity, and reliability thereof;

Communicating or disclosing critical infrastructure information to help prevent, detect, mitigate, or recover from the effects of interference, compromise, or a incapacitation problem related to critical infrastructure or protected systems; and

Voluntarily disseminating critical infrastructure information to its members, State, local, and Federal Governments, or any other entities that may be of assistance in carrying out the purposes specified above.

Infrared [IR] Sensor A sensor available in both active and passive models. An active sensor generates one or more near-IR beams that generate an alarm when interrupted. A passive sensor detects changes in thermal IR radiation from objects located within its field of view. Active sensors consist of transmitter/receiver pairs. The transmitter contains an IR light source such as a gallium-arsenide light-emitting diode [LED] that generates an IR beam. The light source is usually modulated to reduce the sensor's susceptibility to unwanted alarms resulting from sunlight or other IR light sources. The receiver detects changes in the signal power of the received beam. To minimize nuisance alarms from birds or blowing debris, the alarm criteria usually require that a high percentage of the beam be blocked for a specific interval of time.

Insider Compromise A person authorized access to a facility who compromises assets by taking advantage of that accessibility.

Insider Threat An individual who has routine access to and knowledge of the organization either as an employee, contractor, or visitor. The person may be motivated by financial gain, political ideology, or perception of unfair treatment. Insiders, particularly employees, former employees, or contractors, often seek revenge or vent anger over some real or imagined slight. Because of their inside knowledge of operations, these perpetrators have the potential to inflict serious harm to others and to equipment.

Intercom Door/Gate Station Part of an intercom system; typically, where initial communication is received.

Intercom Switcher Part of an intercom system that controls the flow of communications between various stations.

Intercom System An electronic system that allows simplex, half-duplex, or full-duplex audio communications.

Interior Microwave Motion Sensor A sensor that is typically monostatic; the transmitter and the receiver are housed in the same enclosure [transceiver].

International Terrorism There is no universally accepted definition of international terrorism. One definition widely used in United States government circles and incorporated into law defines international terrorism as terrorism involving the citizens or property of more than one country. Terrorism is broadly defined as politically motivated violence perpetrated against noncombatant targets by subnational groups or clandestine agents. For example, kidnapping of U.S. birdwatchers or bombing of U.S.-owned oil pipelines by leftist guerrillas in Colombia would qualify as international terrorism. A terrorist group is defined as a group that practices or that has significant subgroups that practice terrorism [22 U.S.C. 2656f]. One shortfall of this traditional definition is its focus on groups and its exclusion of individual ["lone wolf"] terrorist activity, which has recently risen in frequency and visibility. To these standard definitions, which refer to violence in a traditional form, must be added cyber terrorism. Analysts warn that terrorist acts will include more sophisticated forms of destruction and extortion, such as disabling a national computer infrastructure or penetrating vital commercial computer systems. Finally, the October 12, 2000, bombing of *U.S.S. Cole*, a United States military vessel, raises issues of whether the standard definition would categorize this attack as terrorism, as the *Cole* may not qualify as a "noncombatant."

Though the definition of terrorism may appear essentially a political issue, it can carry significant legal implications. Current definitions of terrorism mostly share one common element: politically motivated behavior. Such definitions do not include violence for financial profit or religious motivation. For example, the high-profile activities of such groups as al Qaeda and Hamas underscore the significance of selective religious ideologies in driving terrorist violence or at least providing a pretext. To illustrate: Osama bin Laden issued a fatwa in 1998 proclaiming in effect that all those who believe in Allah and his prophet Muhammad must kill Americans wherever they find them. Moreover, the growth of international and transnational criminal organizations and the growing range and scale of such operations have resulted in criminal use of large-scale violence with financial profit as the driving motivation.

Complicating matters internationally, nations and organizations have historically been unable to agree on a definition of terrorism, since one person's terrorist is often another person's freedom fighter. To circumvent this political constraint, countries have taken the approach of creating networks of conventions that criminalize specific acts such as kidnapping, detonating bombs, or hijacking airplanes. The International Convention for the Suppression of the Financing of Terrorism, the International Convention for the Safety of Life at Sea and its International Ship and Port Facility Security Code, and the Convention on International Civil Aviation, with its Safeguarding International Civil Aviation Against Acts of Unlawful Interference procedures, come close to a consensus definition by making it a crime to perform designated acts. (Source: *Terrorism and National Security Trends*, Congressional Research Service, Library of Congress [October 2, 2003]).

Intrusion Detection Sensor A device that initiates alarm signals by sensing the stimulus, change, or condition for which it was designed.

Intrusion Detection System [IDS] The combination of components, including sensors, control units, transmission lines, and monitor units, integrated to operate in a specified manner.

Isolated Fenced Perimeter A perimeter with 100 feet or more of space outside the fence that is clear of obstruction, making approach obvious.

Isolation Valve A valve used to isolate a portion of the water system whenever system repair, inspection, or maintenance is required in that segment.

Ionize To split off one or more electrons from an atom, thus leaving it with a positive electric charge. The electrons usually attach to one of the atoms or molecules, giving them a negative charge.

Iridium-192 A gamma-ray-emitting radioisotope used for gamma radiography. Its half-life is 73.83 days.

Isotope A specific element always has the same number of protons in the nucleus. That same element may, however, appear in forms that have different numbers of neutrons in the nucleus. These different forms are referred to as "isotopes" of the element; for example, deuterium [2H] and tritium [3H] are isotopes of ordinary hydrogen [H].

J

Jersey Barrier A protective concrete barrier initially and still used as a highway divider that now also functions as an expedient method for traffic speed control at entrance gates and to keep vehicles away from buildings.

Joint Information Center [JIC] A central point of contact for all news media near the scene of a large-scale disaster. News-media representatives are kept informed of activities and events by public information officers who represent all participating federal, state, and local agencies that are collocated at the JIC.

Joint Information System [JIS] Under the Federal Response Plan [FRP] connection of public-affairs personnel, decision-makers, and news centers by electronic mail, fax, and telephone when a single federal-state-local JIC is not a viable option.

Joint Interagency Intelligence Support Element [JIISE] An interagency intelligence component designed to fuse intelligence information from the various agencies participating in a response to a Weapons of Mass Destruction [WMD] threat or incident within an Federal Bureau of Investigation Joint Operations Center [FBIJOC]. The JIISE is an expanded version of the investigative/intelligence component that is part of the standardized FBI command-post structure. The JIISE manages five functions, including security, collections management, current intelligence, exploitation, and dissemination.

Joint Operations Center [JOC] Established by the Lead Federal Agency [LFA] under the operational control of the federal On-scene Commander or Coordinator [OSC] as a focal point for management and direction of on-site activities, coordination/establishment of state requirements/priorities, and coordination of the overall federal response.

Jurisdiction Typically counties and cities within a state, but states may elect to define differently in order to facilitate their assessment process.

K

Key Resource Publicly or privately controlled resources essential to the minimal operations of the economy and government [6 U.S.C. 101].

L

Laminated Glass Two monolithic glass plies of uniform thickness bonded together with an interlayer material. Many different interlayer materials are used in laminated glass.

Landscaping The use of plantings [shrubs and trees], with or without landforms and/or large boulders, to act as a perimeter barrier against defined threats.

Laser Card A card technology that uses a laser reflected off a card for uniquely identifying it.

Layers of Protection A traditional approach in security engineering using concentric circles extending out from an area to be protected as demarcation points for different security strategies.

Lead Agency The federal department or agency assigned lead responsibility under U.S. law to manage and coordinate the federal response in a specific functional area.

Lead Federal Agency [LFA] The agency designated by the President to lead and coordinate the overall federal response, determined by the type of emergency. In general, an LFA establishes operational structures and procedures to assemble and work with agencies providing direct support to the LFA in order to provide an initial assessment of the situation, develop an action plan, monitor and update operational priorities, and ensure that each agency exercises its concurrent and distinct authorities under U.S. law and supports the LFA in carrying out the President's relevant policy. Specific responsibilities of an LFA vary, according to the agency's unique statutory authorities.

Lethal Dose [50/30] The dose of radiation expected to cause death within 30 days to 50 percent of those exposed without medical treatment. The generally accepted range is from 400–500 rem received over a short period of time.

Level of Protection [LOP] The degree to which an asset is protected against injury or damage from an attack.

Liaison An agency official sent to another agency to facilitate inter-agency communications and coordination.

Life-Cycle Cost Analysis and Repair-Level Analysis The process of identifying equipment spares or repair parts to determine the economic tradeoffs among investment costs. This data is used to formulate maintenance planning and to implement an overall replenishment philosophy.

Limited Area A restricted area within close proximity of a security interest. Uncontrolled movement may permit access to the item. Escorts and other internal restrictions may prevent access to the item. *See also* **Controlled Area** and **Exclusion Area**.

Line of Sight [LOS] Direct observation between two points with the naked eye or hand-held optics.

Line-of-Sight Sensor A pair of devices used as an intrusion detection sensor that monitors any movement through the field between the sensors.

Line-Source Delivery System A delivery system in which the biological agent is dispersed from a moving ground or air vehicle in a line perpendicular to the direction of the prevailing wind. *See also* **Point-Source Delivery System**.

Line Supervision A data-integrity strategy that monitors the communications link for connectivity and tampering. In IDS sensors, line supervision is often referred to as two-state, three-state, or four-state in respect to the number of conditions monitored. The frequency of sampling the link also plays a big part in the supervision of the line.

Liquid Agent A chemical agent that appears to be an oily film or droplets. The color ranges from clear to brownish amber.

Local Government [6 U.S.C. 101]
A county, municipality, city, town, township, local public authority, school district, special district, intrastate district, council of govern-

ments [regardless of whether the council of governments is incorporated as a nonprofit corporation under State law], regional or interstate government entity, or agency or instrumentality of a local government;

An Indian tribe or authorized tribal organization, or Alaska Native village or organizations; and

A rural community, unincorporated town or village, or other public entity, for which an application for assistance is made by a State or political subdivision of a State.

Lock, Barrier, and Access-Delay Technology The proper design, application, and installation of security hardware and locking devices that focus on how to meet the challenge of providing proper security without compromising life-safety issues.

Lone Wolf A disenfranchised, often mentally ill individual who may target his or her victims for their ethnicity, beliefs, or other characteristics or turn on the establishment.

Loss Event An occurrence that produces an operational or financial loss or negative service impact.

M

Magnetic Lock An electromagnetic lock that unlocks a door when power is removed.

Magnetic Stripe Card A card technology that uses a magnetic stripe along the edge with encoded data [sometimes encrypted]. The data is read by moving the card past a magnetic reader head.

Mail-Bomb Delivery A bomb or incendiary device delivered to the target in a letter or package.

Major Application An application that requires special attention to security due to the risk and magnitude of the harm resulting from the loss, misuse, or unauthorized access to or modification of the information in the application. A breach in a major application might comprise many individual application programs and hardware, software, and telecommunications components. Major applications can be either a major software application or a combination of hardware/software where the only purpose of the system is to support a specific mission-related function.

Major Disaster and Emergency Any natural catastrophe [including any hurricane, tornado, storm, high water, winddriven water, tidal wave, tsunami, earthquake, volcanic eruption, landslide, mudslide, snowstorm, or draught], or, regardless of cause, any fire, flood, or explosion, in any part of the United States, which in the determination of the President causes damage of sufficient severity and magnitude to warrant major disaster assistance to supplement the efforts and available resources of States, local governments, and disaster relief organizations in alleviating the damage, loss, hardship, or suffering caused thereby [42 U.S.C. 5122].

Major Event A domestic terrorist attack, major disaster, or other emergency.

Man-Trap An access-control strategy that uses a pair of interlocking doors to prevent tailgating. Only one door can be unlocked at a time.

Mass Care The actions that are taken to protect evacuees and other disaster victims from the effects of the disaster. Activities include providing temporary shelter, food, medical care, clothing, and other essential life-support needs to those people who have been displaced from their homes because of a disaster or threatened disaster.

Mass Notification Capability to provide real-time information to all building occupants or personnel in the immediate vicinity of a building during emergency situations.

Microorganism Any organism, such as bacteria, viruses, and some fungi, that can be seen only with a microscope.

Microwave Motion Sensor A sensor that uses a high-frequency electromagnetic energy to detect an intruder's motion within the protected area. *See also* **Interior Microwave Motion Sensor** and **Sophisticated Microwave Motion Sensor**.

Microwave Sensor An intrusion-detection sensor categorized as either bistatic or monostatic. Bistatic sensors use transmitting and receiving antennas located at opposite ends of the microwave link, whereas monostatic sensors use the same antenna.

Military Installation Army, Navy, Air Force, and Marine Corps base, post, station, or annex [either contractor- or government-operated], hospital, terminal, or other special mission facility, as well as one used primarily for military purposes.

Minimum Essential Infrastructure Resource Element A broad category of resources, all or portions of which constitute the minimal essential infrastructure necessary for a department, agency, or organization to conduct its core mission[s].

Minimum Measure A protective measure that can be applied to all buildings regardless of the identified threat. These measures offer defense or detection opportunities for minimal cost, facilitate future upgrades, and may deter acts of aggression.

Mitigation Those actions taken to reduce the exposure to and impact of an attack or disaster.

Mobile Site A self-contained, transportable shell custom-fitted with specific equipment and telecommunications necessary to provide full recovery capabilities upon notice of a significant disruption.

Motion Detector An intrusion-detection sensor that changes state based on movement in the sensor's field of view.

Moving Vehicle Bomb An explosive-laden car or truck driven into or near a building and detonated.

Mutual-Aid Agreement A prearranged agreement developed between two or more entities to render assistance to the parties of the agreement.

Mycotoxin A toxin produced by fungi.

N

Natural Hazard A naturally occurring event such as a flood, earthquake, tornado, tsunami, coastal storm, landslide, or wildfire that strikes population areas. A natural event is a hazard when it has the potential to harm people or property. The risks of natural hazards may be increased or decreased as a result of human activity; however, they are not inherently human-induced.

Natural Protective Barrier A mountain, desert, cliff, ditch, water obstacle, or other terrain feature that is difficult to traverse.

Nebulizer A device for producing a fine spray or aerosol.

Nerve Agent A substance that interferes with the central nervous system. Exposure is primarily through contact with the liquid [skin and eyes] and secondarily through inhalation of the vapor. Three distinct symptoms associated with nerve agents are: pinpoint pupils, an extreme headache, and severe tightness in the chest. *See also* **G-Series Nerve Agent** and **V-Series Nerve Agent**.

Nonexclusive Zone An area around an asset that has controlled entry but shared or less restrictive access than an exclusive zone.

Nonpersistent Agent An agent that upon release loses its ability to cause casualties after 10 to 15 minutes. It has a high evaporation rate, is lighter than air, and will disperse rapidly. It is considered to be a short-term hazard; however, in small, unventilated areas, the agent will be more persistent.

Nuclear, Biological, or Chemical Weapon Also called a **Weapon of Mass Destruction** [WMD]. A weapon that is characterized by its capability to produce mass casualties.

Nuclear Detonation An explosion resulting from fission and/or fusion reactions in nuclear materials, such as that from a nuclear weapon.

Nuclear Reactor A device in which a controlled, self-sustaining nuclear chain reaction can be maintained with the use of cooling to remove generated heat.

O

On-Scene Coordinator [OSC] The federal official predesignated by the Environmental Protection Agency [EPA] and U.S. Coast Guard to coordinate and direct response and removal under the National Oil and Hazardous Substances Pollution Contingency Plan.

Open-Systems Architecture A term borrowed from the Information Technology [IT] industry to refer to systems capable of interfacing with other systems from any vendor that also use open-system architecture. The opposite would be a proprietary system.

Operator Interface The part of a security-management system that provides the user interface to humans.

Organizational Area of Control An area of control consisting of the policies, procedures, practices, and organization structures designed to provide reasonable assurance that business objectives will be achieved and that undesired events will be prevented or detected and corrected.

Organ-Phosphorous Compound A compound containing the elements phosphorus and carbon, whose physiological effects include inhibition of acetyl cholinesterase. Many pesticides [malathion and parathion] and virtually all nerve agents are organ-phosphorous compounds.

Organism Any individual living thing, whether animal or plant.

Outsider Threat An individual or group that represents a threat to the organization.

P

Parasite Any organism that lives in or on another organism without providing benefit in return.

Passive Infrared [PIR] Motion Sensor A sensor that detects a change in the thermal energy pattern caused by a moving intruder and initiates an alarm when the change in energy satisfies the detector's alarm criteria. These sensors are passive devices because they do not transmit energy; they monitor the energy radiated by the surrounding environment.

Passive Ultrasonic Sensor A sensor that detects acoustical energy in the ultrasonic frequency range, typically between 20 and 30 kilohertz [kHz]. These sensors are used to detect an attempted penetration through rigid barriers [such as metal or masonry walls, ceilings, and floors]. They also detect penetration through windows and vents covered by metal grills, shutters, or bars if these openings are properly sealed against outside sounds.

Passive Vehicle Barrier A vehicle barrier that is permanently deployed and does not require response to be effective.

Patch Panel A concentrated termination point that separates backbone cabling from device cabling for easy maintenance and troubleshooting.

Pathogen Any organism [usually living], such as bacteria, fungi, and viruses, capable of producing serious disease or death.

Pathogenic Agent A biological agent capable of causing serious disease.

Percutaneous Agent An agent that is able to be absorbed by the body through the skin.

Perimeter Barrier A fence, wall, vehicle barrier, landform, or line of vegetation applied along an exterior perimeter used to obscure vision, hinder personnel access, or hinder or prevent intruder access.

Persistent Agent An agent that upon release retains its casualty-producing effects for an extended period of time, usually anywhere from 30 minutes to several days. A persistent agent usually has a low evaporation rate and its vapor is heavier than air; therefore, its vapor cloud tends to hug the ground. It is considered to be a long-term hazard. Although inhalation hazards are still a concern, extreme caution should be taken to avoid skin contact as well.

Physical Destruction Destruction by explosives, mechanical sabotage, or arson to critical assets and interdependent infrastructure.

Physical Security The aspect of security concerned with measures/concepts designed to safeguard personnel; to prevent unauthorized access to equipment, installations, materiel, and documents; and to safeguard them against espionage, sabotage, damage, and theft.

Planter Barrier A passive vehicle barrier, usually constructed of concrete and filled with dirt [and flowers for aesthetics]. Planters, along with bollards, are the usual street furniture used to keep vehicles away from existing buildings. Overall size and depth of installation below grade determine the vehicle-stopping capability of the individual planter.

Plume Airborne material spreading from a particular source; the dispersal of particles, gases, vapors, and aerosols into the atmosphere.

Plutonium-239 [Pu-239] A metallic element used for nuclear weapons. Its half-life is 24,110 years.

Polycarbonate Glazing A plastic glazing material with enhanced resistance to ballistics or blast effects.

Point Sensor A sensor used to protect specific objects within a facility. These sensors [sometimes referred to as proximity sensors] detect an intruder coming in close proximity to, touching, or lifting an object. Several different types are available, including capacitance sensors, pressure mats, and pressure switches. Other types of sensors can also be used for object protection.

Point-Source Delivery System A delivery system in which the biological agent is dispersed from a stationary position. This delivery method results in coverage over a smaller area than with the line-source system. *See also* **Line-Source Delivery System**.

Predetonation Screen A fence that causes an antitank round to detonate or prevents it from arming before it reaches its target.

Preliminary Damage Assessment [PDA] A mechanism used to determine the impact and magnitude of damage and the resulting unmet needs of individuals, businesses, the public sector, and the community as a whole. Information collected is used by the state as a basis for the governor's request for a Presidential declaration and by the Federal Emergency Management Agency [FEMA] to document the recommendation made to the President in response to the governor's request. PDAs are made by at least one state and one federal representative. A local government representative familiar with the extent and location of damage in the community often participates; other state and federal agencies and voluntary relief organizations also may be asked to participate as needed.

Preparedness The existence of plans, procedures, policies, training, and equipment necessary to achieve maximum readiness; the ability to prevent, respond to, and recover from all hazards, disasters, and emergencies, including terrorist attacks and Weapons of Mass Destruction [WMD] incidents. The term "readiness" is used interchangeably with preparedness.

Pressure Mat A mat that generates an alarm when pressure is applied to any part of the mat's surface, such as when someone steps on it. One type of construction uses two layers of copper screening separated by soft-sponge rubber insulation with large holes in it. Another type uses parallel strips of ribbon switches made from two strips of metal separated by an insulating material and spaced several inches apart. When enough pressure is applied to the mat, either the screening or the metal strips make contact, generating an alarm. Pressure mats can be used to detect an intruder approaching a protected object, or they can be placed by doors and windows to detect entry.

Pressure Switch A mechanically activated contact switch or single ribbon switch. Objects that require protection can be placed on top of the switch. When the object is moved, the switch actuates and generates an alarm.

Prevention Activities undertaken by the first-responder community during the early stages of an incident to reduce the likelihood or consequences of threatened or actual terrorist attacks. More general and broader efforts to deter, disrupt, or thwart terrorism are not included in this definition.

Private Nonprofit Facility   Any private nonprofit educational, utility, irrigation, emergency, medical, rehabilitational, and temporary or permanent custodial care facilities [including those for the aged and disabled], other private nonprofit facilities which provide essential services of a governmental nature to the general public, and facilities on Indian reservations as defined by the President [6 U.S.C. 101].

Probability   The chance that a given event will occur.

Process   The series of work activities that perform a function. It could be the making of a product on an assembly line, the development of decision-related information, the transmission of monetary or other transactions, or the movement of people, vehicles, and materials through a system of codes, protocols or other criteria. Loss of process can be partial and temporary or significant and long-term.

Protected System [6 U.S.C. 101]
Any service, physical or computer-based system, process, or procedure that directly or indirectly affects the viability of a facility of critical infrastructure; and

Includes any physical or computer-based system, including a computer, computer system, computer or communications network, or any component hardware or element thereof, software program, processing instructions, or information or data in transmission or storage therein, irrespective of the medium of transmission or storage.

Protection   Any means by which an individual protects his or her body. Measures include masks, self-contained breathing apparatuses, clothing, structures such as buildings, and vehicles.

Primary Asset   An asset that is the ultimate target for compromise by an aggressor.

Primary-Gathering Building   An inhabited building routinely occupied by 50 or more personnel. This designation applies to the entire portion of a building that meets the population-density requirements for an inhabited building.

Probability of Detection $[P_d]$ A measure of an intrusion-detection sensor's performance in detecting an intruder within its detection zone.

Probability of Intercept [P₁] The probability that an act of aggression will be detected and that a response force will intercept the aggressor before the asset can be compromised.

Progressive Collapse A chain-reaction failure of building members to an extent disproportionate to the original localized damage. Such damage may result in upper floors of a building collapsing onto lower floors.

Proprietary Information Information not within the public domain and that the owner has taken some measures to protect. Generally, such information concerns U.S. business and economic resources, activities, research and development, policies, and critical technologies. Although it may be unclassified, the loss of this information could impede the ability of the organization or agency to compete in the world marketplace and could have an adverse effect on the U.S. economy, eventually weakening national security. Commonly referred to as "trade secrets," this information typically is protected under both state and federal laws.

Protection Cost Window The results of cost-analysis efforts that ensure that the cost of recommended protective measures does not cost more than the combined worth of the facilities, systems, and functions being protected.

Protective Barrier A barrier that defines the physical limits of a site, activity, or area by restricting, channeling, or impeding access and forming a continuous obstacle around the object.

Protective Measure An element of a protective system that protects assets against a threat. Protective measures include the integration of people, procedures, and equipment for protection against:

- Malevolent attacks such as terrorism and sabotage
- Criminal activity including felonies and misdemeanors
- Environmental disasters
- Loss of services or injury from system or equipment failures

Protective System An integration of all the protective measures required to protect an asset against the range of threats applicable to it.

Proximity Card A card whose coded pattern is sensed when it is brought within several inches of a reader. Several techniques are used to code cards. One technique uses a number of electrically tuned circuits embedded in the card. Data are encoded by varying resonant frequencies of the tuned circuits. The reader contains a transmitter that continually sweeps through a specified range of frequencies of the tuned circuits. The reader contains a transmitter that continually sweeps through a specified range of frequencies and a receiver that senses the pattern of resonant frequencies contained in the card. Another technique uses an integrated circuit embedded in the card to generate a code that can be magnetically or electrostatically coupled to the reader. The power required to activate embedded circuitry can be provided by a small battery embedded in the card or by magnetically coupling power from the reader.

Proximity Sensor An intrusion-detection sensor that changes state based on the close distance or contact of a human to the sensor. These sensors often measure the change in capacitance as a human body enters the measured field.

Public Facility The following facilities owned by a State or local government [6 U.S.C. 101]:

- Any flood control, navigation, irrigation, reclamation, public power, sewage treatment and collection, water supply and distribution, watershed development, or airport facility;
- Any non-Federal-aid street, road, or highway;
- Any other public building, structure, or system, including those used for educational, recreational, or cultural purposes; and
- Any park

Public Information Officer [PIO] A federal, state, or local government official responsible for preparing and coordinating the dissemination of emergency public information.

Q

Quantitative Relating to, concerning, or based on the amount or number of something measurable or expressed in numerical terms.

R

Rad A unit of absorbed dose of radiation defined as deposition of 100 ergs of energy per gram of tissue. A rad amounts to approximately one ionization per cubic micron.

Radiation High-energy particles or gamma rays that are emitted by an atom as the substance undergoes radioactive decay. Particles can be either charged alpha or beta particles or neutral neutron or gamma rays.

Radiation Sickness The symptoms characterizing the sickness known as radiation injury, resulting from excessive exposure of the whole body to ionizing radiation.

Radiation Waste Disposable radioactive materials resulting from nuclear operations. Wastes are generally classified into two categories, high-level and low-level.

Radio-Frequency [RF] Data Transmission A communications link using radio frequencies to send or receive data.

Radiological Dispersal Device [RDD] A device [weapons or equipment], other than a nuclear explosive device, designed to disseminate radioactive material in order to cause destruction, damage, or injury by means of the radiation produced by the decay of such material.

Radiological Monitoring The process of locating and measuring radiation by means of survey instruments that can detect and measure [as exposure rates] ionizing radiation.

Radioluminescence The luminescence produced by particles emitted during radioactive decay.

Reciprocal Agreement An agreement that allows two organizations to back each other up.

Recovery The long-term activities beyond the initial crisis period and emergency-response phase of a disaster operation that focus on returning all systems in the community to a normal status or to reconstitute these systems to a new condition that is less vulnerable. Also, the ability to operate well enough to meet obligations at an acceptable level and to protect the life and safety of employees and others while transitioning to normalcy.

Regional Operations Center [ROC] The temporary operations facility for the coordination of federal response and recovery activities located at the Federal Emergency Management Agency [FEMA] regional office [or federal regional center] and led by the FEMA regional director or deputy director until the Designated Federal Office [DFO] becomes operational. After the Emergency-Response Team Advance Element [ERT-A] is deployed, the ROC performs a support role for federal staff at the disaster scene.

Report Printer A separate, dedicated printer attached to the electronic security systems used for generating reports utilizing information stored by the central computer.

Request-to-Exit Device A passive infrared motion sensor or push button used to signal an electronic-entry control system that egress is imminent or to unlock a door.

Resolution The level to which video details can be determined in a Closed Circuit Television [CCTV] scene.

Response Deterrence, delay, detection, and assessment are meaningless unless the capability to neutralize an adversary exists. Also referred to as executing the plan and resources identified to perform those duties and services to preserve and protect life and property as well as provide services to the surviving population. The existence and capability of response are operationally dependent on a human function, as are visual assessment and surveillance. As such, the overall technical solution must assist [and not hinder] the security organization in executing its mandate.

Response Force The people who respond to an act of aggression. Depending on the nature of the threat, the response force could consist of guards, special reaction teams, military or civilian police, an explosives ordinance-disposal team, or a fire department. First responders may be a proprietary force, a contract guard service, designated local, state, or federal law enforcement, or a combination of several separate but integral agencies including the Department of Defense [DOD] and elements of the Department of Homeland Security [DHS].

Response Time The length of time from the instant an attack is detected to the instant a security force arrives on site.

Restricted Area An area classified as controlled, limited, or exclusion. *See also* **Controlled Area, Limited Area, Exclusion Area,** and: **Exclusion Zone**.

Retinal Pattern A type of technique based on the premise that the pattern of blood vessels on the human eye's retina is unique to an individual. While the eye is focused on a visual target, a low-intensity infrared [IR] light beam scans a circular area of the retina. The amount of light reflected from the eye is recorded as the beam progresses around the circular path. Reflected light is modulated by the difference in reflectivity between blood-vessel pattern and adjacent tissue. This information is processed and converted to a digital template that is stored as the eye's signature. Users are allowed to wear contact lenses; however, glasses should be removed.

Roentgen Equivalent Man [REM or rem] A unit of absorbed dose that takes into account the relative effectiveness of radiation that harms human health.

Rotating Drum or Rotating Plate Vehicle Barrier An active vehicle barrier used at vehicle entrances to controlled areas based on a drum or plate rotating into the path of the vehicle when signaled.

Route of Exposure [Entry] The path by which a person comes into contact with an agent or organism [e.g., through breathing, digestion, or skin contact].

Routinely Occupied An established or predictable pattern of activity within a building that terrorists could recognize and exploit.

RS-232 Data Institute of Electrical and Electronics Engineers [IEEE] Recommended Standard 232, a point-to-point serial-data protocol with a maximum effective distance of 50 feet.

RS-422 Data Institute of Electrical and Electronics Engineers [IEEE] Recommended Standard 422, a point-to-point serial-data protocol with a maximum effective distance of 4,000 feet.

RS-485 Data Institute of Electrical and Electronics Engineers [IEEE] Recommended Standard 485, a multidrop serial-data protocol with a maximum effective distance of 4,000 meters

S

Sacrificial Roof or Wall A roof or wall that can be lost in a blast without damage to the primary asset.

Safe Haven A secure area within the interior of a facility. A safe haven is a location designed such that it requires more time to penetrate by aggressors than it takes for the response force to reach the protected area to rescue the occupants. It may be a haven from a physical attack or an air-isolated haven from chemical, biological and radiological [CBR] contamination. Also refers to the temporary staging of CBR materials while in transit from one destination to another.

Scramble Keypad A keypad that uses keys on which the numbers change pattern with each use to enhance security by preventing eavesdropping or observation of the entered numbers.

Secondary Asset An asset that supports a primary asset and whose compromise would indirectly affect the operation of the primary asset.

Secondary Hazard A threat whose potential would be realized as the result of a triggering event that of itself would constitute an emergency [e.g., dam failure might be a secondary hazard associated with an earthquake].

Secure-Access Mode The state of an area monitored by an intrusion-detection system in regard to how alarm conditions are reported.

Security Analysis The method of studying the nature of and the relationship between assets, threats, and vulnerabilities.

Security Assessment The process of examining a situation or elements of a program to identify and analyze threats and vulnerabilities to determine the potential for loss and identifying cost-effective protective measures and residual risks. Also, the process of analyzing security requirements through a systematic approach that addresses vulnerability and leads to recommendations to be adopted by governing authorities. It [1] identifies risk and early-warning signals that enable the assessment of vulnerabilities and [2] uses a comprehensive methodology to outline the preparation, performance, reporting, and follow-up criteria for ensuring security-posture effectiveness of facilities, systems, functions, and resources in accordance with generally accepted security standards and practices.

Security Console Specialized furniture, racking, and related apparatus used to house the security equipment required in a control center.

Security Engineering The process of identifying practical, risk-managed short- and long-term solutions to reduce and/or mitigate dynamic man-made hazards by integrating multiple factors, including construction, equipment, manpower, procedures, and human interface.

Security-Engineering Design Process The process through which assets requiring protection are identified, the threat to and vulnerability of those assets is determined, and a protective system is designed to protect the assets.

Security-Management-System Database A database that is transferred to various nodes or panels throughout the system for faster data processing and protection against communications-link downtime.

Security Management System Distributed Processing In a Security Management System, a method of data processing at various nodes or panels throughout the system for faster data processing and protection against communications links downtime.

Security Risk A potential occurrence that would be detrimental to plans or programs. Risk is measured as the combined effect of the likelihood of the occurrence and a measured or assessed consequence given to that occurrence. Also, an event that has a potentially negative impact, with the possibility that such an event will occur and adversely affect an entity's assets and operations as well as the achievement of its mission and strategic objectives. As applied to the homeland-security context, risk is most prominently manifested as a "catastrophic" or "extreme" event related to terrorism.

Security-Risk Analysis The identification of consequences that are likely to occur and the indicators of the start of the problem.

Security-Risk Assumption The acknowledgement of the existence of the risk and a decision to accept the consequences if failure occurs. The management decision that declares a particular terrorist threat and program vulnerability so low and the probability of an attack against a particular asset so low that damage or destruction of a critical asset including the potential of injury or loss of life is acceptable.

Security-Risk Control The process of continually sensing the condition of a program and developing options and fallback positions to permit alternative lower risk solutions without violating the integrity of the overall security program.

Security-Risk Management A continuous process of managing, through a series of mitigating actions that permeate an entity's activities, the likelihood of an adverse event happening and having a negative impact. In general, security risk is managed as a portfolio, addressing entity-wide security risk within the entire scope of activities. Security risk management addresses "inherent" or preaction risk [i.e., risk that would exist absent any mitigating action] as well as "residual" or postaction risk [i.e., the risk that remains even after mitigating actions have been taken].

Security-Risk-Management Framework A concept based on the proposition that a threat to a vulnerable asset results in punitive risk consisting of the following components:

- **Internal [or Implementing] Environment:** The institutional "driver" of security-risk management, serving as the foundation of all elements of the security-risk-management process. The internal environment includes an entity's organizational and management structure and processes that provide the framework to plan, execute, control, and monitor an entity's activities, including security-risk management. Within the organizational and management structure, an operational unit that is independent of other operational [business] units is responsible for implementing the entity's security-risk management-function. This is typically the corporate security organization. The unit is supported by and directly accountable to an entity's senior management.

  For its part, senior management:

  1. Defines the entity's risk tolerance [i.e., how much risk is an entity willing to assume in order to accomplish its mission and related objectives]

  2. Establishes the entity's security-risk-management philosophy and culture, how an entity's values and attitudes view risk and how its activities and practices are managed to deal with security risk

  The security organization:

  1. Designs and implements the entity's security-risk-management process

  2. Coordinates internal and external evaluation of the process and helps implement any corrective action

- **Threat [Event] Assessment**: Threat is defined as a potential intent to cause harm or damage to an asset [e.g., natural environment, people, manmade infrastructures, and activities and operations]. Threat assessment consists of the identification of adverse events that can affect an entity. Threats might be present at the global, national, regional, or local level, and their sources include terrorists, criminal enterprises, and individuals. Threat information emanates from "open" sources and intelligence [both strategic and tactical]. Intelligence information is characterized as "reported" [or raw] and "finished" [fully fused and analyzed]. As applied to homeland-security terrorism risk, adverse-event scenarios consist of six stages, as illustrated:

STAGE	DESCRIPTION
Intent	The terrorist develops malice and intent to do harm.
Target acquisition	The terrorist chooses specific target[s] among assets.
Planning	The terrorist researches the targets and various attack options.
Preparation	Full commitment stage—the terrorist prepares to launch the attack.
Execution	The terrorist carries out the attack.
"Grace period"	Depending on the nature and success of the attack, there could be a time lag between the attack and its impact.

- **Criticality Assessment:** An asset's relative importance. The identification and evaluation of an entity's assets based on a variety of factors, including the importance of its mission or function, the extent to which people are at risk, or the significance of a structure or system in terms of, for example, national security, economic security, or public safety. Criticality assessments are important because they provide, in combination with the framework's other assessments, the basis for prioritizing which assets require greater or special protection relative to finite resources.

- **Vulnerability Assessment:** Vulnerability is defined as the inherent state [either physical, technical, or operational] of an asset that can be exploited by an adversary to cause harm or damage. Vulnerability assessments identify these inherent states and the extent of their susceptibility to exploitation relative to the existence of any protective measures. As applied to the global supply chain, a vulnerability assessment might involve, first, establishing a comprehensive understanding of the business and commercial aspects of the chain [as a complex system with multiple interacting participants]; and, second, "mapping" the chain and identifying vulnerability points that could be exploited.

- **Security-Risk Assessment** A qualitative and/or quantitative determination of the likelihood [probability] of the occurrence of an adverse event and the severity, or impact, of its consequences. Security-risk assessments include scenarios under which two or more risks interact, creating greater or lesser impacts.

- **Security-Risk Characterization** Involves designating risk as, for example, low, medium, or high [other scales, such as numeric, are also used]. Is a function of the probability of an adverse event occurring and the severity of its consequences. Risk characterization is the crucial link between assessments of risk and the implementation of mitigation actions, given that not all risks can be addressed, because resources are inherently scarce. Accordingly, risk characterization forms the basis for deciding which actions are best suited to mitigate the assessed risk.

- **Mitigation Evaluation** The identification of mitigation alternatives to assess the effectiveness of the alternatives. The alternatives should be evaluated for their likely effect on risk and cost.

- **Mitigation Selection** Involves a management decision on which mitigation alternatives should be implemented, taking into account risk, costs, and efficacy. Selection is based upon preconsidered criteria. There are as of yet no clearly preferred selection criteria, although potential factors might include risk reduction, net benefits, equality of treatment, or other stated values. Mitigation selection does not necessarily involve prioritizing all resources to the highest risk area but rather attempts to balance overall risk and available resources.

- **Risk Mitigation** The implementation of mitigation actions, in priority order and commensurate with assessed risk. Depending on its risk tolerance, an entity may choose **not to take any action to mitigate risk [this is characterized as risk acceptance].** If the entity does choose to take action, such action falls into three categories:

  1. **Risk Avoidance** Existing activities that expose the entity to risk
  2. **Risk Reduction** Implementing actions that reduce likelihood or impact or risk
  3. **Risk Sharing** Implementing actions that reduce the likelihood or impact by transferring or sharing risk

     In each category, the entity implements actions as part of an integrated "systems" approach, with built-in redundancy to help address residual risk [the risk that remains after actions have been implemented]. The systems approach consists of taking actions in personnel [e.g., training, deployment], processes [e.g., operational procedures], technology [e.g., software or hardware], infrastructure [e.g., institutional or operational, such as port configurations], and governance [e.g., management and internal control and assurance]. In selecting actions, the entity assesses their costs and benefits, in which the amount of risk reduction is weighed against the cost involved, and identifies potential financing operations for the actions chosen.

- **Monitoring and Evaluation of Risk Mitigation:** Monitoring entails the assessment of the functioning of actions against strategic objectives and performance measures to make necessary changes. Includes, where and when appropriate: peer review, testing, and validation; an evaluation of the impact of the actions on future options; and identification of unintended consequences that in turn would need to be mitigated. Monitoring and evaluation help ensure that the entire security-risk-management process remains current and relevant and reflects changes in:

1. The effectiveness of the actions

2. The risk environment in which the entity operates

The security-risk-management process should be repeated periodically, restarting the "loop" of assessment, mitigation, and monitoring and evaluation.

Security Strategy Mission, vision, value statements, and the policies and procedures that support the organization.

Semi-Isolated Fenced Perimeter A fence line where approach areas are clear of obstruction for 60 to 100 feet outside of the fence and where the general public or other personnel seldom have reason to be.

Senior FEMA Official [SFO] The official appointed by the Director of the Federal Management Agency [FEMA] or his representative who is responsible for deploying to the Joint Operations Center [JOC] to serve as the senior interagency consequence-management representative on the command group and to manage and coordinate activities taken by the consequence-management group.

Serial Interface An integration strategy for data transfer where components are connected in series.

Shielded Wire Wire with a conductive wrap used to mitigate electromagnetic emanations.

Shielding Materials [lead, concrete, etc.] used to block or attenuate radiation for protection of equipment, materials, or people.

Situational Crime Prevention A crime-prevention strategy based on reducing the opportunities for crime by increasing the effort required to commit a crime, increasing the risks associated with committing the crime, and reducing the target appeal or vulnerability [whether property or person]. This opportunity reduction is achieved by management and uses policies such as procedures and training as well as physical approaches such as alteration of the build environment.

Smart Card A card is embedded with a microprocessor, memory, communication circuitry, and a battery. The card contains edge contacts that enable a reader to communicate with the microprocessor. Entry control information and other data may be stored in the microprocessor's memory.

Software-Level Integration An integration strategy that uses software to interface systems. An example of this would be digital video displayed in the same computer-application window and linked to events of a security-management system.

Sophisticated Microwave Motion Sensor A sensor that may be equipped with electronic range gating. This feature allows the sensor to ignore the signals reflected beyond the settable detection range. Range gating may be used to effectively minimize unwanted alarms from activity outside the protected area.

Special Nuclear Material [SNM] Plutonium and uranium enriched in the isotopes uranium-233 or uranium-235.

Specific Threat Known or postulated aggressor activity focused on targeting a particular asset.

Spore A reproductive form that some microorganisms can take to become resistant to environmental conditions, such as extreme heat or cold while in a "resting stage."

State and Local Government When used in a geographical sense, has the same meaning given to those terms in section 2 of the Homeland Security Act of 2002 [6 U.S.C. 101], Public Law 107-296.

Strain-Sensitive Cable A transducer that is uniformly sensitive along its entire length. It generates an analog voltage when subject to mechanical distortions or stress resulting from fence motion. Strain-sensitive cables are sensitive to both low and high frequencies. Because the cable acts like a microphone, some manufacturers offer an option that allows the operator to listen to fence noises causing the alarm. Operators can then determine whether the noises are naturally occurring sounds from wind or rain or are from an actual intrusion attempt.

Standoff Distance A distance maintained between a building or portion thereof and the potential location of an explosive detonation or other threat.

Standoff Weapon A weapon such as an antitank weapon or mortar that is launched from a distance at a target.

State Coordinating Officer [SCO] The person appointed by the governor to coordinate state, commonwealth, or territorial response and recovery activities with Federal Response Plan [FRP]-related activities of the federal government in cooperation with the Federal Coordinating Officer [FCO].

State Liaison A Federal Emergency Management Agency [FEMA] official assigned to a particular state, who handles initial coordination with the state in the early stages of an emergency.

State Any State of the United States, the District of Columbia, Puerto Rico, the Virgin Islands, Guam, American Samoa, the Commonwealth of the Northern Mariana Islands, and any possession of the United States [6 U.S.C. 101].

Stationary Vehicle Bomb An explosive-laden car or truck stopped or parked near a building.

Storm Surge A dome of sea water created by the strong winds and low barometric pressure in a hurricane that causes severe coastal flooding as the hurricane strikes land.

Strain-Sensitive Cable A transducer that is uniformly sensitive along its entire length and generates an analog voltage when subjected to mechanical distortions or stress resulting from fence motion. These cables are typically attached to a chain-link fence about halfway between the bottom and top of the fence fabric with plastic ties.

Structural Protective Barrier A man-made device [e.g., fence, wall, floor, roof, grill, bar, roadblock, sign, or other construction] used to restrict, channel, or impede access.

Structural Vibration Sensor A sensor that detects low-frequency energy generated in an attempted penetration of a physical barrier [such as a wall or a ceiling] by hammering, drilling, cutting, detonating explosives, or employing other forcible methods of entry. A piezoelectric transducer senses mechanical energy and converts it into electrical signals proportional in magnitude to the vibrations.

Superstructure The supporting elements of a building above the foundation.

Supplies-Bomb Delivery A bomb or incendiary device concealed and delivered to a supply or material handling points such as a loading dock.

System A generic term used to refer to either a major application or a general support structure.

System-Development Life Cycle The scope of activities associated with a system, encompassing the system's initiation, development and acquisition, implementation, operation and maintenance, and ultimately its disposal, which instigates another system initiation.

System Event An event that occurs normally in the operation of a security-management system. Examples include access-control operations and changes of state in intrusion-detection sensors.

System Functional Specification A design-guidance document that specifies system performance objectives. It tells planners and designers how the security system is to protect a given resource against a particular vulnerability and what level of protection is to be achieved. The functional specification does not tell planners or designers how the performance objectives will be accomplished nor does it provide a list of equipment. Rather, the functional specification contains design criteria and instructions to be used to deliver a highly reliable security system that meets or exceeds stated needs and expectations.

System Integration The final step in the design process. It matches all the pieces together and makes individual items of equipment and different parts of a subsystem work as a total system. The approach provides a cost-effective combination of equipment, personnel, and administrative procedures. Each piece of the system enhances the others, resulting in a balanced approach to optimum security.

System Software Controls that limit and monitor access to the powerful programs and sensitive files that control the computer hardware and secure applications supported by the system.

T

Tactic A specific method of achieving the aggressor's goals to injure personnel, destroy assets, or steal materiel or information.

Tamper Switch An intrusion-detection sensor that monitors an equipment enclosure for breaches.

Tangle-Foot Wire Barbed wire or tape suspended on short metal or wooden pickets outside a perimeter fence to create an obstacle to approach.

Taut-Wire Sensor A sensor combining a physical taut-wire barrier with an intrusion-detection sensor network. The taut-wire sensor consists of a column of uniformly spaced horizontal wires up to several hundred feet in length, securely anchored at each end and stretched taut. Each is individually tensioned and attached to a detector located in a sensor post. Two types of detectors are commonly used, mechanical switches and strain gauges.

Tear-Gas [Riot-Control] Agent An agent that produces irritating or disabling effects that rapidly disappear within minutes after exposure ceases. Examples are bromobenzylcyanide [CA], chloroacetophenone [CN, commercially known as Mace], chloropicrin [PS], CNB [CN in benzene and carbon tetrachloride], CNC [CN in chloroform], CNS [CN and chloropicrin in chloroform, CR [dibenz-pb,f]-1,4-oxazepine, a tear gas], CS [tear gas], and capsaicin [pepper spray].

Technical Assistance The provisioning of direct assistance to state and local jurisdictions to improve capabilities for program development, planning, and operational performances related to responses to WMD terrorist incidents. Also referred to as consultant technical-assistance services.

Technological Hazard An incident that can arise from human activities such as manufacture, transportation, storage, and use of hazardous materials. For the sake of simplicity, it is assumed that technological emergencies are accidental and that their consequences are unintended.

Tempest An unclassified short name referring to investigations and studies of compromising emanations. It is sometimes used synonymously for the term "compromising emanations" [e.g., tempest tests, tempest inspections].

Terrorism
Any activity that [6 U.S.C. 101]:
Involves an act that is dangerous to human life or potentially destructive of critical infrastructure or key resources; and is a violation of the criminal laws of the United States or of any State or other subdivision of the United States; and
Appears to be intended to intimidate or coerce a civilian population; to influence the policy of a government by intimidation or coercion; or to affect the conduct of a government by mass destruction, assassination, or kidnapping.

Thermally Tempered Glass [TTG] Glass that is heat-treated to have a higher tensile strength and resistance to blast pressures, although with a greater susceptibility to airborne debris.

Threat Any indication, circumstance, or event with the potential to cause loss of or damage to an asset or an indication of something impending.

Threat Analysis A continual process of compiling and examining all available information concerning potential threats and human-caused hazards. A common method to evaluate terrorist groups is to review the factors of existence, capability, intentions, history, and targeting.

Time-Critical The loss of any service function that poses an imminent threat to the survival of the organization.

Time-Date Stamp Data inserted into a Closed Circuit Television [CCTV] video signal with the time and date of the video as it was created.

Time-Sensitive The loss of any service function that poses a near-term threat to the survival of the organization.

TNT Equivalent Weight The weight of TNT [trinitrotoluene], which has an equivalent energetic output to that of a different weight of another explosive compound.

Tornado A local atmospheric storm, generally of short duration, formed by winds rotating at very high speeds, usually in a counterclockwise direction. The vortex, up to several hundred yards wide, is visible to the observer as a whirlpool-like column of winds rotating about a hollow cavity or funnel. Winds may reach 300 miles per hour or higher.

Toxic-Free Area An area within a facility in which the air supply is free of toxic chemical or biological agents.

Toxicity A measure of the harmful effects produced by a given amount of a toxin on a living organism. The relative toxicity of an agent can be expressed in milligrams of toxin needed per kilogram of body weight to kill experimental animals.

Toxin A poisonous substance produced by living organisms.

Triple-Standard Concertina [TSC] Wire A type of fence that uses three rolls of stacked concertina. One roll will be stacked on top of two other rolls that run parallel to each other while resting on the ground, forming a pyramid.

Tsunami A sea wave produced by an undersea earthquake. Such sea waves can reach a height of 80 feet and can devastate coastal cities and low-lying coastal areas.

Twisted-Pair Wire A type of wire that uses pairs of wires twisted together to mitigate electromagnetic interference.

Two-Person Rule A security strategy that requires two people to be present in or gain access to a secured area to prevent unobserved access by any individual.

U

Undesirable Event Any action or plan that threatens, damages, or destroys assets or disrupts service operations. Within this report it is used interchangeably with threat, domestic and international terrorism, sabotage, economic espionage, criminal activity, unlawful act, loss event, event, and major emergency or disaster.

United States The fifty States, the District of Columbia, Puerto Rico, the Virgin Islands, Guam, American Samoa, the Commonwealth of the Northern Mariana Islands, any possession of the United States, and any waters within the jurisdiction of the United States [6 U.S.C. 101].

Unlawful Act   A felony under United States federal or state law.

Unobstructed Space   Space around an inhabited building without obstruction large enough to conceal explosive devices 150 mm [6 inches] or greater in height.

Unshielded Wire   Wire that does not have a conductive wrap.

Uranium 235 [U-235]   Naturally occurring element found at 0.72 percent enrichment. U-235 is used as a reactor fuel or for weapons; however, weapons typically use U-235 enriched to 90 percent. Its half-life is 7.04×10^8 years.

V

Vaccine   A preparation of killed or weakened microorganism products used to artificially induce immunity against a disease.

Vandal   An individual or group committing a crime of opportunity—a spontaneous action without provoking causes.

Vapor Agent   A gaseous form of a chemical agent. If heavier than air, the cloud will be close to the ground. If lighter than air, the cloud will rise and disperse more quickly.

Vault   A reinforced room for securing items.

Vector   An agent, such as an insect or rat, capable of transferring a pathogen from one organism to another.

Venom   A poison produced in the glands of some animals [e.g., snakes, scorpions, or bees].

Vertical Rod   Typical door hardware often used with a crash bar to lock a door by inserting rods vertically from the door into the doorframe.

Vibration Sensor   An intrusion-detection sensor that changes state when vibration is present.

Video Intercom System   An intercom system that also incorporates a small CCTV system for verification.

Video Motion-Detection Sensor   A sensor that generates an alarm when an intruder enters a selected portion of a CCTV camera's field of view. The sensor processes and compares successive images between the images against predefined alarm criteria. There are two categories of video motion detectors, analog and digital. Analog detectors generate an alarm in response to changes in a picture's contrast. Digital devices convert selected portions of the analog-video signal into digital data that are compared with data converted previously; if differences exceed preset limits, an alarm is generated. The signal processor usually provides an adjustable window that can be positioned anywhere on the video image. Available adjustments permit changing horizontal and vertical window size, window position, and window sensitivity. More sophisticated units provide several adjustable windows that can be individually sized and positioned. Multiple windows permit concentrating on several specific areas of an image while ignoring others. For example, in a scene containing six doorways leading into a long hallway, the sensor can be set to monitor two critical doorways.

Video Multiplexer A device used to connect multiple video signals to a single location for viewing and/or recording.

Virus An infectious microorganism that exists as a particle rather than as a complete cell. Particle sizes range from 20 to 400 nanometers [onebillionth of a meter]. Viruses are not capable of reproducing outside of a host cell.

Visual Display A display or monitor used to inform the operator visually of the status of the electronic security system.

Visual Surveillance An ocular or photographic device [such as binoculars and cameras with telephoto lenses] to monitor facility or installation operations or to see assets.

Voice Recognition A biometric technology that is based on nuances of the human voice.

Volatility A measure of how readily a substance will vaporize.

Volumetric Motion Sensor A sensor designed to detect intruder motion within the interior of a protect volume. Volumetric sensors may be active or passive.

Vomiting Agent An agent that produces nausea and vomiting effects; can also cause coughing, sneezing, pain in the nose and throat, nasal discharge, and tears. Examples are adamsite [DM], diphenylchloroarsine [DA], and diphenylcyanoarsine [DC].

V-Series Nerve Agent A chemical agent of moderate to high toxicity developed in the 1950s. It is generally persistent. Examples are VE [phosphonothioic acid, ethyl-, S-[2-[diethylamino] ethyl] O-ethyl-ester], VG [phosphorothioic acid, S-[2-[diethylamino] ethyl] O, O-diethyl ester], VM [phosphonothioic acid, methyl-S-[2-diethylamino] ethyl] O-ethyl ester], VS [phosphonothioic acid, ethyl, S-[2-[bis[1-methylethyl]amino] ethyl] O-ethyl ester], and VX [phosphonothioic acid, methyl-S-[2-[bis[1-methylethyl]amino]ethyl] O-ethyl ester].

Voluntary In the case of any submittal of critical infrastructure information to a covered Federal agency and in the absence of such agency's exercise of legal authority to compile access to or submission of such information and may be accomplished by a single entity or an Information Sharing and Analysis Organization on behalf of itself or its members [6 U.S.C. 101].

Vulnerability Any weakness that can be exploited by an aggressor or, in a nonterrorist threat environment, make an asset susceptible to hazard damage.

W

Warm Site An environmentally conditioned workspace that is partially equipped with information technology [IT] and telecommunications equipment to support relocated operations in the event of a significant disruption.

Warning The alerting of emergency-response personnel and the public to the threat of extraordinary danger and the related effects that specific hazards may cause.

Watch Indication in a defined area that conditions are favorable for the specified type of severe weather [e.g. flash-flood watch, severe-thunderstorm watch, tornado watch, tropical-storm watch].

Waterborne Contamination Chemical, biological, or radiological agent introduced into a water supply.

Weapons-Grade Material Nuclear material considered most suitable for a nuclear weapon. It usually connotes uranium enriched to above 90 percent uranium-235 or plutonium with greater than about 90 percent plutonium-239.

Weapons of Mass Destruction [WMD] Any device, material, or substance used in a manner, in a quantity or type, or under circumstances showing an intent to cause death or serious injury to persons or significant damage to property. An explosive, incendiary, or poison-gas bomb, grenade, or rocket having a propellant charge of more than 4 ounces, a missile having an explosive incendiary charge of more than 0.25 ounce, or a mine or device similar to the above; poison gas; weapon involving a disease organism; or weapon that is designed to release radiation or radioactivity at a level dangerous to human life.

Weigand Protocol A security-industry-standard data protocol for card readers.

X

X-ray An invisible, highly penetrating electromagnetic radiation of much shorter wavelength [higher frequency] than visible light. Very similar to a gamma ray.

Z

Zoom The ability of a CCTV camera to close and focus or open and widen the field of view.

INDEX

Page numbers in italics refer to exhibits